Environmental Analytical Chemistry

JOIN US ON THE INTERNET VIA WWW, GOPHER, FTP OR EMAIL:

WWW: http://www.thomson.com
GOPHER: gopher.thomson.com
FTP: ftp.thomson.com
EMAIL: findit@kiosk.thomson.com

A service of

Environmental Analytical Chemistry

Edited by

F.W. Fifield
Principal Lecturer in Analytical Chemistry

and

P.J. Haines
Consultant Chemist
formerly Principal Lecturer in Physical Chemistry

School of Applied Chemistry
Kingston University
Kingston upon Thames
Surrey

BLACKIE ACADEMIC & PROFESSIONAL
An Imprint of Chapman & Hall

London · Weinheim · New York · Tokyo · Melbourne · Madras

Published by
Blackie Academic and Professional, an imprint of Chapman & Hall
2–6 Boundary Row, London SE1 8HN, UK

Chapman & Hall, 2–6 Boundary Row, London SE1 8HN, UK

Chapman & Hall GmbH, Pappelallee 3, 69469 Weinheim, Germany

Chapman & Hall USA, 115 Fifth Avenue, Fourth Floor, New York, NY 10003, USA

Chapman & Hall Japan, ITP-Japan, Kyowa Building, 3F, 2-2-1 Hirakawacho, Chiyoda-ku, Tokyo 102, Japan

DA Book (Aust.) Pty Ltd, 648 Whitehorse Road, Mitcham 3132, Victoria, Australia

Chapman & Hall India, R. Seshadri, 32 Second Main Road, CIT East, Madras 600 035, India

First edition 1995
Reprinted 1996 (twice), 1997

© 1995 Chapman & Hall

Typeset in 10/12 pt Times by Pure Tech India Ltd., Pondicherry, India
Printed in Great Britain by Clays Ltd, St Ives plc

ISBN   0 7514 0052 1 (PB)

A catalogue record for this book is available from the British Library

Library of Congress Catalog Card Number: 95-76277

∞ Printed on permanent acid-free text paper, manufactured in accordance with ANSI/NISO Z39.48–1992 and ANSI/NISO Z39.48–1984 (Permanence of Paper)

Preface

In the final quarter of the twentieth century the health of our global environment has become a matter of wide concern. This concern has stimulated a wide ranging and intensive search for an understanding of the way in which the natural environment functions, and the way in which the human race is bringing about environmental changes. These studies are heavily dependent on observation and quantitative measurements. It is clearly impossible to discuss sensibly the greenhouse effect without measurement of atmospheric carbon dioxide level; eutrophication of a lake without measurement of dissolved oxygen and nutrients; or heavy metal pollution without measurement of metal concentrations in soils and water. Many different parameters are studied, and diverse types of measurements made. These measurements must be designed and executed so as to be both relevant to the problem being studied, and reliable in themselve, i.e. they must be valid. Only by a proper understanding of the capabilities and limitations of the measurement methods themselves, and the context within which they are being applied can such validity be ensured. Increasingly, measurements are also made to demonstrate compliance with a framework of environmental legislation and regulations.

Many environmental studies are concerned with the amounts and distribution of chemical species. Very many different species with diverse properties are of interest. These species may be elements or compounds, simple or complex in structure, organic or inorganic, natural or artificial in origin, and occur at vastly different levels of concentration. Measurements in this context fall very much into the realm of analytical chemistry, which is the central theme of this book. The challenges of chemical analysis in the environmental context are immense, demanding high levels of expertise and skill across a wide range of analytical methodology.

In compiling this book we have firstly, in chapter 1, addressed the fundamental questions of the general properties of analytical measurements and in chapter 2, their limitations, validity and interpretation. Secondly, we have recognised the importance of a knowledge of fundamental chemical principles to anyone seeking to develop a proper understanding of chemical analysis. Readers with strong chemical backgrounds can ignore chapter 3, but it is the nature of environmental science to involve people from diverse initial subject areas. We anticipate, therefore, that a significant number of our readers will find reference to chapter 3 helpful at various times. It has been accepted by us from the outset that it is not sensible to attempt a fully comprehensive coverage of environmental analytical chemistry, and we have concentrated on those techniques, methods and applications which seem to us to be the most important.

Chapter 12 on the use of biological indicators in environmental assessments could be considered to be outside the strict realms of analytical chemistry. However, in practice the two areas are often so closely interdependent in use, that we have taken the view that it should be included. In order to sustain the appropriate level of expertise across the diversity of subject matter, chapters have been written by specialists and significant numbers of the examples used throughout the text are drawn from the personal experiences of the authors.

We are grateful for the many discussions that we have had over the years with colleagues, discussions which have helped shape our approach to environmental analytical chemistry, and, from time to time, to correct misconceptions. The hard work and forbearance of the publishers are also fully acknowledged.

F.W. Fifield
P.J. Haines

Contributors

Dr E.M. Buckley   School of Applied Chemistry, Kingston University, Penrhyn Road, Kingston upon Thames, Surrey KT1 2EE

Dr G.L. Christie   School of Applied Chemistry, Kingston University, Penrhyn Road, Kingston upon Thames, Surrey KT1 2EE

Mr F.W. Fifield   School of Applied Chemistry, Kingston University, Penrhyn Road, Kingston upon Thames, Surrey KT1 2EE

Mr P.J. Haines   38 Oakland Avenue, Farnham, Surrey GU9 9DX

Dr C.K. Laird   Kingston Analytical Services, School of Applied Chemistry, Kingston University, Penrhyn Road, Kingston upon Thames, Surrey KT1 2EE

Dr R. Manly   School of Life Sciences, Kingston University, Penrhyn Road, Kingston upon Thames, Surrey KT1 2EE

Dr N.I. Ward   Department of Chemistry, The University of Surrey, Guildford, Surrey GU2 5XH

Dr C.J. Welch   School of Applied Chemistry, Kingston University, Penrhyn Road, Kingston upon Thames, Surrey KT1 2EE

Contents

Part I General principles and techniques   **1**

1 Introduction   **3**
   F.W. FIFIELD and P.J. HAINES

1.1 Environmental science and analytical chemistry   3
1.2 Analytical chemistry   4
1.3 Overall analytical processes   6
     1.3.1 Defining the aims of an analytical programme   6
     1.3.2 Selection of an analytical method   7
     1.3.3 Sampling, sample handling and pretreatment   7
     1.3.4 Analytical measurements   8
     1.3.5 Method validation   9
     1.3.6 Data assessment and interpretation   9
     1.3.7 Safety   9

2 Analytical environmental data: assessment and interpretation   **10**
   F.W. FIFIELD

2.1 Introduction   10
2.2 Basic concepts and definitions   10
     2.2.1   True result   11
     2.2.2   Population   11
     2.2.3   Statistical sample   11
     2.2.4   Error   11
     2.2.5   Accuracy   11
     2.2.6   Precision   12
     2.2.7   The range or spread of data   12
     2.2.8   Distribution curves   12
     2.2.9   The mean   13
     2.2.10 The mode   13
     2.2.11 The median   14
     2.2.12 Degrees of freedom   14
     2.2.13 Standard deviation   14
     2.2.14 The variance   15
     2.2.15 Confidence levels and confidence limits   15
2.3 The nature and origin of errors   15
     2.3.1   Types of errors   16
2.4 Frequency distributions   18
     2.4.1   Normal and Student t distributions   19
     2.4.2   Other distributions   20
2.5 Assessment and interpretation of analytical results   20
     2.5.1   Introduction   20
     2.5.2   Data reliability   21
     2.5.3   Precision comparisons   22
     2.5.4   The assessment and comparison of means   24
     2.5.5   Graphical methods   26
     2.5.6   Detection limits   31
Further reading   31

3   Chemical principles   **32**
     P.J. HAINES

3.1   Introduction   32
      3.1.1   Periodicity   34
      3.1.2   Atomic spectra   35
      3.1.3   Nature of the electron   36
      3.1.4   Population of energy levels   40
3.2   Atomic orbitals and chemical bonds   40
      3.2.1   Ionic bonds   40
      3.2.2   Covalent molecules   41
      3.2.3   Polyatomic molecules   45
      3.2.4   Metal compounds and complexes   47
3.3   Molecular energy levels   48
      3.3.1   Energy of assemblies of molecules   50
3.4   Enthalpies of formation and reaction   51
3.5   Entropy and free energy   52
3.6   Free energy and equilibrium   54
3.7   The effects of temperature   55
3.8   Application to equilibria   55
      3.8.1   Phase equilibria   55
      3.8.2   Ions in solution   56
      3.8.3   Solubility   56
      3.8.4   Acid–base equilibria   57
      3.8.5   Oxidation–reduction equilibria   59
      3.8.6   Electrochemical reactions   59
      3.8.7   Complexation   61
3.9   Reaction kinetics   61
3.10 Examples of reaction kinetics   63
      3.10.1 Radioactive reactions   63
      3.10.2 Ionic reactions   63
      3.10.3 Solid-state reactions   64
      3.10.4 Photochemical reactions   64
3.11 Summary   65
Further reading   65

4   Titrimetry and gravimetry   **66**
     F.W. FIFIELD

4.1   Introduction   66
4.2   Titrimetry   66
      4.2.1   Introduction   66
      4.2.2   Acid–base titrations   68
      4.2.3   Complexometric titrations   71
      4.2.4   Redox titrations   76
4.3   Gravimetry   77
      4.3.1   Principles   77
      4.3.2   Gravimetric procedures   78
Further reading   80

5   Separation techniques   **81**
     F.W. FIFIELD

5.1   Introduction   81
5.2   Solvent extraction   81
      5.2.1   Introduction   81
      5.2.2   Solvent extraction of analytes from environmental samples   81
      5.2.3   Separation of mixtures by solvent extraction   82

5.3   Chromatography   86
      5.3.1   Introduction   86
      5.3.2   Characteristics of chromatograms   87
      5.3.3   High-performance liquid chromatography   90
      5.3.4   Ion-exchange chromatography   93
      5.3.5   Thin-layer and paper chromatography   96
      5.3.6   Gas chromatography   98
5.4   Other separation techniques   102
      5.4.1   Supercritical fluid chromatography (SFC)   102
      5.4.2   Gel permeation chromatography (GPC)   102
      5.4.3   Electrophoresis   103
      5.4.4   Distillation and volatilisation   103
      5.4.5   Precipitation   104
Further reading   104

6 General principles of spectrometry   105
P.J. HAINES

6.1   Introduction   105
6.2   Energy levels   105
6.3   Types of transition   105
      6.3.1   Lasers   106
6.4   Molecular dissociation   108
6.5   Electromagnetic radiation   109
6.6   The electromagnetic spectrum   111
6.7   Interaction of species with electromagnetic radiation   113
6.8   Absorption laws   115
6.9   Spectrometric instrumentation   116
      6.9.1   Single-beam spectrometer   116
      6.9.2   Double-beam spectrometer   118
      6.9.3   Fourier-transform instruments   119
Further reading   120

7 Atomic spectrometry   121
F.W. FIFIELD

7.1   Introduction   121
7.2   Flame emission spectrometry   123
      7.2.1   The chemical flame   123
7.3   Plasma spectrometry   124
      7.3.1   Introduction   124
      7.3.2   Inductively coupled plasma-atomic emission spectrometry 125
      7.3.3   Inductively coupled plasma-mass spectrometry   128
7.4   X-ray emission techniques   129
      7.4.1   Introduction   129
      7.4.2   Electron probe microanalysis   133
      7.4.3   X-ray fluorescence spectrometry   135
7.5   Atomic absorption spectrometry   137
      7.5.1   Introduction   137
      7.5.2   Sharp-line radiation   137
      7.5.3   AAS measurements   138
      7.5.4   Flame AAS   139
      7.5.5   Electrothermal AAS   140
      7.5.6   AAS using mercury vapour or volatile hydrides   142
      7.5.7   Use of atomic spectrometry   142
Further reading   144

8   Molecular spectrometry   **145**
P.J. HAINES

8.1   Introduction   145
8.2   Ultraviolet and visible spectrophotometry   145
      8.2.1   Instrumentation   145
      8.2.2   Band spectra   146
      8.2.3   Polyatomic organic molecules   147
      8.2.4   Solvent effects   148
      8.2.5   Metal complexes   149
      8.2.6   Applications   149
      8.2.7   UV fluorescence methods   149
      8.2.8   Combined separation and UV techniques   151
8.3   Infrared spectrometry   152
      8.3.1   Sampling   152
      8.3.2   Infrared absorption   154
      8.3.3   Polyatomic molecules   156
      8.3.4   Combinations of infrared and separation techniques   158
      8.3.5   Applications of infrared spectrometry in environmental analysis 159
8.4   Nuclear magnetic resonance spectrometry   160
      8.4.1   Instrumentation   162
      8.4.2   Solvents for NMR work   163
      8.4.3   The chemical shift   163
      8.4.4   The peak area   165
      8.4.5   Spin–spin coupling   166
      8.4.6   Applications of NMR   170
8.5   Mass spectrometry   170
      8.5.1   Instrumentation   170
      8.5.2   Isotopic composition and accurate masses   173
      8.5.3   Nitrogen rule   173
      8.5.4   Fragmentation   174
      8.5.5   Applications of mass spectrometry   176
Further reading   179

9   Measurement of ionising radiations and radionuclides   **180**
F.W. FIFIELD

9.1   Introduction   180
9.2   Ionising radiations and radioactivity   180
      9.2.1   Alpha radiation (α)   181
      9.2.2   Beta radiation (β^- or β^+)   181
      9.2.3   Gamma radiation (γ)   184
      9.2.4   Internal conversion   185
      9.2.5   Radioactive decay   186
      9.2.6   Units of radioactivity and radiation measurement   187
9.3   The detection and measurement of radiation   187
      9.3.1   Gas ionisation detectors   188
      9.3.2   Semiconductor detectors   189
      9.3.3   Sodium iodide detectors   191
      9.3.4   Organic scintillators   192
      9.3.5   Liquid scintillation counting   192
      9.3.6   Detection by films   194
      9.3.7   Concentration and separation of radionuclides   195
Further reading   195

10   Electrochemical techniques   **196**
      E.M. BUCKLEY

10.1   Introduction   196
10.2   Electrochemical principles   196
10.3   Potentiometric techniques   197
        10.3.1   Introduction and theory   197
        10.3.2   Practical considerations and applications   200
        10.3.3   Potentiometric titrations   209
        10.3.4   Current developments   210
10.4   Voltammetric and controlled potential techniques   210
        10.4.1   Introduction   210
        10.4.2   Theory   210
        10.4.3   Practical considerations and applications   211
        10.4.4   Techniques   215
10.5   Electrochemical detection in flowing streams   219
        10.5.1   Introduction   219
        10.5.2   Potentiometric measurements in flowing streams   219
        10.5.3   Voltammetric detection in flowing streams   220
10.6   Other electroanalytical techniques   221
        10.6.1   Introduction   221
        10.6.2   Conductometry   222
        10.6.3   Coulometry   222
        10.6.4   Electrogravimetry   223
Further reading   223

11   Thermal methods of analysis   **224**
      P.J. HAINES

11.1   Introduction   224
11.2   Definitions   224
11.3   General apparatus   225
11.4   Factors affecting thermal analysis results   226
        11.4.1   The sample   226
        11.4.2   The crucible   227
        11.4.3   The rate of heating (dT/dt)   227
        11.4.4   The atmosphere   227
        11.4.5   The mass of the sample   228
11.5   Thermogravimetry   228
        11.5.1   Apparatus   228
        11.5.2   Applications of thermogravimetry   230
11.6   Differential thermal analysis (DTA) and differential scanning calorimetry (DSC)   233
        11.6.1   Differential thermal analysis (DTA)   233
        11.6.2   Differential scanning calorimetry (DSC)   233
        11.6.3   Apparatus   234
        11.6.4   Applications   236
11.7   Thermomechanical analysis (TMA) and dynamic mechanical analysis (DMA)   239
        11.7.1   Apparatus   239
        11.7.2   Applications   240
11.8   Simultaneous techniques and product analysis   242
        11.8.1   Apparatus   242
        11.8.2   Evolved gas analysis (EGA)   242
        11.8.3   Analysis of products and reactions   244
11.9   Environmental applications of thermal methods   246
        11.9.1   Geological materials   246
        11.9.2   Recycling   246
        11.9.3   Residues   247
        11.9.4   Vaporisation studies   247

11.9.5	Flue gas treatment	247
11.9.6	Purity	248
11.10	Summary	248
Further reading		248

12   Biological indicators   249
R. MANLY

12.1	Introduction	249
12.2	Monitoring community structure	251
	12.2.1 Diversity indices	252
	12.2.2 Similarity indices	253
	12.2.3 Species abundance patterns	253
	12.2.4 Multivariate analyses	254
12.3	Bioindicator methods	255
	12.3.1 Biotic indices	255
	12.3.2 Pollutant mapping	260
	12.3.3 Morphological and histological indicators	261
	12.3.4 Detector and sentinel organisms	264
	12.3.5 Comparative methods	264
12.4	Microbiological monitoring	265
	12.4.1 Intrusive microorganisms and faecal contamination	266
12.5	Bioaccumulators	268
12.6	Bioassays	272

Part II   Specific applications   277

13   Speciation   279
G.L. CHRISTIE

13.1	The importance of speciation	279
13.2	Definition of speciation	279
13.3	The determination of trace metal speciation	279
	13.3.1 Computer modelling	279
	13.3.2 Experimental determination of speciation	284
13.4	Concluding remarks	289
Further reading		290

14   The analysis of atmospheric samples   291
C.K. LAIRD

14.1	Introduction	291
14.2	Atmospheric analyses	292
	14.2.1 Measurements of atmospheric composition	292
	14.2.2 Emission measurements	292
	14.2.3 Indoor and workplace atmospheres	292
14.3	Techniques for gas analysis	293
	14.3.1 Gas chromatography	293
	14.3.2 Spectrometric methods	293
	14.3.3 Electrochemical sensors	299
	14.3.4 Chemical methods and detector tubes	301
14.4	Sampling	302
	14.4.1 Ambient air	302
	14.4.2 Emissions	303
14.5	Calibration of gas analysers	305

14.6 Analysis of the major air pollutants and oxygen   306
    14.6.1 Carbon monoxide   306
    14.6.2 Nitrogen oxides   307
    14.6.3 Ozone   308
    14.6.4 Sulphur dioxide   309
    14.6.5 Oxygen   310
14.7 Analysis of some secondary air pollutants   313
    14.7.1 Methane, non-methane hydrocarbons and non-methane organics   313
    14.7.2 Chlorofluorocarbons   314
    14.7.3 Organic nitrates   314
    14.7.4 Nitrous oxide   314
    14.7.5 Combustible gases   314
14.8 Sampling and analysis of particles and aerosols   315
    14.8.1 Emissions sampling   315
    14.8.2 Particle and aerosol characterisation and collection   317
    14.8.3 Particle and aerosol composition   318
Further reading   318

15   Trace elements   320
N.I. WARD

15.1 Trace elements in the environment   320
15.2 Natural levels and chemical forms   321
15.3 Trace element contamination and pollution   328
    15.3.1 Air particulates   328
    15.3.2 Natural waters   330
    15.3.3 Soils and sediments   332
    15.3.4 Plants   333
    15.3.5 Animals and humans   334
15.4 Sampling and sample preparation   336
    15.4.1 Atmospheric samples   336
    15.4.2 Water samples   337
    15.4.3 Soils and sediments   339
    15.4.4 Plants   340
    15.4.5 Biological tissues and fluids   341
    15.4.6 Sample digestion methods   342
15.5 Methods of analysis   344
15.6 Selected important examples   348
    15.6.1 Lead   248
    15.6.2 Aluminium   350

16   Environmental radiation and radioactivity   352
F.W. FIFIELD

16.1 Introduction   352
16.2 The hazards of ionising radiations and their assessment   353
16.3 Natural sources of radiation   354
    16.3.1 Radionuclides of geological origin   354
    16.3.2 Radionuclides resulting from cosmic rays   360
16.4 Artificial sources of radiation   360
    16.4.1 Radiation and radioactivity in research and medicine   361
    16.4.2 Nuclear power   361
16.5 Radiogenic dating   364
    16.5.1 Geological dating   365
    16.5.2 Carbon dating   366
    16.5.3 Tritium dating   366
    16.5.4 Lead-210 dating   367
    16.5.5 Thermoluminescence or electron spin resonance dating   367
Further reading   368

17   Contaminated landsites   369
F.W. FIFIELD

17.1   Introduction   369
17.2   The nature of contaminated sites   370
        17.2.1   Origins of contamination   370
        17.2.2   Physical characteristics of landsites   371
17.3   Assessments   372
        17.3.1   Investigational plan   373
Further reading   375

18   The analysis of water   376
F.W. FIFIELD

18.1   Introduction   376
18.2   pH, acidity and alkalinity   377
18.3   Dissolved oxygen and oxygen demand   379
18.4   Total organic carbon   381
18.5   Metals   381
18.6   Dissolved salts   381
18.7   Trace organics   383
18.8   Radioactivity and radionuclides in water   384
18.9   Water surveys and sampling   384
Further reading   385

19   The determination of trace amounts of organic compounds   386
C.J. WELCH

19.1   Introduction   386
19.2   Sample preparation   388
        19.2.1   Macretion, dissolution and extraction   389
        19.2.2   Partition   390
        19.2.3   Concentration of the analyte   393
19.3   Chromatography   395
        19.3.1   Gas chromatography   395
        19.3.2   High-performance liquid chromatography   396
        19.3.3   Quantitation   396
19.4   Screening analysis   397
19.5   GC applications: pesticide analysis   398
19.6   HPLC: trace analysis   401
19.7   HPLC applications   401
        19.7.1   Aromatic hydrocarbons   401
19.8   GC/HPLC applications   403
        19.8.1   Trace organic analysis in water   403
        19.8.2   Analysis of organic materials in soil and sediments   405
        19.8.3   Analysis of amines   406
References   407

Glossary   409

Index   417

Part I
General principles and techniques

Introduction 1
F.W. Fifield and P.J. Haines

Environmental science encompasses the study of the whole human environment and in doing so makes use of all scientific disciplines. In a vast number of studies the need emerges for information on the composition of the parts of the environment concerned. It is not possible for example to study the natural transport of substances as in the water cycle, or other natural processes without measurement of the substances being transported. The assessment of the depletion of the ozone layer requires measurements of the amount of ozone present just as any investigation of the greenhouse effect demands measurements on the concentrations of greenhouse gases. Heavy metals and pesticides are substances whose use and environmental concentrations are now controlled by legislation in many countries, which cannot be enforced unless the concentration can be measured.

In some cases it is necessary to know the exact chemical structure of compounds. Organic compounds, which differ structurally only in minor and subtle ways, can have very different physiological effects and pose very different levels of health hazards. Heavy metals such as lead and mercury are dangerous pollutants in their inorganic form. If, however, they are converted into organo-metallics by either natural or anthropogenic action they are conferred additionally with lipid solubility. As a result, their mobility in the human body is substantially enhanced and they become much more hazardous.

Two aspects of environmental science which frequently arise are requirements for the surveillance or monitoring of a particular area or ecosystem. Although related, and often utilising the same measurements, the two concepts are different. Surveillance implies the assessment and continued observation of a particular system to ensure that it is 'healthy' and remains so. This is obviously a difficult problem because every parameter cannot be identified and continuously checked. On the precept that the health of the system is measured by the health of the living organisms that it supports, biological monitoring is often employed. This approach, which is developed more fully in chapter 12, selects certain monitoring organisms to represent the overall system and keeps a check on their health. This use of biological monitoring leads on to the general principles of monitoring. Individual

1.1 Environmental science and analytical chemistry

parameters are selected and regularly assessed by appropriate measurements. For example, in order to comply with legislation on the release of pollutants into the environment, specified elements or compounds will need to be measured on a regular basis, either in environmental samples, or for instance in factory discharges. Similarly, in order that food and drinking water are known to be suitable for consumption, the monitoring of specified qualities and particular components will be needed.

Some of the most newsworthy aspects of environmental science are what might be termed 'pollution incidents'. These generally involve the sudden release of pollutants from a point source such as an oil tanker, mine workings or factory. Extensive measurements will be required for the assessment of the short and longer term effects of an incident, as well as in the forensic task of providing an unambiguous identification of the source of pollution.

A high proportion of the measurements needed in the circumstances outlined above fall within the realm of *analytical chemistry*.

1.2 Analytical chemistry

Analytical chemistry employs physico-chemical principles to make measurements of chemical species i.e. atoms, ions, molecules and free radicals. The material collected and subjected to analysis is known as the *sample*. It consists of two parts, the *analyte* or *analytes* which are the substances to be determined, and the rest of the sample which comprises the *matrix*. The matrix can exert considerable influence over the way in which the analysis is carried out and the quality of the results obtained. Where the aim is to identify or detect the presence of a species *qualitative analysis* is carried out. The detailed determination of molecular structure is known as *structural analysis* and is regarded as a specialised form of qualitative analysis. In circumstances where the amount of a substance present is measured, *quantitative analysis* is used. It is easy to define different types of analysis, but in practice the distinctions are less clear cut. For example, a qualitative test may involve the use of a reagent which develops a colour in the presence of the analyte. However, below a certain concentration or *detection limit* the colour will not be detectable. Thus, the qualitative test is quantitative also in the sense that it can only show that the analyte concentration does not exceed a certain detection limit. Zero concentration of an analyte can never be demonstrated in practice and does not exist in principle. It should be noted that the situation with regard to living organisms is different and it is sometimes possible to say for a sample that no organisms of a particular type are present. The framework within which a zero result is reported must be very carefully defined.

Returning to the example of the use of a colour reaction as an analytical tool, it is self-evident that the intensity of the colour developed is related to the amount of analyte present. If proportionality of the response of a test is reproducible over a range of analyte concentrations, the basis for quantitative analysis may exist. This is expressed by equation (1.1).

$$I = kC \tag{1.1}$$

where I is the magnitude of the measurement, C, the amount or concentration of the analyte and k, a proportionality factor. A value for k may be calculated from first principles, but because the theoretical relationships involved are often complex and in any case k is frequently matrix dependent, it is almost invariably obtained via a *calibration process*. In this, measurements are made on mixtures of accurately known composition with a matrix as closely similar to that of the sample as possible. Calibration data are often presented and used in the form of a graphical plot of response against amount of analyte. This is known as a *calibration curve* or *graph* (Figure 1.1).

Environmental science presents some very difficult problems for analytical chemists.

1. The range of analyte types is very wide. They vary from simple inorganic species at one end of the scale to complex bio-molecules at the other.
2. Sample matrices are frequently complex and unknown. They may be solid, liquid or gaseous.
3. Analytes may have to be measured at very low concentrations, e.g. a few ppb (μg kg^{-1}).
4. Environmental programmes often generate very large numbers of samples and place a premium on automation.
5. Some analyses can only be made at, or very close to the sampling site. Readily portable and robust analytical equipment or remote sensing devices are thus required.

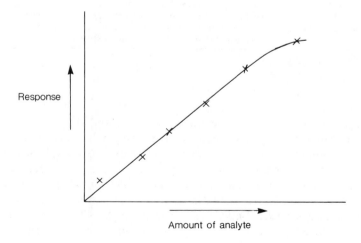

Figure 1.1 A typical calibration curve.

1.3 Overall analytical processes

The practical sequence of operations which is used to obtain an analytical result from a sample is known as the *analytical method*. Its constituent parts are often based upon different chemical principles, e.g. precipitation, distillation, titration. These are called *techniques*. An illustrative list of some groups of techniques appears in Table 1.1. If relevant and reliable analytical information is to be obtained in an environmental investigation more than a mastery of analytical techniques and methods is required. It is essential to have a unified view of the whole process from the definition of the aims of the investigation through to the final interpretation and presentation of the results. It is convenient to do this in a number of stages, i.e.

1. definition of the aims;
2. selection and development of an appropriate method;
3. sampling plan, sample collection, sample handling and pretreatment;
4. separations;
5. final measurements;
6. method validation;
7. assessment and interpretation of results;
8. safety.

Table 1.1 A classification of some analytical techniques

Type	Properties measured
Gravimetric	Weight of pure analyte or of a stoichiometric compound of it
Volumetric	Volume of a solution containing a known amount of a reagent reacting with the analyte
Spectrometric	Wavelengths and intensity of electromagnetic radiation emitted or absorbed by the analyte
Mass spectrometric	Abundance of atomic ions or molecular fragments derived from the analyte
Electrochemical	Electrical properties of analyte solutions
Chromatographic	Physico-chemical properties of analytes following separation
Nuclear or radiochemical	Energy and intensity of nuclear radiations emitted by the analyte
Thermal	Physico-chemical properties of the analyte as heat is applied to it

1.3.1 Defining the aims of an analytical programme

This stage is clearly the fundamental one in designing an overall analytical procedure. Sometimes it is straightforward as in the case of monitoring a system to comply with specified legal requirements. Indeed, it is increasingly common to find a standard overall procedure also specified. In other circumstances the problem can be much more difficult. This is especially so in investigatory or research situations. It may not even be obvious what measurements are needed let alone the levels of sensitivity precision and accuracy required. To define the analytical requirement, unquestionably requires full consultation between the environmental scientists and the analytical chemists.

1.3.2 Selection of an analytical method

Having defined the aims of the analysis it is possible to proceed to the selection of an appropriate method. A whole range of factors need to be taken into account in this selection process. It is as important to avoid 'over analysing' as it is to avoid producing data which are inadequate in quality or quantity. Some of the important factors are listed below. These are not in any strict order of significance as this may change from one situation to another.

1. what is the analyte?;
2. detection limits;
3. precision;
4. accuracy;
5. turn-round time;
6. are analyses needed 'on-site'?;
7. number of samples and any consequent automation;
8. cost.

Even where methods specified by regulatory bodies are not to be used, generally accepted ones for many analytes are well reported. It is clearly appropriate to use such methods although they may need some modification in order to fit the particular circumstances of the investigation.

1.3.3 Sampling, sample handling and pretreatment

Samples must be collected according to a plan which ensures that they are representations of the original aims. Throughout subsequent handling and treatment they must remain so. Many variables need to be taken into account when producing a sampling plan. The general pattern is best illustrated by considering some typical examples. In a land survey for a particular analyte, both lateral and vertical variations in concentration need to be assessed. A routinely used approach is to divide the study area with a regular grid pattern, taking samples at the centre of each grid square and at predetermined depths. The dimensions of the grid will clearly be related to the size of the area to be covered. For relatively small sites such as industrial ones, 10 m squares may be appropriate. On the geochemical scale, 1 km squares are more likely with a more detailed study to follow on if an initial large scale survey indicates local areas of interest. In completing the picture, other attributes of the site might be examined, such as groundwater, streams or vegetation. Finally temporal variations and the effects of the seasons need to be borne in mind. For example, nitrate concentrations will depend on the stage of the agricultural cycle and the degree of leaching by rainfall.

In the survey of a large body of water such as a lake, the same spatial parameters need to be considered and a similar pattern of grid and depth sampling employed. Given the capacity of a liquid for ready mixing, it is unlikely in most cases that samples as closely spaced as 10 m would be collected. Depth studies might also be over rather greater distances. It may well be that the dilution effected by the water means that analytes are present at very low concentrations. The use of pollution indicators such as sediments or filter-feeding shellfish may be advantageous in such circumstances. Seasonal effects here would be exemplified by *seasonal overturn* which occurs in many lakes. In this process thermal effects acting on the water bring about mixing of the upper and lower strata of water twice yearly, in spring and autumn.

The sampling of vapours and gaseous materials presents its own special problems. Physically, the samples are difficult to handle, and the high mobility of substances leads to rapid dispersion. Analytes may thus need to be trapped, for example by cooling to a liquid form, or sorption on to a solid surface.

Having collected samples, their handling and storage prior to analysis must be given careful consideration. Losses of the analyte or contamination can occur by simple mechanical means, evaporation or absorption from external sources. Decomposition is also a possibility and is a particular problem with organic and biological samples. Refrigerated storage is one obvious possible solution. The difficulties outlined above are of especial concern where sampling is being carried out at a site remote from full laboratory facilities. Measurement on-site with portable field equipment must then be effected where necessary.

Sample pretreatments constitute the stage in the analytical process in which the sample is prepared for measurement of the analyte. There is no general standard procedure and considerable variations will be encountered. The analytes in water for example may well need to be concentrated prior to measurements. Simple evaporation, solvent extraction or other chemical separation procedures may be required. Solid samples will usually need to be solubilised. Aggressive chemical reagents, including strong acids or alkalis, are often used for this. Analyte losses must be carefully guarded against at this stage, as must contamination.

In ideal circumstances direct measurements can be made on analytes without significant matrix effects or interference. Sometimes this is possible especially where spectrometric methods are employed to distinguish the analyte signal from others. Often, however, this cannot be achieved and separations are needed. The principles of the most important ones are discussed in chapter 5.

1.3.4 *Analytical measurements*

The making of an analytical measurement is often the simplest and most straightforward stage in the overall process. This is especially so in modern laboratories where instrumental and automated procedures now dominate. Many of the important measurement techniques are reviewed in later chapters. It is timely to recall here that unless the earlier stages in the analytical process have been properly executed, the results obtained will be meaningless or, perhaps worse, misleading.

1.3.5   Method validation

It is axiomatic that the experimental results obtained must give a good indication of the true result. The calibrations carried out in setting up the method go part of the way to ensuring this. Matrix effects have previously been mentioned and these are difficult to allow for fully in calibration standards unless the matrix is very simple. Analysis of validated standards to check the performance of a method is almost always essential. Validated standards used should have a composition which is closely similar to the sample both in terms of its analyte concentration and matrix. As a result of extensive previous analysis by a variety of methods, the analyte content is known within narrow limits. Standards used can be 'in-house' ones which have been established within the laboratory and these are valuable for the routine checking of method performance. However, reference must also always be made to standards which are accepted nationally and internationally. They can be obtained from bodies like the National Bureau of Standards (NBS) in The United States and The Bureau of Certified Reference Materials (BCR) in Europe.

1.3.6   Data assessment and interpretation

Finally, the data yielded by an analytical programme are used to try to fulfil the original aims defined at the beginning. A two stage process is required for this. The reliability of the data must first be assessed. As discussed in chapter 2, errors are implicit in all measurements and correct procedures must be used to ensure that data with excessive errors which endanger the validity of the conclusions are not used. It is then possible to attempt to interpret the data in the light of the defined aims. A blend of environmental, analytical and statistical expertise will be required to do this soundly. Careful thought should also be given to data presentation, so that it is readily digested and conclusions are clearly demonstrated.

1.3.7   Safety

Obviously a normal safety assessment for the programme must be made. It should, however, be borne in mind that the initial interest in environmental analytes often derives fom their hazardous properties, and that special precautions may well be needed.

2 Analytical environmental data: assessment and interpretation

F.W. Fifield

2.1 Introduction

The essence of the use of analytical chemistry in environmental science is to make measurements in order to help attain a specified aim. As has been emphasised in chapter 1, measurement programmes must be planned with great care to ensure that they are directed properly towards that aim. It is self-evident that the data collected as a result of the measurements must be as reliable as possible. In order to ensure this, a sound understanding of the techniques and methods being used and their limitation in a particular application is needed. Much of the rest of this book is directed towards providing just such an understanding. It is also necessary to subject the data acquired to regular statistical assessment to detect unreliable results. Finally, conclusions drawn from the experimental data must be soundly based. This requires further use of statistical methods in quantifying the confidence which may be placed in the conclusions.

The key elements in the obtaining and interpreting environmental analytical data are:

1. an understanding of environmental processes as a basis for planning measurement programmes;
2. a sound knowledge of the techniques and methods in use, especially with regard to their limitations;
3. an appreciation of the characteristics of experimental data together with appropriate statistical techniques for its assessment and interpretation.

2.2 Basic concepts and definitions

Below are brief explanations of some basic concepts and definitions of some important terms and quantities to which reference is made later. Some subjects are expanded upon in subsequent sections.

2.2.1   True result

For any measurement there is an ultimately correct value which is known as the *true result*. In reality, all measurements are subject to some degree of uncertainty and the true result is never known. It is estimated with varying degrees of efficiency by experimental measurement. When a calibration standard is used in assessing a method, it is assumed to represent the true result for the measurement. However, it is clear that this also is only an approximation as there will be uncertainty associated with its initial preparation.

2.2.2   Population

In making measurements to obtain an exactly correct value for a true result an infinite number would be needed. This data set is known as the *population*.

2.2.3   Statistical sample

It is obviously impossible to have access to the complete population of results and in practice a much smaller number of measurements is made as a sample of the population. From this sample, the true result and the characteristics of the whole population are estimated. Often the sample is small, five or less, and care has to be used in the interpretation of the sample data, and in predicting the characteristics of the population. Care must be exercised in distinguishing between the term sample used in this context and the *analytical sample* which is the physical subject of chemical analysis.

2.2.4   Error

Error is simply the difference between an experimental result and the true result, and may be attributed to different sources. It can only be estimated, because of the unknown value of the true result and the variability of measurements. The numerical difference is known as the *absolute error* and is quoted in the same units as the measurement. More useful for estimating the significance of an error is the *relative error* where the error is expressed as a proportion of the overall result. Errors are discussed more fully in section 2.3.

2.2.5   Accuracy

The closeness of an experimental result to the true result is known as its *accuracy*, which is defined in terms of error and similarly can only be estimated. As the aim of a measurement is to estimate the true result, accuracy is a most important quantity.

2.2.6   Precision

Even when measurements are carried out under closely standardised conditions, some variability amongst replicate measurements will occur. This variability, which can derive from many sources is called the *precision* of the measurement. It is most frequently quantified in terms of the *standard deviation*. Distinction is made between repeatability and reproducibility. The former refers to variations where replicate measurements are made under closely controlled circumstances in one place and in a short timescale. The latter applies when time and place and other circumstances may differ substantially, but with the method remaining ostensibly the same. The detailed distribution of the variable data is also an important characteristic.

2.2.7   The range or spread of data

The numerical difference between the highest and lowest values in a data set is known as its *range* or *spread*. It constitutes one estimate of precision but reference to Figure 2.1 will show its limitations. Two data sets with similar ranges can have very different overall characteristics. A more sophisticated assessment of data distribution is thus required although some useful non-parametric statistical methods which make no assumptions, are often based on the use of ranges. Figure 2.1 illustrates the concepts of true result, error, accuracy, precision and range.

2.2.8   Distribution curves

If a series of measurements is made on the same sample by as near identical procedures as possible, a variation will be seen in the results obtained. The variation stems from the uncertainties implicit in the measurement procedures. The results represent a set of *replicates*. When a plot of numerical values against frequency of occurrence is made a *frequency distribution curve* is obtained. The characteristics of this curve will depend upon the measurement used, how it is applied and the size of the data set. In the study of environmental problems, the investigation of the distribution of environmental constituents is often important and different types of distribution curves may be obtained. The subject of distribution curves is addressed more fully in section 2.4

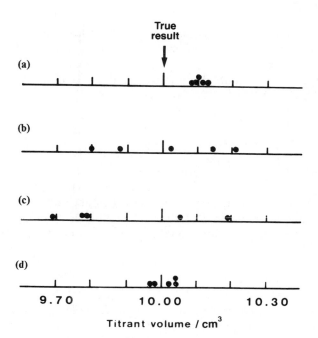

Figure 2.1 Graphical representation of accuracy and precision: (a) precise but inaccurate; (b) accurate but imprecise; (c) inaccurate and imprecise; and (d) accurate and precise. Source: Miller, J.C. and Miller, J.N. (1988) *Statistics for Analytical Chemistry*, 2nd edition, Ellis Horwood, Chichester.

2.2.9   The mean

The arithmetic average of a set of data is known as its *mean* ($\bar{x}$) and is defined by

$$\bar{x} = \sum_i x_i/n \tag{2.1}$$

where x_i represents the individual results and n the number of results. When the distribution of data is symmetrical, the mean is the best statistic to represent the central tendency of the whole data set.

2.2.10   The mode

If data are unsymmetrically distributed they may be best represented by the most frequently occurring value, i.e. the one corresponding to the peak of the distribution curve. This is known as *the mode*.

2.2.11   The median

Where no assumptions can be made about the distribution of a data set it is better to use the *median* to represent the central tendency. The median is calculated by placing the data in ascending order of magnitude when it is the $\frac{1}{2}(n + 1)$th value if n, the number of data, is odd or the midpoint between the $\frac{1}{2}$ and $(\frac{1}{2}n + 1)$th if even.

2.2.12   Degrees of freedom

A *degree of freedom* is an independent variable. In this context the number of degrees of freedom of a replicate set of data is equal to the total number of data. However, for a statistic derived from the data the number of degrees of freedom may be reduced. Rules for calculating the number of degrees of freedom are quoted on an individual basis in the text where appropriate.

2.2.13   Standard deviation

As has been indicated above, the *standard deviation* (σ) is an important statistic used in expressing the characteristic of a data set and hence also the performance of an analytical method. Essentially it represents the distribution of the data around its mean. For a single data set it is expressed by equation (2.2):

$$\sigma = \sqrt{\sum_i (x_i - \mu)^2/n} \qquad (2.2)$$

where x_i, is an individual result, μ is the true mean of the data set and n is the number of data. In practice μ is not known and the statistic is more correctly calculated from equation (2.3):

$$\sigma_{n-1} = \sqrt{\sum_i (x_i - \bar{x})/(n - 1)} \qquad (2.3)$$

where $\bar{x}$ is the experimental mean calculated from the experimental data. It will be seen that as n increases σ_{n-1} approaches σ_n and for practical purposes may be assumed to be the same when $n > 30$. The standard deviation is the most important precision indicator. It is often quoted in the form of the *relative standard deviation* (RSD) which is expressed by equation (2.4):

$$\text{RSD} = 100\sigma_{n-1}/\bar{x} \qquad (2.4)$$

The units are percent. The term *coefficient of variation* (CV) is also used synonymously with RSD.

A better estimate of the standard deviation may often be obtained by the pooling of results from more than one set of data. Thus, σ may be calculated from K sets of data:

$$\sigma = \sqrt{\frac{\sum_{i=1}^{N_1} (x_i - \bar{x}_1)^2 + \sum_{i=1}^{N_2} (x_i - \bar{x}_2)^2 + \ldots + \sum_{i=1}^{N_K} (x_i - \bar{x}_K)^2}{M - K}} \qquad (2.5)$$

where $M = N_1 + N_2 + \ldots + N_K$. One degree of freedom is lost with each set pooled. In the special circumstances where two data sets are concerned, and whose precisions are known to be similar (see F-test, section 2.5.3), the pooled value can be computed from the individual σ values according to equation (2.6).

$$\sigma^2 = \{(n_1 - 1)\sigma_1^2 + (n_2 - 1)\sigma_2^2\}/(n_1 + n_2 - 2) \qquad (2.6)$$

2.2.14   The variance

σ^2, *the variance* is an important quantity particularly because variances are additive and the overall variance for a process may be estimated by summing the individual variances for its constituent parts, as expressed in equation (2.7)

$$\sigma_{\text{overall}}^2 = \sigma_1^2 + \sigma_2^2 + \ldots + \sigma_n^2 \qquad (2.7)$$

This is an important relationship when considering the sources of the overall variability in an analytical process.

2.2.15   Confidence levels and confidence limits

Because of the variability associated with measurements, conclusions can only be drawn with various degrees of confidence, typically at 90%, 95% or 99%, depending upon the circumstances. These are known as *confidence levels*. On the other hand *a confidence interval* may be defined around a specified value, e.g. the mean, within which there is a known probability of the true result falling. The *confidence limits* define the interval quantitatively.

2.3 The nature and origin of errors

The concept of errors representing the differences between experimental results and true results has been introduced in section 2.2 With regard to experimental results it is essential to be able to:

1. identify the nature of the errors present;
2. assess their magnitude;

3. minimise and, where possible, correct for them;
4. express results so that the limitations imposed by errors are clearly stated.

In order to achieve this an understanding of the nature of the errors and their origin is essential.

2.3.1   Types of error

Gross errors.   A measurement may be invalidated by a major event such as the complete failure of an instrument, serious loss from a sample or totally inaccurate recording of a result. Errors of this type are known as *gross errors*. When they occur there is no alternative to the complete repetition of the experiment.

Determinate errors.   Errors may be *determinate* in that they can be determined, e.g. by calibration, and corrections made. This type of error is also known as *bias* or *systematic* and can be traced to sources such as:

1. sampling and sample handling;
2. the analytical method;
3. faulty instrumentation;
4. mistakes by operators.

Statistical methods and carefully designed experiments may be needed to pinpoint the source. Determinate errors may be *constant* or *proportional*. It is important to appreciate the nature of these because of their significantly different effects on analytical results. Constant errors have a fixed value and would be typified by an instrumental fault or a consistent mis-reading by an operator such as assessing the colour change of an indicator in a titration. As the error has a fixed value, it follows that it will decrease in significance as the magnitude of the measurement increases. For example, an error of 0.1 cm^3 on a titration of 10 cm^3 is one of 1% but on 100 cm^3 is reduced to 0.1%. Thus by taking a larger volume of sample, the effect of the error can be reduced. This is not so for a proportional error, which increases in proportion to the magnitude of the measurement. A simple example would be the weighing of an hygroscopic powder which absorbs water in direct relation to its surface area. Figure 2.2 illustrates the effects of constant and proportional errors on experimental results. From the practical point of view of series of measurements, using increasing amounts of sample will show what type of determinate error is present.

Indeterminate errors.   *Statistical* or *random* are other terms used to specify indeterminate errors. They are generated by the small variations and

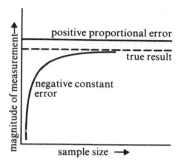

Figure 2.2 The effects of constant and proportional errors on a measurement of concentration. Source: Fifield, F.W. and Kealey, D. (1990) *Principles and Practice of Analytical Chemistry*, 3rd edition, Blackie Academic & Professional, Glasgow.

fluctuations in the various stages of a method, e.g. small uncertainties in weighing and volume measurements, small variations in temperature or humidity, fluctuations in electrical supplies and many other factors. Such errors have equal chances of being positive or negative and to some extent self-cancellation of errors occurs.

A *normal error curve* (Figure 2.3) reflects the distribution of errors in replicate measurements, and has Gaussian characteristics. Precision is determined by, and is a measurement of indeterminate error. It is clear that unlike determinate errors, indeterminate errors by their very nature cannot be determined, even in principle. The uncertainty generated with respect to

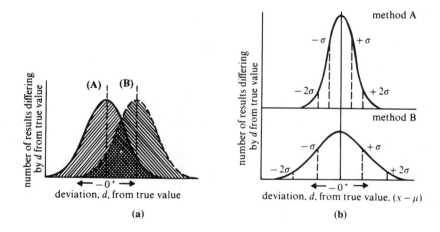

Figure 2.3 Normal error curves. (a) Curve (A) shows a normal distribution about the true value. Curve (B) shows the effect of a determinate error on the normal distribution. (b) Curves showing the result of the analysis of a sample by two methods of differing precision. Method A is the more precise or reliable. Source: Fifield, F.W. and Kealey, D. (1990) *Principles and Practice of Analytical Chemistry*, 3rd edition, Blackie Academic & Professional, Glasgow.

a particular measurement may be estimated and expressed quantitatively by assessing a replicate set of measurements. This uncertainty, usually expressed as the standard deviation or variance is a most important consideration in the interpretation of results. A significant aspect is the distinction of the uncertainty associated with the measurement from the uncertainty associated with the sampling. The two are conveniently related by their variances as shown in equation (2.8)

$$\sigma^2_{overall} = \sigma^2_{measurement} + \sigma^2_{sampling} \qquad (2.8)$$

Close control over the methodology of measurement and sampling can reduce, but not eliminate the uncertainty. Furthermore, these variations need to be distinguished from actual environmental variations which may exist with regard to an environmental constituent that is being measured. The *analysis of variance* (ANOVA) is an elegant statistical technique which can be used. It is discussed a little more fully in section 2.5.3.

It should be noted however, that indeterminate errors do not always have a strictly symmetrical distribution. For example, in titrations there is a greater tendency for inexperienced operators to 'overshoot' the end point rather than undershoot it. Careful attention to the detail of a method can usually overcome such difficulties but it is important to recognise their potential existence, and where appropriate test for them (section 2.4).

Type 1 and type 2 errors.   Type 1 and type 2 errors are concepts specifically associated with statistical decision making and are discussed in section 2.5.2.

2.4 Frequency distributions   The relationship between a particular value for a parameter and its probability has been earlier identified as its distribution. In practical circumstances a plot of the actual value against the frequency of occurrence is usually produced. Where small numbers of data are involved a *histogram* is often used but for larger data sets this translates into a smooth *distribution curve*, whose characteristics may be formulated as mathematical equations. In environmental science a number of different distributions will be met and it is important for an environmental scientist to achieve some familiarity with them although the assistance of a specialist statistician is frequently needed. An appreciation will be needed in the understanding of environmental phenomena as well as for the assessment and interpretation of analytical results. Distributions may have very different characteristics as illustrated schematically in Figure 2.4. Typical named distributions which may be met are:

1.   normal (Gaussian),
2.   Student *t*,
3.   *F*,
4.   log-normal,

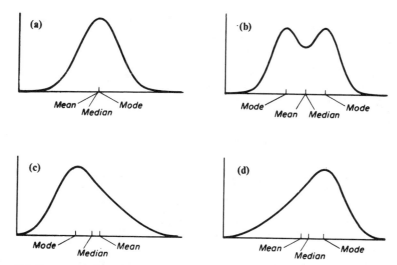

Figure 2.4 Frequency distributions showing measures of central tendency. Values of the variable are along the abscissa (horizontal axis) and the frequencies are along the ordinate (vertical axis). Distributions (a) and (b) are symmetrical, (c) is positively skewed and (d) is negatively skewed. Distributions (a), (c) and (d) are unimodal, and distribution (b) is bimodal. Source: Zar, J.H. (1984) *Biostatistical Analysis*, 2nd edition, Prentice Hall, Englewood Cliffs, NJ.

5. binomial,
6. Poisson,
7. chi-squared.

2.4.1 Normal and Student t distributions

Of particular importance in dealing with analytical data and methods are the *normal* (Gaussian) and *Student t* distributions. It is generally assumed that replicate analytical measurements follow a normal distribution although significant exceptions do exist (section 2.3.1). This follows from the indeterminate errors in the measurements which usually have equal chances of being positive or negative. The normal distribution curve is expressed by equation (2.9):

$$y = \frac{\exp[-(x - \mu)^2/2\sigma^2]}{\sigma\sqrt{2\pi}} \qquad (2.9)$$

where μ is the mean, σ is the standard deviation and π has its usual value. It will be seen that the *peakedness* or *kurtosis* of the curve is determined by σ. A typical normal distribution is illustrated by the normal error curve in Figure 2.3 and again in Figure 2.4.

It is often impractical to repeat a measurement more than a few times, i.e. five or less. In these circumstances the data are more correctly part of a

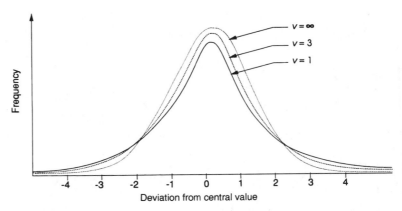

Figure 2.5 The t distribution for various degrees of freedom, ν. For $\nu = \infty$, the t distribution is identical to the normal distribution.

t distribution, illustrated in Figure 2.5. The key difference from the normal distribution is the tendency for larger deviations from the mean to occur. σ and t are statistics that appear frequently in methods for the assessment of results.

2.4.2 Other distributions

Space does not allow for accounts of the intimacies of other distributions and more specialist sources will need to be consulted where this becomes necessary. Much of the routine treatment of analytical results can be managed in terms of the normal and t distributions although statistics such as F and chi-squared (χ^2) are used from time to time.

2.5 Assessment and interpretation of analytical results

2.5.1 Introduction

Many varied and often sophisticated techniques exist for the assessment and interpretation of analytical data. With the development and ready availability of microcomputers these techniques are accessible on a routine and relatively labour free basis. The rather ill-defined term, *chemometrics*, has been introduced to describe collectively the more advanced ones. If full advantage is to be taken of the techniques available, two fundamental points must be borne in mind. Firstly, as has been emphasised previously, the measurement programme must be carefully and properly designed and executed to fulfil the aims of the investigation. Secondly, the data obtained must be as reliable as possible and shown to be so prior to drawing inferences or firm conclusions from it. The second of these aspects is the main concern of this section. An informed, structured and disciplined approach is needed.

2.5.2   Data reliability

It has been intimated previously that analytical results usually follow normal statistics. If this is in any doubt then it must be checked. A commonly used test is the chi-squared test, in which the χ^2 statistic is calculated and compared with tabulated critical values at a nominated confidence level. If the critical value is exceeded by the experimental value, then the data cannot be regarded as following a normal distribution and non-parametric methods may have to be used. The computation of χ^2 is rather complex and is not usefully illustrated here but the test should be readily available in statistical computer programmes. Further explanation can be found in the recommended further reading texts.

Outliers. Where a normal distribution exists, from time to time results which are significantly higher or lower than the mean will be obtained. For a large data set, with positive and negative deviations being equally likely one will be balanced by another, leaving the mean as a good measure of the central tendency of the results. For small data sets, however, one significantly different result or *outlier* may produce an unacceptable perturbation in the value of the mean. An early step in the examination of a set of results for their reliability is to look for, and where appropriate, reject outliers. It may be that the outlier is so vastly different from the rest of the set, perhaps as a result of a gross error, that it obviously should be rejected. On the other hand, as possible outliers lie closer and closer to the main set, the decision needs to be taken on more quantitative grounds, at a specified level of confidence. A number of methods for assessing outliers exist, but the simplest and most commonly used is based upon the use of the *rejection quotient* (Q) and is known as the *Q-test*. For a set of data, arranged in increasing order of magnitude, the potential outlier will be the highest or the lowest value:

$$x_1 \, x_2 \, x_3 \, x_4 \ldots x_n: \quad Q = \frac{x_2 - x_1}{x_n - x_1} \quad \text{or} \quad \frac{x_n - x_{n-1}}{x_n - x_1} \tag{2.10}$$

The experimental value of Q calculated in this way is then compared with tabulated critical values for a specified level of confidence. Ninety percent confidence is typically selected and the critical values are given in Table 2.1. If the experimental value exceeds the critical value, the datum is rejected. The application of the Q-test is exemplified below.

Table 2.1 Critical values of Q at the 90% confidence level

Number of results	Q_{crit} (90% confidence)
2	–
3	0.94
4	0.76
5	0.64
6	0.56
7	0.51
8	0.47
9	0.44
10	0.41

Example 2.1.   Five replicate measurements of an analyte gave the following results which have been arranged in rank order.

$$1.13, \quad 1.14, \quad 1.17, \quad 1.19, \quad 1.29 \, \%$$

On inspection it seems that 1.29 may be an outlier. Q is calculated from equation (2.10)

$$Q = \frac{1.29 - 1.19}{1.29 - 1.13} = 0.625$$

Q critical (90% confidence for five results) is 0.54 from Table 2.1. As this is not exceeded by the experimental value, the questioned result is retained.

When decisions over the inclusion or exclusion of data are being made on a probability basis, an incorrect decision will always be possible. Thus data which should rightly be included may be rejected in what is called a *type 1 error*. Similarly, data which should be excluded may be retained in a *type 2 error*. Changing the confidence levels will change the balance of probability of the two types of error, i.e. by reducing the possibility of type 1 errors, the possibility of type 2 is increased. A 90% level of confidence for including all relevant data is often selected as a compromise.

2.5.3   Precision comparisons

A frequent requirement is to assess and compare the precisions of different data sets. This may be required in the comparison of one method of analysis with another or the performance of one laboratory with another. When a number of factors contribute to the overall variability, it may be important to identify which one is the least precise. For these purposes the use of the statistic F is valuable. F is the ratio of the variances of the two sets of results as expressed in equation (2.11):

$$F = \frac{\sigma_x^2}{\sigma_y^2} \tag{2.11}$$

For convenience, the largest variance is used as the numerator ensuring that F-values are always greater than 1. The *F-test* is simple to apply, with the experimental value of F being compared with tabulated critical values at a specified confidence level (Table 2.2). If the experimental value exceeds the tabulated value then the two precisions are deemed to be different at that level of confidence. When using the table of critical values, $(n - 1)$ degrees of freedom for both numerator and denominator should be employed. Example 2.2 illustrates the application of the simple F-test.

Table 2.2 Critical values for F at the 5% level

Degrees of freedom (denominator)	Degrees of freedom (numerator)						
	3	4	5	6	12	20	∞
3	9.28	9.12	9.01	8.94	8.74	8.64	8.53
4	6.59	6.39	6.26	6.16	5.91	5.80	5.63
5	5.41	5.19	5.05	4.95	4.68	4.56	4.36
6	4.76	4.53	4.39	4.28	4.00	3.87	3.67
12	3.49	3.26	3.11	3.00	2.69	2.54	2.30
20	3.10	2.87	2.71	2.60	2.28	2.12	1.84
∞	2.60	2.37	2.21	2.10	1.75	1.57	1.00

Note: At the 5% level, the critical value will be exceeded by the ratio of variances of samples from the same population about 1 in 20 times on a probability basis alone.

Example 2.2   A second laboratory analysed the same sample for which results are reported in Example 2.1, using the same method. They obtained the following results.

$$1.11, \quad 1.12, \quad 1.14, \quad 1.19, \quad 1.25 \quad 1.34\,\%$$

No result was considered to be an outlier. Precisions were compared by means of the F-test (equation (2.11)):

$$F = \frac{(0.0893)^2}{(0.0639)^2} = 1.95$$

For 5 degrees of freedom for the numerator and 4 for the denominator, F, critical from Table 2.2 is 6.26. As this is not exceeded by F, experimental, the precisions are deemed to be similar.

A more sophisticated and very powerful technique based upon the use of F is the *analysis of variance* (ANOVA). It constitutes an elegant method of apportioning and comparing sources of variability. A simple example (one-way ANOVA) would be a situation where replicate samples, i.e. of soil had been taken to get a good estimate of an environmental component. If each sample were then subjected to replicate analysis a good estimate of the concentration of the component at the sampling site could be obtained from the overall mean of these results. If the overall variability is high, it is important to know the contributions to variability from differences between samples on the one hand, and from differences between replicate measurements on the other. In terms of ANOVA these would be assessed as *between batch* and *within batch* variations, respectively. More specialist texts need to be consulted for a fuller account. The computational power of microcomputers makes the results of this valuable technique readily available.

2.5.4   *The assessment and comparison of means*

It has been established earlier that the mean is usually the best measure of the central tendency of a replicate set of results. As such, it is important to have methods available for the assessment of experimental means, especially for those of small data sets which exhibit a t distribution rather than a normal one. There are three particular requirements to consider:

1. the relationship between the experimental mean for the data set and the true mean of the population from which it is drawn;
2. comparisons between experimental means and target values or accepted results;
3. comparisons between experimental means.

The confidence interval about a mean. An effective way of relating an experimental mean to the true mean is to define a *confidence interval* around it within which there is a specified confidence of finding the true mean, e.g. 90% The confidence interval is defined in equation (2.12):

$$\bar{x} \pm t\sigma_{n-1}/\sqrt{n} \qquad\qquad (2.12)$$

where $\bar{x}$ is the experimental mean, t is a statistical factor derived from the normal error curve and tabulated at different levels of confidence in Table 2.3.

Table 2.3   Values of t for various levels of confidence

Degrees of freedom	Confidence level (%)				
	80	90	95	99	99.9
1	3.08	6.31	12.7	63.7	637
2	1.89	2.92	4.30	9.92	31.6
3	1.64	2.35	3.18	5.84	12.9
4	1.53	2.13	2.78	4.60	8.60
5	1.48	2.02	2.57	4.03	6.86
6	1.44	1.94	2.45	3.71	5.96
7	1.42	1.90	2.36	3.50	5.40
8	1.40	1.86	2.31	3.36	5.04
9	1.38	1.83	2.26	3.25	4.78
10	1.37	1.81	2.23	3.17	4.59
11	1.36	1.80	2.20	3.11	4.44
12	1.36	1.78	2.18	3.06	4.32
13	1.35	1.77	2.16	3.01	4.22
14	1.34	1.76	2.14	2.98	4.14
∞	1.29	1.64	1.96	2.58	3.29

Example 2.3. The calculation of a confidence interval (90%) can be illustrated using the data of Example 2.1, i.e.

$$1.18 \pm \frac{2.13 \times 6.39 \times 10^{-2}}{5}$$

$$\pm 0.08$$

t values are found by looking in the column for the appropriate confidence level at $(n - 1)$ degrees of freedom.

Comparison of a mean with an accepted value. Typical circumstances in which this type of comparison is needed are those where a method is being validated by the analysis of a reference sample, or where a substance, whose concentration must be controlled, is being monitored. A version of the t-test is employed which indicates whether the numerical difference between the experimental mean and the target value is real or whether it is just a function of the inherent variability of the measurements. A value of t is computed using equation (2.13) which is then compared with the tabulated values in order to assess the significance of the difference between the mean $\bar{x}$, and the target value μ. Again $(n - 1)$ degrees of freedom are used.

$$t = [(\bar{x} - \mu)/\sigma_{n-1}]\sqrt{n} \tag{2.13}$$

Example 2.4. It is convenient again to use the data in Example 2.1. If the mean of 1.18 is compared with a target value of 1.25, firstly, t is evaluated using equation (2.13), i.e.

$$t = \frac{1.25 - 1.18}{0.0639} \sqrt{5}$$

$$= 2.45$$

From Table 2.3 it will be seen that this experimental value of t exceeds the critical value for 4 degrees of freedom at the 90% confidence level. Thus there is greater than 90% confidence so that the difference is real and the method of analysis would appear to be giving low results.

Comparison of two experimental means. The third aspect concerning the assessment of means arises when two experimental means need to be compared, for example in the comparison of two methods of analysis used on one sample, or the results of the analysis of two different samples analysed by the same method. In this case a value for t is obtained from equation (2.14), and used to assess the numerical difference in the same way as has been demonstrated in Example 2.4.

$$t = [(\bar{x} - \bar{y})/\sigma_p]\sqrt{mn/(m + n)} \tag{2.14}$$

where $\bar{x}$ is the mean of n results in the first set and $\bar{y}$ of m in the second. σ_p is the pooled standard deviation which can be evaluated from equation (2.6). It is essential to note that this version of the t-test can only be used if the F-test has been previously applied to show that the precisions of the two data sets are similar.

Example 2.5. A further replicate set of data was obtained on the same sample and by the same method as in Example 2.1, but by a different operator. The new operator obtained a mean of 1.24% from 6 valid data. It is necessary to know the significance of the difference between the two means. The standard deviation for the data reported by the second operator was 0.0645 which is not significantly different from that of the first operator.

Firstly, the pooled standard deviation is calculated from equation (2.6):

$$\sigma_p^2 = \frac{(5 - 1) \times 0.0639^2 + (6 - 1) \times 0.0645^2}{(5 + 6 - 2)}$$

$$\sigma_p = 0.0642$$

Secondly, calculate t from equation (2.14):

$$t = \frac{(1.24 - 1.18)}{0.0642} \sqrt{\frac{6.5}{6 + 5}} = 6.70$$

This value of t is compared with the tabulated values for $(n + m - 2)$ degrees of freedom, and it will be seen that it exceeds the tabulated value for 99.9% confidence. Hence the difference is highly significant and the two operators have reported different results.

2.5.5   Graphical methods

Graphs are valuable and important in both the presentation and interpretation of analytical data. In the form of *control charts* they play a direct role in decision making, and as *calibration curves*, are routinely used in quantitative analysis. There are three stages in the preparation of graphs in this context.

1. Establishing the existence of an inter-relationship or *correlation* between two or more variables.
2. Determining the nature of the relationship between the variables and the range over which it can be applied. For example, calibration curves ideally have a straight line form, but often show curvature at high and low concentrations of analytes. The straight line region is often known as the *working range*.
3. Plotting the best line through the experimental points, i.e. getting the *best fit*.

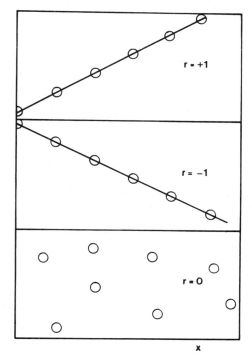

Figure 2.6 The product-moment correlation coefficient, r. Source: Miller, J.C. and Miller, J.N. (1988) *Statistics for Analytical Chemistry*, 2nd edition, Ellis Horwood, Chichester.

The mathematical techniques available for dealing with these matters are varied, and where multivariate systems are concerned, complex. It should be noted, however, that modern computers make many of these complex methods readily available for use. In this text only the relatively simple aspects of bivariate systems are considered.

Correlation. Simple straight line relationships are frequently sought and expected. In these circumstances the calculation of the product-moment correlation coefficient r (Pearson's) is often used. It is calculated from equation (2.15), and has values between 1 which represents a strong positive correlation, 0, where none is discernible and -1, for a strong negative correlation (Figure 2.6). Numerical values of r are related to confidence levels in Table 2.4.

$$r = \frac{\sum_i \{(x_i - \bar{x})(y_i - \bar{y})\}}{\sqrt{\left[\sum_i (x_i - \bar{x})^2\right]\left[\sum_i (y_i - \bar{y})^2\right]}} \qquad (2.15)$$

Table 2.4 Critical values of the correlation coefficient r for different confidence levels and different numbers of data.

n	95%	99%	n	95%	99%
5	0.75	0.87	18	0.44	0.56
6	0.71	0.83	20	0.42	0.54
7	0.67	0.80	25	0.38	0.49
8	0.63	0.77	30	0.35	0.45
9	0.60	0.74	40	0.30	0.39
10	0.58	0.71	50	0.27	0.35
12	0.53	0.66	60	0.25	0.33
14	0.50	0.62	80	0.22	0.23
16	0.47	0.59	100	0.20	0.25

The use of r without a visual inspection of the raw data can lead to fallacious conclusions, a point that is well illustrated in Figure 2.7. The use of scatter plots alongside correlation coefficients is strongly advised.

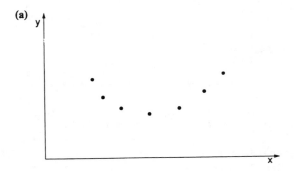

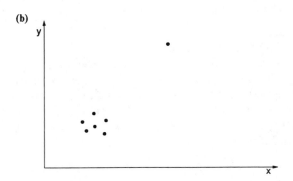

Figure 2.7 Illustration of the risks of using a correlation coefficient without inspection of the data. (a) A relationship clearly exists despite $r = 0$ and (b) a relationship is unlikely despite $r = 0.6$.

Graph plotting.   The process of fitting the best line to experimental data is often known as regression Where a straight line relationship has been established, the method of least squares is widely used. Equation (2.16) represents a straight line:

$$y = bx + a \qquad (2.16)$$

The values of a, the intercept on the y-axis, and b the slope, can be calculated from equations (2.17) and (2.18), respectively:

$$a = \bar{y} - b\bar{x} \qquad (2.17)$$

$$b = \frac{\sum_i \{(x_i - \bar{x})(y_i - \bar{y})\}}{\sum_i (x_i - \bar{x})^2} \qquad (2.18)$$

Where relationships are not straight lines, polynomial curve fitting routines are used. It remains important to have some idea of the form of the relationship so that the more convoluted patterns produced by computers can be sensibly discarded.

Control charts.   Useful forms of graphs employed extensively in some areas of analytical science are *control charts*. When analysis is being used to monitor a particular parameter in a system, it is important to know to what extent the parameter is changing in value and as closely as possible if a significant change has begun to occur. It may also be important to establish the cause of a change and when and what corrective action is needed. Two types of control chart are in widespread use.

The first type is the *Shewhart chart*. These are employed in pairs, known as *averages* and *ranges charts*. In the former, the averages of small runs of results, typically 5, are plotted against the sequential number of the observation (Figure 2.8a). Any deviation from the expected, or target value, can then be observed. In order to quantify the confidence with which a deviation is being seen, the chart is marked with warning and action limits, usually defined by $\pm 2\sigma\sqrt{n}$ and $\pm 3\sigma\sqrt{n}$ corresponding to 95% and 99.7%, respectively. n refers to the number of results in the run.

A ranges chart is constructed in a similar way by plotting the ranges of the runs of results against observation number (Figure 2.8b). Shewhart charts are often used in the quality assurance of analytical results where reference samples are measured at regular intervals in order to monitor the performance of an analytical method. Movement of the average may indicate that a bias is developing in the method, whereas an increase in range indicates loss of precision, perhaps linked to wear on a device such as an auto-injector for a chromatograph.

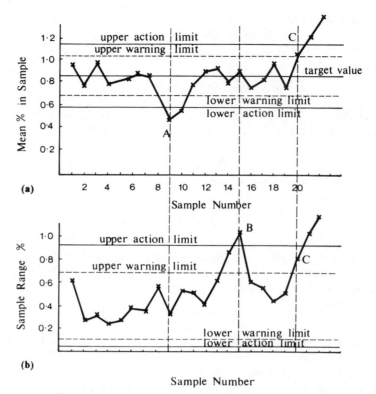

Figure 2.8 A typical pair of Shewhart charts. (a) Averages chart and (b) ranges chart. Point A shows a lack of control of averages only, point B of ranges only and point C of both together. Source: Fifield, F.W. and Kealey, D. (1990) *Principles and Practice of Analytical Chemistry*, 3rd edition, Blackie Academic & Professional, Glasgow.

A second form of control chart which has valuable uses is the *cusum chart*. In constructing this, the *cumulative sum* of the deviations of the observations from the target value is continuously computed and plotted as a function of the observation number. When the system average remains stable, positive and negative deviations will tend to cancel each other, and the cusum plot will run horizontally close to zero. If the system average begins to change, the plot will move increasingly above or below the zero line (Figure 2.9). The deviation will become apparent quickly and this rapid response is a feature of cusum charts and their use. In order to provide confidence levels for discerning a deviation, a *v-mask* can be used (Figure 2.9). When the points fall between the arms of the v the system is under control, but when they fall outside it is out of control. Clearly the values of θ and d need to be defined to provide appropriate confidence levels. In its simplest form a v-mask is engraved on transparent plastic. An example of the use of a cusum plot would be in monitoring the concentration of an effluent from an industrial plant or the quality of the water in a river.

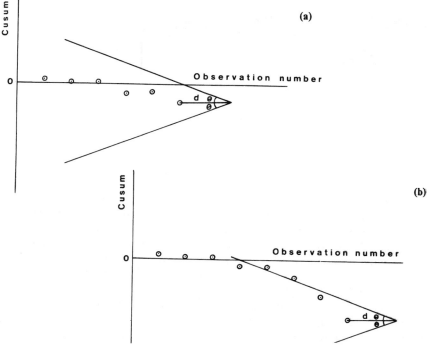

Figure 2.9 Illustration of cusum charts and v-masks: (a) system under control; (b) system going out of control. Source: Miller, J.C. and Miller, J.N. (1988) *Statistics for Analytical Chemistry*, 2nd edition, Ellis Horwood, Chichester.

2.5.6 Detection limits

The concept of the *detection limit* (DL) or *limit of detection* (LOD) has already been introduced in section 1.2. It is not difficult to appreciate the idea of the minimum working limit for a method of analysis. Defining it in a quantitive manner is a more complex matter. In essence, the requirement is to define the point at which a signal from an analyte is distinguishable and measurable above the background signal. This has to be done taking into account the inherent variability associated with the measurement. The standard deviation is extensively used to quantify this variation. If a good estimate of the standard deviation of the background is available, a point 2σ or 3σ above mean background ($\bar{x}$) can be used with confidence levels of 90% and 95.5% respectively. Where the number of data available is small, a more sophisticated expression, equation (2.16), is appropiately used. In some other circumstances a *determination limit* defined by 6σ, may be quoted.

$$DL = 2\bar{x}_B + 4.66\sigma \qquad \text{(95\% confidence)} \qquad (2.16)$$

Müller, J.C. and Müller, J.N. (1994) *Statistics for Analytical Chemistry*, 3rd edition, Ellis Horwood, Chichester, UK.

Zar, J.H. (1984) *Biostatistical Analysis*, Prentice Hall, Englewood Cliffs, NJ.

Further reading

3 Chemical principles

P.J. Haines

3.1 Introduction

In order to understand fully the chemical nature of the analyses we have to conduct, and also the chemical nature of the processes which are involved in the environment, it is necessary to consider the basic principles of chemistry. These have evolved as a result of observations, experiments and theories made since the Egyptians started their empirical technological processes to make better metals and alloys, to make glass, and to use dyes, soap and medical products.

The ancients knew about nine elements: carbon, copper, gold, iron, lead, mercury, silver, sulphur and tin, all of which may be found naturally as the elements. Chemists experimenting with minerals and organic materials both improved the processes of separation such as filtration and distillation which we still use today and also increased the number of elements which were known and the number and nature of the compounds which they formed.

Although the ancient Greeks used the term 'atom' to denote the small particles of which matter was made, it was the work of Boyle, Black and Dalton and others in the 18th and 19th centuries which showed the nature of elements and the differences between them.

The Law of Constant Composition states that "No matter how a compound is formed, it contains the same elements in the same proportions by weight".

John Dalton in 1803 propounded the Law of Multiple Proportions that "When two elements A and B combine to form more than one compound, the different weights of A which combine with a fixed weight of B are in simple proportion". By measuring experimentally the combining weights of elements, Dalton was enabled to propose his Atomic Theory:

1. All elements are made up of atoms which are homogeneous and indivisible.
2. The atoms of a given element are all alike, but different from those of any other element.
3. When elements combine to form compounds, their atoms join together in simple small whole numbers to form 'molecules'.

While modern thinking has contradicted almost all of these proposals, it is doubtful whether such rapid progress in chemistry could have been made without Dalton's ideas.

In analytical chemistry, the use of distinct atomic masses, now based on the International Union of Pure and Applied Chemistry (IUPAC) recommendations of 1973 and related to the ^{12}C isotope having an atomic mass of 12 (i.e. $A_r(^{12}C) = 12$), is a direct consequence of the second proposal, and is most useful in every type of method.

The 'splitting' of the atom into smaller particles was started with the discovery of X-rays, cathode rays and of the electron, all produced by electric discharges through gases. Thomson's measurement of the charge/mass ratio of the electron and the measurement of electronic charge by Millikan provided a great stimulus to the theoretical understanding of the electrical nature of matter.

The discovery and study of radioactivity by Becquerel and the Curies in the 1890s demonstrated the reactivity and stability of the atomic nucleus. The production of α-, β- and γ-rays by radioactive processes has proved to be both a boon and a bane! The electrical properties of these 'emanations' and analyses of the radioactive decay sequence have helped with many investigations. Nuclear reactions, however, have also contributed to environmental problems.

The need for a complementary positive particle to give neutral atoms with electrons was solved with the discovery of the proton, again from discharge experiments. The fact that the same element could have atoms of differing masses but with the same chemical properties led Soddy to propose that isotopes existed, that is atoms of the same element with different masses, for example, lead gives three stable isotopes, depending on its radioactive source:

From uranium, the lead isotope of mass 206 is obtained, but thorium gives Pb-208, and other sources can produce Pb-207. In contrast to Dalton's ideas, elementary atoms *can* be different! The reason for these differences was shown eventually to be due to the different numbers of neutrons present in each nucleus.

Rutherford's work on the scattering of α-particles by thin metallic films suggested that atoms had a small dense core, or nucleus, with a positive charge and a mass, A, made up of the total masses of protons and neutrons contained. The nucleus must be surrounded by a number of electrons to balance the sum of protonic charges (Table 3.1).

Table 3.1 Properties of fundamental particles

Particle	Symbol	Charge/C	Mass/amu
Proton	^1p	$+1.602 \times 10^{-19}$	1.00757
Neutron	^1n	0	1.00893
Electron	e	-1.602×10^{-19}	0.0005486

Measurement of the masses of atoms was made easier with the construction of the mass spectrometer (see chapters 7 and 8) which showed that many elements consisted of mixtures of isotopes in characteristic proportions. One such element is chlorine which naturally has 75.8% Cl-35 and 24.2% Cl-37, giving a relative atomic mass $A_r(Cl) = 35.45$.

3.1.1 Periodicity

From early experimental work, it was clear that elements formed 'families' with similar properties, but with different atomic masses. Thus, calcium, strontium and barium are all found together as carbonates of formula MCO_3, copper, silver and gold are all found as metals, and chlorine, bromine and iodine are all reactive volatile elements, readily forming the halide salts, such as NaCl, NaBr and NaI.

When the elements were arranged according to their atomic masses (A_r), a periodic behaviour was noted by Newlands, Mendeleyev and others, and a Periodic Table was constructed by Mendeleyev, which allowed predictions of new elements, of radioactive decay series and of similarities in chemical behaviour. The discovery of the noble gases added a new group to the periodic table.

Some difficulties still existed, since certain elements were clearly out of sequence when arranged by mass. For example, argon has an $A_r = 39.95$, and potassium has an $A_r = 39.10$, but they clearly belong chemically with the noble gases and the alkali metals, respectively. The periodic table (Figure 3.1) groups elements of the same 'combining power' or 'valency' as illustrated above.

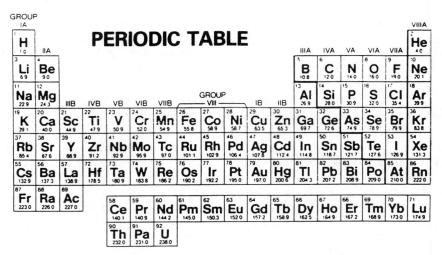

Figure 3.1 Periodic table in the 'long' form. Each box shows the symbol for the element, its atomic number (top left) and relative atomic mass (bottom centre).

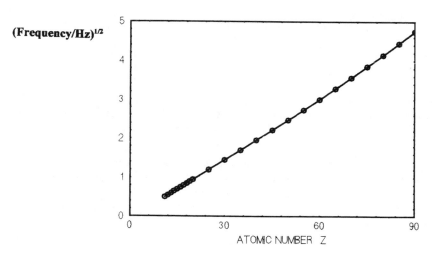

Figure 3.2 Plot of (frequency)$^{1/2}$ against atomic number (Moseley Law Plot).

The concept of atomic number (Z) was suggested to denote the number of the element in the periodic classification, and represents the number of protons (or electrons) in the atom. Moseley had shown from plots of the characteristic X-ray lines that a relationship existed between the frequency of a characteristic X-ray line and the atomic number (Figure 3.2):

$$\nu^{1/2} = a(Z - b) \qquad (3.1)$$

This gave an experimental basis for using atomic number, and provides a means of identifying elements by their X-ray spectra.

3.1.2 Atomic spectra

Experimental observation of the spectra of elements shows that they generally give discrete line spectra when excited in a flame or an arc. When a continuous light source, such as an incandescent filament lamp, is shone through the element, it absorbs discrete lines. Without further elaboration, this gives us a method of analysing for elements, as discussed in chapter 7. This behaviour suggests that atoms may only have certain discrete energy levels and can increase or decrease their energy by jumps between these definite levels. If the difference in energy is ΔE between the upper energy level, E_U and the lower energy level E_L then the frequency, ν, of radiation absorbed or emitted is given by:

$$\Delta E = E_U - E_L = h\nu \qquad (3.2)$$

where h is Planck's constant, 6.626×10^{-34} J s.

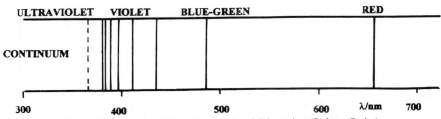

Figure 3.3 Spectrum of hydrogen in the visible region (Balmer Series).

It is also a most important guide to atomic structure. The simplest of atoms, hydrogen, consists of a single proton and a single electron. It also produces the simplest spectrum shown in Figure 3.3.

The lines may be allocated into series whose frequency is accurately given by formula:

$$\nu = R \times c \times (1/n_2^2 - 1/n_1^2) \tag{3.3}$$

where R is called the Rydberg constant, 1.097×10^7 m^{-1}; c is the velocity of light, 2.998×10^8 m s^{-1}; and n is called a quantum number. For the Balmer series, which occurs in the visible region of the spectrum, $n_2 = 2$, and n_1 can take values of 3, 4, 5 etc. Other series have $n_2 = 1$ (Lyman, far ultraviolet), 3 (Paschen, far infrared), 4 (Brackett) and 5 (Pfund).

The explanation of these series requires the formulation of the quantum theory and of the wave nature of the electron, which are very large subjects indeed. For further information the appropriate sections in the recommended textbooks should be consulted. A very brief description of the principles is given below.

3.1.3   Nature of the electron

To explain the behaviour of an electron it is sometimes convenient to describe it as a wave, for example when considering electron diffraction, and sometimes as a particle, for example in classical charge/mass experiments. In reality, it may have the properties of both wave and particle.

The Heisenberg Uncertainty Principle places a limit on the accuracy with which either the velocity (or momentum) and the position of a particle may be known. It may be stated that the product of uncertainty in position, Δx, and the uncertainty in momentum, Δp, may not be less than the quantity $h/2\pi$, where h is Planck's constant, 6.626×10^{-34} J s.

$$\Delta p \cdot \Delta x \geq h/2\pi \tag{3.4}$$

These concepts gave rise to the *wave equation* describing the hydrogen atom. The wave functions, ψ, are solutions of this equation, and $\psi_{(x, y, z)}^2$ gives the probability of finding the electron at any point x, y, z.

$$-\frac{h^2}{8\pi^2 m}\nabla^2\psi - \frac{e^2}{4\pi\varepsilon_0 r}\psi = E\psi \qquad (3.5)$$

In this expression, e and m are the charge and mass of the electron; r is the distance from the nucleus; ε_0 is the permittivity of free space; and E is the total energy.

It is referred to as the time-independent Schrodinger Wave Equation. For other atoms the equation and its solutions are much more complex.

This wave equation has a number of allowed solutions. Each solution corresponds to an allowed energy level of the electron, and an allowed probability distribution. For example, the lowest energy, or *ground state* is

$$\psi_1 = (1/(\pi a_0^3)^{1/2})\exp(-r/a_0) \qquad (3.6)$$

where $a_0 = 52.9$ pm, the atomic length unit.

Other solutions to the wave equation are found to be restricted by quantum numbers. These numbers, which arose firstly from the older Bohr theory explaining atomic spectra, are now found to arise automatically as a requirement in solving the Schrodinger equation. To each stationary energy state (or eigenvalue) there belongs a set of integer quantum numbers, n, l and m (Table 3.2).

Table 3.2 The quantum numbers n, l and m

Quantum number	Name	Allowed values
n	Principal	1, 2, 3, 4, 5, . . .
l	Azimuthal	$(n-1)$, $(n-2)$, . . . 2, 1, 0
m	Magnetic	$+1$, $+(l-1)$, $+(l-2)$, . . . , 1, 0, -1, . . . $-(l-2)$, $-(l-1)$, -1

Thus, when n is 1, l and m can only be 0, and only one value is possible.

When $n = 3$ and $l = 2$, m may be 2, 1, 0, -1, -2. There are thus five different orbitals. When $n = 3$, $l = 1$, there are three, and when $n = 3$, only one. These different orbitals are given symbols derived from the original observations of atomic spectra (Table 3.3):

s (sharp)   for $l = 0$
p (principal)   for $l = 1$
d (diffuse)   for $l = 2$
f (fundamental) for $l = 3$ and so on.

There is also a fourth quantum number, describing the spin of each electron in the atom. This is the spin quantum number, m_S, which may take values of $\pm 1/2$ only.

To understand the build up of the Periodic Table, we use Pauli's principle that:

No two electrons may have all four quantum numbers the same.

Table 3.3 Atomic orbitals with n values up to 4

n	l	m	Symbol	Levels
1	0	0	1s	1
2	0	0	2s	4
	1	+1, 0, −1	2p	
3	0	0	3s	9
	1	+1, 0, −1	3p	
	2	+2, +1, 0, −1, −2	3d	
4	0	0	4s	16
	1	+1, 0, −1	4p	
	2	+2, +1, 0, −1, −2	4d	
	3	3, 2, 1, 0, −1, −2, −3	4f	

and Hund's Rules that:

Electrons tend to avoid as far as possible being in the same orbital *and* electrons in orbitals of the same energy have parallel spins.

Filling the orbitals in order of increasing energy means filling the 1s orbital first, then 2s, then 2p and so on (Figure 3.4).

The energy of these orbitals is known from the measurement of X-ray and atomic spectra, and the periodicity now appears as a natural consequence of atoms having similar electronic structures, for example the noble gases having full shells:

He   2
Ne   2, 8
Ar   2, 8, 8
Kr   2, 8, 18, 8

and the halogens having 1 less than a full shell:

F   2, 7
Cl   2, 8, 7

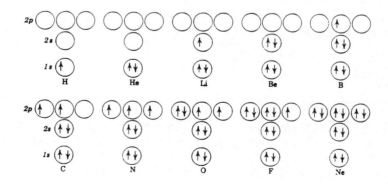

Figure 3.4 The 'Aufbau Principle': the build-up of the periodic table. Source: Alberty, R.A. and Daniels, F. (1980). *Physical Chemistry*, 5th edition, Wiley, New York.

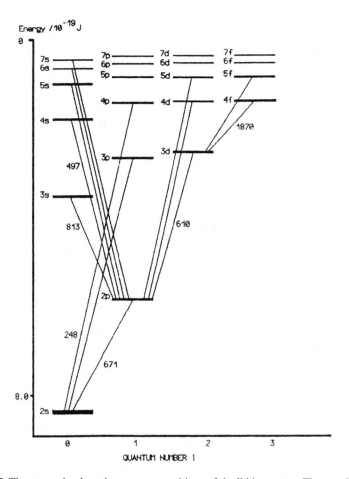

Figure 3.5 The energy levels and spectrum transitions of the lithium atom. The wavelengths in nanometres of some of the transitions are given beside the connecting lines.

    Br   2, 8, 18, 7
    I   2, 8, 18, 18, 7
    At   2, 8, 18, 32, 18, 7

The quantum theory also restricts the transitions which may take place. Although the principal quantum number, n, may change by any integer value, the quantum number, l, must change by $+1$ or -1 only.

This means that the atomic and X-ray spectra we observe involve jumps between discrete, quantised energy levels. This is shown in the spectral diagram for lithium, but with the doublet structure excluded (Figure 3.5).

3.1.4 Population of energy levels

The probability of finding a particular species in an energy level depends on several factors, particularly the absolute temperature, T. Boltzmann showed that for two single levels of energy, E_L and E_U, the population ratio should be:

$$N_U/N_L = \exp(-(E_U - E_L)/kT) \tag{3.7}$$

where $k = R/N_A = 1.380 \times 10^{-23}$ J K^{-1} and is called the Boltzmann constant.

If the levels are degenerate, that is, if g levels have the same energy, E, then the formula becomes:

$$N_U/N_L = (g_U/g_L) \cdot \exp(-(E_U - E_L)/kT) \tag{3.8}$$

This shows that at higher temperatures, the more energetic levels may be populated, and also that the population of the upper level relative to the lower falls off as the energy difference $(E_U - E_L)$ increases.

3.2 Atomic orbitals and chemical bonds

3.2.1 Ionic bonds

If sufficient energy is available, the atom may lose one (or more) of its outermost electrons. For example,

$$Na(g) - e = Na^+(g) \qquad I = 496 \text{ kJ mol}^{-1} \tag{3.9}$$

This is generally reported as the Ionisation Potential, I, in electron volts per atom, or kJ per mol (1 eV = 96.49 kJ mol^{-1}).

Conversely, the energy needed to add an electron to an atom, the Electron Affinity, E, corresponds to reactions such as

$$Cl(g) + e = Cl^-(g), \qquad E = -349 \text{ kJ mol}^{-1} \tag{3.10}$$

It would seem unlikely on energetic grounds that ionic compounds could ever be made. However, the electrostatic attractions between ions in the regular array on a crystal lattice contributes a *very* large stabilising energy, the crystal lattice energy, U:

$$Na^+(g) + Cl^-(g) = Na^+ \cdot Cl^- \text{ (solid)}, \qquad U = -764 \text{ kJ mol}^{-1} \tag{3.11}$$

This means that, despite the energy needed to sublime sodium metal, and to dissociate chlorine molecules to atoms, sodium and chlorine readily form the very stable ionic salt, sodium chloride.

In general, binary ionic compounds are formed between metals from Groups IA and IIA of the periodic table, and non-metals from Groups VI and VII. More particular rules apply with complex cations and complex anions or oxyanions.

The dissociation and separation of ions in solvents, and their reduction electrochemically provides the important techniques of electrochemical analysis discussed in chapter 10.

3.2.2   Covalent molecules

If atoms do not lose or gain electrons completely, they may still form bonds by sharing their electrons. Such bonds are called covalent bonds and require that two electrons are shared. The electrons may be contributed by each of the atoms forming the bond, or one atom may contribute *both* electrons when a dative bond is formed. The simplest way of representing these is by 'dot and cross diagrams' as shown in Figure 3.7a.

The best way of obtaining an understanding of the shapes of covalent molecules is to consider their relationship to the atomic orbitals from which they are formed. The shapes of atomic orbitals, predicted by solving the Schrodinger wave equation, are shown in Figure 3.6.

In order to combine these orbitals to join separate atoms into molecules, certain rules must be followed.

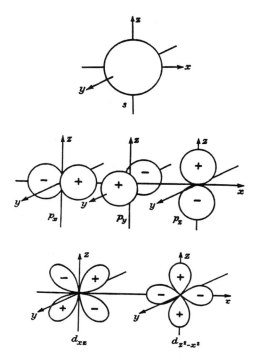

Figure 3.6 Shapes of atomic orbitals. Source: Coulson, C.A. (1952) *Valence*, Oxford University Press, Oxford.

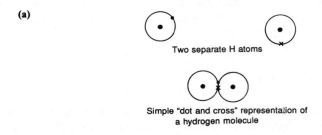

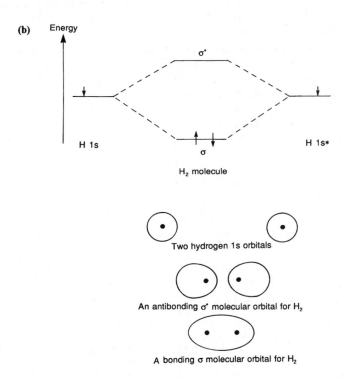

Figure 3.7 Representations of the bonding between two H atoms: (a) diagram of the hydrogen molecule; (b) combining two H ls orbitals.

1. The atomic orbitals must be of similar energy.
2. The orbitals must overlap.
3. The number of molecular orbitals formed must equal the original number of atomic orbitals.

The simplest case to consider is the hydrogen molecule ion H_2^+ (Figure 3.7).

The productive overlap in the σ_S case gives a bonding molecular orbital of lower energy, E_B, whereas the destructive overlap of the σ_S^* case produces an antibonding molecular orbital of higher energy, E_A. The fact that the overlap region lies chiefly between the atoms means that this is a σ (sigma) bond.

Let us consider a more complex case for combining p orbitals, such as the fluorine molecule (Figure 3.8). The 1s orbitals are now too far apart to take part in the bonding, but the 2s and 2p orbitals can overlap.

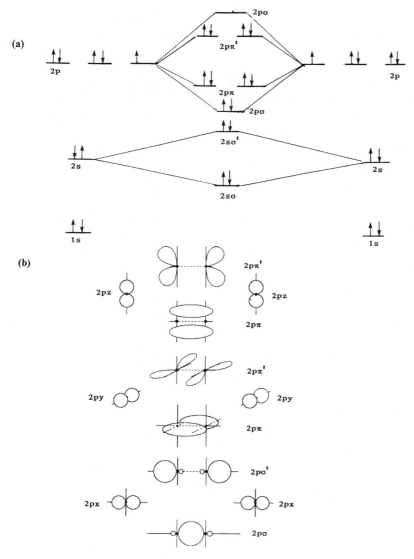

Figure 3.8 (a) The atomic orbitals of fluorine atoms and their combination to give the fluorine molecule F_2 (b) (Note: some texts place the $2p\pi$ levels lower in energy than the $2p\sigma$).

The 2s orbitals would form both σ_s and σ_s^* orbitals. The $2p_z$ orbitals on each fluorine would overlap along the line joining the atoms and form σ_p and σ_p^* orbitals, whereas the $2p_x$ and $2p_y$ would overlap to the side of the bond and form π_p and π_p^* molecular orbitals of equal energy, as shown.

Filling the orbitals with the seven electrons available from each fluorine atom, we fill both bonding and antibonding s orbitals for the 2s levels. The next two electrons fill the $2p\sigma$ level, leaving eight more, which fill both the $2p\pi$ and $2p\pi^*$ levels, leaving only the $2p\sigma^*$ without electrons. There are therefore four filled bonding and three filled antibonding levels, leaving a net bonding of one. Thus we can write the F_2 molecule with a single F–F bond.

One notable and important case is oxygen which has one less electron per atom. The energy levels are (approximately) the same, but here the $2p\pi^*$ have only one electron each, and these have the same spin and are 'unpaired'. This has two consequences. First there are two more bonding electrons, so we have a double bond joining the oxygen atoms, and, secondly, oxygen molecules are paramagnetic due to the unpaired spins (Figure 3.9).

For heteronuclear diatomics, such as NO, the atoms will have different energies and different numbers of electrons but the same principles are followed, giving the diagram shown in Figure 3.10. Note that NO has an unpaired electron, too!

Experimental verification of the energetic structure of these molecules has been provided by photoelectron spectroscopy.

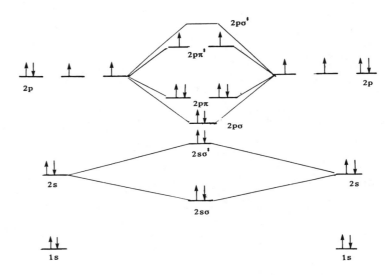

Figure 3.9 The energy levels for the oxygen atoms and for the oxygen O_2 molecule.

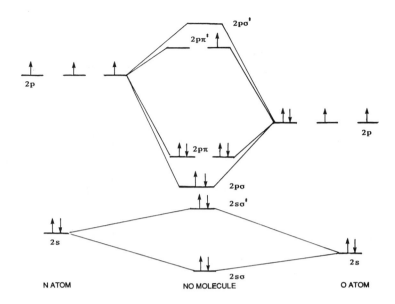

Figure 3.10 The energy levels for the nitrogen and oxygen atoms and for the NO molecule.

3.2.3 Polyatomic molecules

The extension of the molecular orbital theory to more complex molecules is possible, though certain assumptions must be made. The reader should consult theory texts for more detailed information. The shapes of molecules, and of groups within molecules, may be predicted by use of the Valence Shell Electron Pair Repulsion (or VSEPR) Theory.

Any molecule made up of single bonds may be regarded as being constructed of sets of electron pairs. Each single bond involves two electrons, and any extra electrons will remain as non-bonded or lone pairs. Only two rules are needed:

1. The pairs arrange themselves to maximise their distance apart.
2. Lone pairs repel more than bonding pairs.

The theory predicts that molecules with two bonding pairs only must be linear. With three pairs, triangular, with four, tetrahedral and so on.

Structures involving lone pairs are slightly distorted from these. Thus, ammonia, NH_3, where the central nitrogen has five bonding electrons, uses three of these to form bonds with the hydrogens which each contribute one electron to the bonding pairs. There are two non-bonding electrons left, so there are a total of four pairs and the basic structure is tetrahedral, slightly distorted by the lone pair effect.

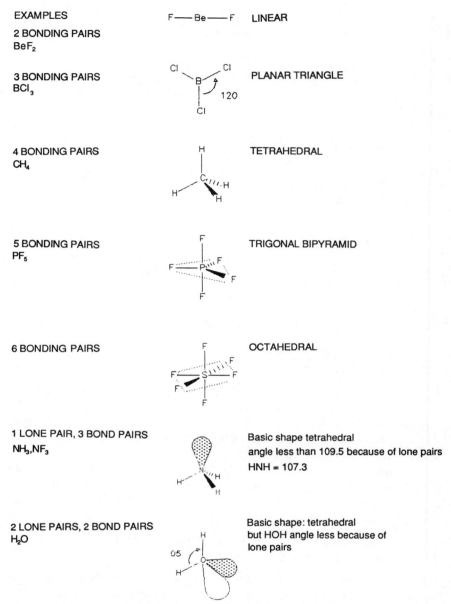

EXAMPLES

2 BONDING PAIRS
BeF$_2$

F——Be——F   LINEAR

3 BONDING PAIRS
BCl$_3$

PLANAR TRIANGLE

4 BONDING PAIRS
CH$_4$

TETRAHEDRAL

5 BONDING PAIRS
PF$_5$

TRIGONAL BIPYRAMID

6 BONDING PAIRS

OCTAHEDRAL

1 LONE PAIR, 3 BOND PAIRS
NH$_3$,NF$_3$

Basic shape tetrahedral
angle less than 109.5 because of lone pairs
HNH = 107.3

2 LONE PAIRS, 2 BOND PAIRS
H$_2$O

Basic shape: tetrahedral
but HOH angle less because of
lone pairs

Figure 3.11 Simple valence shell electron pair repulsion (VSEPR) structures.

Figure 3.11 gives some further examples, including molecules with multiple bonds which are treated as a single '*link*' as though they were just a bonding pair.

In more complex molecules, especially organic species, it is convenient to

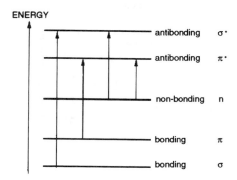

ENERGY

antibonding $\sigma^{\cdot}$

antibonding $\pi^{\cdot}$

non-bonding   n

bonding π

bonding σ

Figure 3.12 Organic electronic energy levels of molecular orbitals showing possible transitions.

consider the energy levels very simply as σ, π or n (for non-bonding) plus their complementary antibonding levels, π^* and σ^*. These generally have energies in the order shown in Figure 3.12, and transitions between the levels give rise to organic electronic spectra.

3.2.4   Metal compounds and complexes

Metal atoms may bond to other species in a variety of ways, ranging from ionic bonds to form salts or ion pairs to covalent or dative bonding to ligands to form complex ions or compounds. As an example, the metal cadmium is important as an environmental pollutant. It may be present as the ionic sulphide CdS, as the ionic species $CdBr^+$ or $CdBr_4^{2-}$, or as organo-metallic compounds like $(C_2H_5)_2 \cdot Cd$ or as many different complexes, such as $Cd(NH_3)_2Cl_2$. This ability to form many different species in natural reactions must be studied carefully and speciation is discussed fully in chapter 14.

Transition metal complexes. When we consider the periodicity of the elements, we find that there are certain elements which exist in different oxidation states, for example, iron has two series of compounds based on iron(II) or 'ferrous, Fe^{2+}' and iron(III) or 'ferric, Fe^{3+}'. This behaviour is due to the electronic structure of the atoms, especially the d-electrons and the elements are termed transition elements.

The electronic structure of the iron atom is:

Fe: ($Z = 26$): 2, 8, 8, 8,   that is $1s^2$, $2s^2$, $2p^6$, $3s^2$, $3p^6$, $3d^6$, $4s^2$

The energies of the 3d and 4s levels are almost the same, and it is relatively easy to promote electrons from one to the other. Also, it is fairly easy to

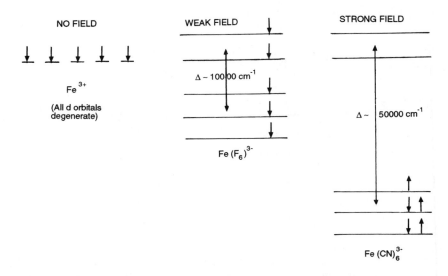

Figure 3.13 Effect of octahedral ligand field on d orbitals.

produce multiply charged ions. Fe^{2+} has two and Fe^{3+} three electrons less than the neutral iron atom. These electrons would occupy the 3d orbitals singly with parallel spins. Because of the shapes of the d orbitals (shown in Figure 3.6), introduction of coordinating groups or 'ligands', which have a field due to their electron pairs will exert a 'ligand field' on the d orbitals. This field interacts to different extents on the different d orbitals, depending on the geometry of the complex formed.

For an octahedral complex, where the ligands attach along the principal axes, x, y, z, the d_{z^2} and $d_{x^2-y^2}$ orbitals point directly at the ligands and are raised in energy, whereas the d_{xy}, d_{xz} and d_{yz} point *between* the ligands and are lowered in energy. Thus, in the presence of coordinating ligands, the energy levels of the d electrons are no longer the same, and the coordination produces a ligand field stabilisation energy (Figure 3.13).

In tetrahedral, square planar and other complexes, the splitting is different. Transitions between these d levels produce the characteristic colours of transition metals and allow the use of complexes in the colorimetric determination of metals.

3.3 Molecular energy levels

The quantum theory may be applied to any type of energy level, and the energy of each level is defined by its particular set of quantum numbers.

When single atoms are considered, the energy required to change their electronic configuration is generally very high. By contrast, when single atoms of a gas such as helium move around in space, the energy needed to raise them from one position to another, referred to as 'translation' is very small. The quantised energy levels for this type of movement are so closely spaced that they form an almost continuous band.

In a gas, the atoms or molecules move rapidly around in space, and their kinetic energy and collisions give rise to the pressure and temperature-dependent kinetic energy of a gas. The work of Boyle, Gay-Lussac and Graham and others gave us the gas laws.

For an ideal gas

$$pV = nRT = 1/3 \ Nm\langle v^2 \rangle \qquad (3.12)$$

where p is the pressure in Pascals (N m^{-2}); V the volume of the vessel in m^3; n the moles of gas present; R the molar gas constant, 8.314 J K^{-1} mol^{-1}; T the temperature in K; N the number of particles (atoms, molecules); m the mass of each particle in kg; and $\langle v^2 \rangle$ the mean square speed of the particles (m s^{-1})^2. These equations provide most useful explanations of the behaviour of gases at low pressures, although they must be modified for easily liquefied gases and for higher pressures.

The product of pressure and volume pV depends on the amount of gas present, n, and on the temperature, T, as discovered by Boyle and Charles. The average speed of molecules depends upon their mass and on temperature, and the root mean square speed is given for a gas with molar mass M by

$$\sqrt{\langle v^2 \rangle} = \sqrt{(3RT/M)} \qquad (3.13)$$

This allows us to predict the diffusion rates of gases.

Molecules have translational energy levels, but since they can now rotate about their centre of mass, and vibrate stretching or bending their bonds, there are two further sets of energy level to consider. All these energy levels are defined by quantum numbers, and contribute to the total energy of the molecule (Figure 3.14).

Translation in one dimension:

$$E_{\text{trans}} = n^2 \cdot (h^2/(8ma^2)) \qquad (3.14)$$

where a is the length of container; n the translational quantum number; and m the particle mass.

Rotation of a diatomic rotor:

$$E_{\text{rot}} = J(J + 1) \cdot (h^2/(8\pi^2 I) \qquad (3.15)$$

where I is the moment of inertia of molecule; and J the rotational quantum number.

Vibration of a diatomic molecule:

$$E_{\text{vib}} = (v + 1/2) \cdot h\nu \qquad (3.16)$$

where ν is the fundamental vibration frequency; and v the vibrational quantum number.

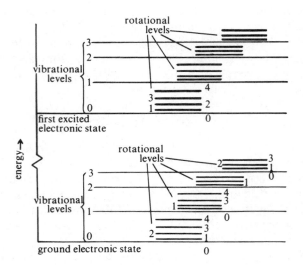

Figure 3.14 Electronic, vibrational and rotational molecular energy levels, not to scale (only the first two electronic levels are shown). Source: Fifield, F.W. and Kealey, D. (1990) *Principles and Practice of Analytical Chemistry*, 3rd edition, Blackie Academic & Professional, Glasgow.

Transitions between rotational levels occur with the absorption of energy in the radiofrequency or microwave spectral region, whereas vibrational transitions, and (rotation + vibration) transitions occur in the infrared region, and electronic, and electronic plus vibration plus rotation transitions involve energy changes in the ultraviolet and visible region.

3.3.1 Energy of assemblies of molecules

If we wish to consider the energy of a body, it is difficult to do this on an 'absolute' basis. Often what we observe is the change in energy. When we perform mechanical work on a gas, or heat it or cause it to expand, the energy change is between the initial state and the final state. To relate this to the molecular changes occurring is very difficult.

We have seen that the various types of atomic and molecular energy, electronic, vibrational, rotational and translational, may be related to molecular parameters and to definite quantum numbers. In a system with a fixed, large number of particles, such as a 1 mol of gas, with about 6×10^{23} molecules, the number of molecules occupying each energy level may be found from the Boltzmann equation which shows how the distribution varies with temperature, T.

The total thermal energy of a mol of molecules, U, above the energy they would have at 0 K. U_0 is the sum of all terms for energy, E_i, of each level and its population, N_i

$$U - U_0 = \Sigma N_i(E_i - E_0) \tag{3.17}$$

A quantity called the 'molecular partition function', z, may be used to help to work out the energies:

$$z = \Sigma g_i \exp(- (E_i - E_0)/kT) \tag{3.18}$$

where z can incorporate all the different types of energy level.

It can be shown that, for 1 mol:

$$U - U_0 = RT^2(\mathrm{d} \ln(z)/\mathrm{d}T) \tag{3.19}$$

The heat energy needed to raise the temperature of the system by 1 K is the heat capacity. If the volume of the system is held constant, this is written C_V, and is the rate of change of U with temperature. Differentiating the above equation with respect to temperature gives C_V.

Since the total energy of the system should be conserved, we should be able to relate any change in the internal energy, U, with the heat absorbed by the system, q, and the work done on the system, w, by the First Law of Thermodynamics:

$$\Delta U = q + w \tag{3.20}$$

At constant pressure, an extra amount of heat is needed to expand the system, so we have C_p, which may be related to C_V and the properties of the system. For 1 mol of an ideal gas:

$$C_p - C_V = R \tag{3.21}$$

The heat content of the system at constant pressure, called the enthalpy H and related to U by the equation

$$H = U + pV \tag{3.22}$$

where p is the pressure, and V the volume of the system.

For 1 mol of an ideal gas, $pV = RT$, so we can find H from the molecular energies and the partition function.

As stressed above, it is usually the changes in properties and energy which interest us, and so we may choose standard conditions to which these changes refer. A very useful one is the formation of compounds. If we start with the elements in their most stable form at a particular temperature and pressure, say 298 K and 1 atmosphere, then the energy needed to form them from themselves is zero! These standard conditions are often used when tabulating values.

For any reaction between elements, producing a new compound, the change in energy represents the enthalpy of formation ΔH_f of that compound. This too is generally reported under the standard conditions stated above. For example,

3.4 Enthalpies of formation and reaction

$$H_2(\text{gas, 1 atm}) + 1/2\ O_2(\text{gas, 1 atm}) = H_2O(\text{liquid, 1 atm})$$

$$\Delta H_f^{\ominus} = -285.8 \text{ kJ mol}^{-1} \tag{3.23}$$

$$C(\text{graphite, 1 atm}) + O_2(\text{gas, 1 atm}) = CO_2(\text{gas, 1 atm})$$

$$\Delta H_f^{\ominus} = -393.5 \text{ kJ mol}^{-1} \tag{3.24}$$

Since each compound in a reaction equation could, theoretically, be returned to its elements, which could then be re-reacted to for the products required, it may be deduced that, for any reaction:

$$\Delta H(\text{reaction}) = \Sigma \Delta H_f(\text{products}) - \Sigma \Delta H_f(\text{reactants}) \tag{3.25}$$

For example,

$$CH_4 + 2O_2 = CO_2 + 2H_2O$$

$$\Delta H^{\ominus} = \Delta H_f^{\ominus}(CO_2) + 2\Delta H_f^{\ominus}(H_2O) - \Delta H_f^{\ominus}(CH_4) - 2\Delta H_f^{\ominus}(O_2) \tag{3.26}$$

$$= -393.5 - 2 \times 285.8 - (-74.8) - 0$$

$$= -890.3 \text{ kJ}$$

Note that the oxygen has zero enthalpy of formation!

Enthalpy changes for reactions, for physical changes such as melting or vaporisation, and for interactions such as adsorption or solution may be measured and related to each other by this type of procedure.

3.5 Entropy and free energy

It might seem reasonable to suppose that all spontaneous changes would take place with a decrease in enthalpy, because they are going to a more stable state. We find many instances where the change occurs, but the enthalpy does not alter. Probably the most obvious example is the diffusion of gases, which is very important in environmental systems.

Suppose we have two gases, A and B, which are in a container, but held apart by a barrier. They are at the same pressure and temperature. The gases do not react together or with the container. When the barrier is removed, the gases mix spontaneously because of the motions of the molecules! It can be shown that the enthalpy does not change, so we need to suggest a different 'driving force' to explain this change.

The changes which tend to occur do so in such a way that the disorder of the system and its surroundings increases.

This is one way of stating the Second Law of Thermodynamics. A property, S, called the entropy measures this disorder, and Boltzmann proposed an equation,

$$S = k \ln W \tag{3.27}$$

where W is the number of arrangements that the particles can adopt for the specific macroscopic state.

We can also relate the entropy to the partition function and molecular energy levels. For 1 mol of an ideal gas,

$$S = (U - U_0)/T + R \ln(z/N_A) + R \tag{3.28}$$

For a change such as the melting of a solid, the entropy change is also related to the heat, q, absorbed reversibly by the system

$$\Delta S(\text{melting}) = q_p(\text{melting})/T_m = \Delta H(\text{melting})/T_m \tag{3.29}$$

It is, however, rather a nuisance to have to consider *both* the system *and* the surroundings when trying to decide whether a reaction will proceed spontaneously or not.

The Gibbs Free Energy, G, or Gibbs' Function combines the properties of H and of S into one function. It is defined as

$$G = H - TS \tag{3.30}$$

The Gibbs Free Energy may be regarded as the best indication of the spontaneity of a reaction. If the free energy of a system does not change when a small (differential) change is applied, then the system is at equilibrium. If the free energy decreases, so that the system could do work, then the reaction will occur spontaneously. If the free energy increases, then the reaction is unlikely, or going away from equilibrium. This is summarised in Figure 3.15.

Free energy may be related to all sorts of work, but is especially useful with the work done by electrical cells. If the cell has an electromotive force (emf), E, then

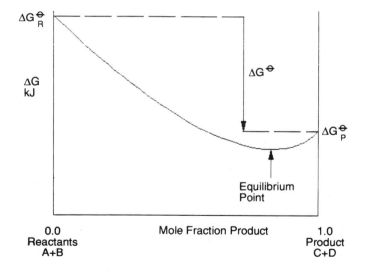

Figure 3.15 Free energy changes in a reaction and equilibrium.

$$\Delta G = -nFE \tag{3.31}$$

where n is the number of electrons transferred in the cell reaction, and F the Faraday = 1 mol of electronic charge = 96500°C.

3.6 Free energy and equilibrium

In 1864, Guldberg and Waage stated in the Law of Mass Action that "the rate at which a substance reacts is proportional to its active mass and the rate of a chemical reaction depends on the product of the active masses of the reactants".

For most reactions, there exists a set of conditions where a final stage is reached, and no more change will take place (unless the conditions are altered!). This condition of balance between the forward and backward tendencies is the position of equilibrium and will depend on the 'active masses' of the reactants and products.

It can be shown that the free energy per mole (or the chemical potential) of a compound is given by

$$G_{m, A} = G^{\ominus}_{m, A} + RT \ln a_A \tag{3.32}$$

where a is called the activity of A and is the pressure or concentration corrected for non-ideality. When $a = 1$, the second term becomes zero, and thus the standard free energy per mole is $G^{\ominus}_m$.

Consider a general change,

$$A + B \rightleftharpoons C + D \tag{3.33}$$

and let all reactants be ideal gases, for simplicity. We can write the free energy change, ΔG as

$$\Delta G = G_C + G_D - G_A = G_B \tag{3.34}$$

$$= \Delta G^{\ominus} + RT \ln(p_C \cdot p_D / p_A \cdot p_B) \tag{3.35}$$

When equilibrium is reached, the pressures of A, B, C and D will have altered to their equilibrium values, and ΔG must be zero, so that:

$$\Delta G^{\ominus} = -RT \ln \{p_C \cdot p_D / p_A \cdot p_B\}_{eq} \tag{3.36}$$

$$= -RT \ln K_P \tag{3.37}$$

This is called the Standard Reaction Isotherm Equation and K_p is called the equilibrium constant in terms of pressures for this reaction. $\Delta G^{\ominus}$ is the standard free energy change for the reaction and, like the standard enthalpy change may be calculated from tabulated $\Delta G^{\ominus}_f$ values.

Similar arguments can be used to derive relationships in terms of concentrations, c or most generally in terms of activities, a. For example, for the dissociation of a weak acid such as ethanoic acid (acetic acid),

$$HO_2C \cdot CH_3 \rightleftharpoons H^+ + CH_3 \cdot CO_2^- \qquad (3.38)$$

$$K_a = a_{H^+} \cdot a_{CH_3CO_2^-}/a_{HO_2C \cdot CH_3} = 1.8 \times 10^{-5} \qquad (3.39)$$

$$\Delta G^{\ominus} = - RT \ln K_a = + 27.07 \text{ kJ mol}^{-1} \qquad (3.40)$$

The value of K is fixed for a particular temperature, but the variation of equilibrium constants with temperature is important for both physical processes like solubility, and for chemical reactions. The variation of K with temperature depends upon the standard enthalpy change, $\Delta H^{\ominus}$:

3.7 The effects of temperature

$$\ln K = - \Delta H^{\ominus}/RT + \text{constant} \qquad (3.41)$$

This is called the van't Hoff equation or the reaction isochore.

Some examples of the types of equilibrium which may be involved in chemistry, especially environmental chemistry should now be considered.

3.8 Application to equilibria

3.8.1 Phase equilibria

Matter may exist in different states, for example, gaseous, liquid or solid. There may be several different crystalline forms of a solid, for example silica, SiO_2, has at least three different crystalline forms. Each different homogeneous state, limited by its boundary, is called a phase. Equilibria may exist between phases of the same material or of different chemical components. For example, the vaporisation of a pure liquid such as water, H_2O (liquid) $\rightleftharpoons H_2O$(vapour), depends on the temperature and pressure, and the equilibrium constant is the vapour pressure characteristic of the substance. The solubility of a gas in a liquid is a similar case:

$$O_2(\text{gas}) + H_2O(\text{liquid}) \rightleftharpoons O_2(\text{solution in water}) \qquad (3.42)$$

The solubility depends on the gas pressure, the temperature and the nature of the gas and liquid.

Of particular interest in separation techniques is the distribution of a solute between two phases. Considering the equilibrium which will be set up between a solute A which dissolves unequally in two liquids or a liquid and a gas or a liquid and a solid:

$$A(\text{liquid 1}) \rightleftharpoons A(\text{liquid 2}) \qquad (3.43)$$

The equilibrium (or distribution) constant K_D is given by

$$K_D = [\text{A in liquid 2}]/[\text{A in liquid 1}] \qquad (3.44)$$

where the terms in square brackets denote concentrations.

3.8.2 Ions in solution

If the material being studied can form ions, for example sodium chloride $Na^+ \cdot Cl^-$, it will behave in very particular ways in solution. Firstly, in order to overcome the electrostatic attraction between the ions the solvent must be polar. That is, it must have molecules within which there is a separation of equal, but opposite charges. Water, H_2O, is a good example, with the hydrogen atoms being more positive and the oxygen more negative. Non-polar materials such as tetrachloromethane, CCl_4 show little polar character. The water molecules lower the force between the oppositely charged ions, and also they may attach themselves to the ions to hydrate (or for solvents in general, solvate) them.

3.8.3 Solubility

The solubility of solids in liquids is most important in analysis when considering precipitation, and in environmental work when looking at the composition of the hydrosphere. A general equation for solution of salts in water may be written

$$MX(\text{solid}) \rightleftharpoons M^{P+}(\text{aq}) + X^{P-}(\text{aq}) \tag{3.45}$$

where the description 'aq' refers to aqueous, or water solutions. Unless the solutions are very dilute indeed, it is most important to realise that the ions M^{P+} and X^{P-} will interact. They may come together as ion pairs, in some cases, but in all cases they will be held by their electrostatic forces and the polar forces of the solvent in an ionic atmosphere surrounding each ion of opposite charge. This will generally lower the activity of the ion, and instead of the concentration, we must use the activity:

$$a = c\gamma \tag{3.46}$$

where γ is the activity coefficient.

Various theories can be used to predict the value of γ, for example the Debye–Hückel theory, but these generally only apply in very dilute solutions.

The solubility of a solid such as sodium chloride, NaCl, in water has the equilibrium:

$$NaCl(\text{solid}) \overset{H_2O}{\rightleftharpoons} Na^+(\text{aq}) + Cl^-(\text{aq}) \tag{3.47}$$

The pure solid NaCl is in its standard state, so $a = 1$. The concentrations of the ions are the same, and at 25°C are found to be about 36 g per 100 g of water, that is $(36 \times 10/58.5)$ mol per dm^3 or 6.15 M. The activity coefficient γ at this concentration is about 0.99, so that the equilibrium constant is

$$K = (a_{Na^+} \cdot a_{Cl^-}/a_{NaCl}) \qquad (3.48)$$

$$= (6.15 \times 0.99)^2/1 = 37.1$$

and

$$\Delta G^{\ominus} = -8.314 \times 298 \times \ln(37.1) = -8954 \text{ J mol}^{-1} \qquad (3.49)$$

This is a fairly large negative value, and indicates that the solubility equilibrium goes largely to the 'product' side under these conditions. It is also found that the ions in salts like NaCl are completely separated in aqueous solution, and are referred to as strong electrolytes.

Consider a further example. How might the solubility of silver chloride be discovered?

$$AgCl(\text{solid}) = Ag^+(aq) + Cl^-(aq) \qquad (3.50)$$

Literature values for the standard free energies of the species at 298 K are:

	$\Delta G^{\ominus}(\text{kJ mol}^{-1})$
AgCl(solid)	−109.8
$Ag^+(aq)$	+77.1
$Cl^-(aq)$	−131.2

Therefore, for this reaction,

$$\Delta G^{\ominus} = (77.1 - 131.2 - (-109.8)) \qquad (3.52)$$

$$= +55.7 \text{ kJ mol}^{-1}$$

Thus,

$$K = 1.73 \times 10^{-10} \qquad (3.53)$$

and the solubility of silver chloride can be calculated as 1.31×10^{-5} mol dm^{-3}. This shows how very insoluble silver chloride is!

Unfortunately, data are not always available to do every calculation!

Because of the polar properties of the water molecule, it has a great tendency to solvate, or to bond to ions. This helps the solubility in many cases, and the composition of sea water, as well as rain and river water reflects the solubility of salts and of gases such as oxygen into the aqueous mixture.

The solubility of non-polar organic materials in non-polar solvents may be studied in similar fashion. The general rule that 'like dissolves like' would predict that a solvent of similar properties would dissolve a material most readily. Thus non-polar hydrocarbons such as oils or waxes dissolve most easily in non-polar solvents like hexane or benzene, but are almost insoluble in water.

3.8.4 Acid–base equilibria

Since environmental systems will chiefly involve water, it is most important we consider the effects of certain solutions in water. The molecule of water itself can dissociate partially, and is thus a weak electrolyte:

$$H_2O\,(\text{liquid}) \rightleftharpoons H^+(\text{aq}) + OH^-(\text{aq}) \qquad (3.54)$$

Measurement and calculation show that the equilibrium constant or ionisation constant at 25°C for this reaction, taking the activity of pure water as 1, is

$$K_W = 10^{-14} \qquad (3.55)$$

This is also called the ionic product of water.

Since in neutral water there are equal concentrations of hydrogen ions, H^+ and hydroxyl ions OH^-, their concentrations must be 10^{-7} mol dm^{-3}.

Because there is an equilibrium or balance, increasing the concentration of hydrogen ions always decreases the concentration of hydroxyl ions in an aqueous system. If $a_{H^+} = 10^{-5}$, then a_{OH^-} must be $10^{-14}/10^{-5} = 10^{-9}$

The nuisance of having to write everything in terms of large negative exponents is avoided if a pX scale is used, defined for hydrogen ions as the pH scale:

$$pH = -\log_{10}(a_{H^+}) \qquad (3.56)$$

The examples above mean that in neutral solution pH = 7, and if pH = 5 then pOH = pK_W − pH = 14 − 5 = 9.

Anything which puts in an excess of hydrogen ions over this value is called an acid, whereas anything which decreases the hydrogen ions, or increases the hydroxyl ions is alkaline.

Strictly, all the possible equilibria in solution must be considered, both the fully dissociated acids like HCl, the weak partially dissociated acids like ethanoic acid, CH_3CO_2H, and all species including the water. This makes calculation complex, and we must often simplify. This is important for speciation as discussed in chapter 14.

The weak acid ethanoic acid has a dissociation constant $K_a = 1.8 \times 10^{-5}$, and if water contains it at a total concentration 0.001, then

$$K_a = a_{H^+} \cdot a_{CH_3CO_2^-}/(0.001 - a_{H^+}) = 1.8 \times 10^{-5} \qquad (3.57)$$

Since the hydrogen and ethanoate ions have the same concentration, we can solve this equation for the activity of hydrogen ion

$$a_{H^+} = 1.25 \times 10^{-4} \quad \text{or} \quad pH = 3.9 \qquad (3.58)$$

For a weak base, such as ammonia solution, the dissociation forms more hydroxyl ions:

$$NH_3 \cdot H_2O = NH_4^+ + OH^- \qquad (3.59)$$

$$K_b = a_{NH_4^+} \cdot a_{OH^-}/a_{NH_3H_2O} = 1.78 \times 10^{-5} \qquad (3.60)$$

A solution which is 0.005 M in ammonia will give an activity of hydroxyl ions of 2.90×10^{-4} M. Thus pOH = 3.54 and pH = 10.46. This solution is thus alkaline.

The solution of carbon dioxide gas in water will produce the weak carbonic acid H_2CO_3, and the solution of sulphur dioxide will give sulphurous acid H_2SO_3, but this may be transformed into the stronger sulphuric acid H_2SO_4 by oxidation. Oxides of nitrogen can catalyse this process and also produce nitric acid HNO_3.

The importance of acids and bases in analysis and in the environment is enormous. The reactions which take place between any acid and any base serve as one basic method for analysing for them. For example, to determine the concentration of ethanoic acid in an aqueous solution, such as a fermented and oxidised plant material, we may titrate with an alkali like sodium hydroxide solution, adding small amounts until the solution becomes neutral:

$$HO_2C \cdot CH_3(aq) + Na^+(aq) + OH^-(aq) = Na^+ + CH_3CO_2^- + H_2) \quad (3.61)$$

3.8.5 Oxidation–reduction equilibria

Whenever there is a transfer of electrons, a process of oxidation takes place for one species, accompanied by a process of reduction for another. The simplest processes might involve hydrogen and an oxide:

$$H_2(gas) + CuO(solid) \rightleftharpoons Cu(solid) + H_2O(liquid) \quad (3.62)$$

The hydrogen is oxidised to water, the copper(II) oxide reduced to copper metal. Electrons are added to the copper to convert from ions to metal, and removed from the hydrogen.

Also important is the effect of oxygen itself:

$$3/2 O_2(gas) + 2Fe(metal) \rightleftharpoons Fe_2O_3(solid) \quad (3.63)$$

These reactions, too, show the equilibrium behaviour and in aqueous media, the transfer of electrons may be measured using electrochemical methods.

3.8.6 Electrochemical reactions

Consider the reduction of silver chloride by hydrogen:

$$1/2 H_2(gas) + AgCl(solid) \rightleftharpoons H^+(aq) + Cl^-(aq) + Ag(solid) \quad (3.64)$$

Hydrogen is oxidised to hydrochloric acid in aqueous solution, and silver chloride reduced to metallic silver, and *one* electron is transferred for each atom of silver produced.

This reaction has a $\Delta G^\ominus$ of -21.5 kJ mol^{-1} at 25°C. The reactions may be separated out into two 'half-cell reactions':

$$1/2 H_2(gas) - e = H^+ \quad (3.65)$$

$$AgCl(solid) + e = Ag(solid) + Cl^- \qquad (3.66)$$

A convention has been adopted that the first reaction under standard conditions has a $\Delta G^{\ominus}$ value of 0, so that the standard free energy change for the silver reaction is -21.5 kJ mol^{-1}.

These reactions can be studied electrochemically, using electrodes which respond specifically to hydrogen ions and to chloride ions, as described in chapter 10. Since the electrical work done may be related to the free energy change, the electrical potential difference for the cell reaction, is

$$\Delta G = -nFE \qquad (3.67)$$

and under standard conditions:

$$\Delta G^{\ominus} = -nFE^{\ominus} \qquad (3.68)$$

By convention, for the cell reaction,

$$1/2\,H_2(gas, 1\ atm) + AgCl(s) = H^+ + Cl^- + Ag \qquad (3.69)$$

the eletrochemical cell would be written

$$Pt, H_2 | H^+ | Cl^- | AgCl, Ag \qquad (3.70)$$

and the potential difference between the electrodes, or the cell electromotive force (emf) under standard conditions is

$$-\Delta G^{\ominus}/1F = 0.222\ V \qquad (3.71)$$

This value is the standard reduction electrode potential of the silver–silver chloride electrode,

$$AgCl(solid) + e = Ag(solid) + Cl^- \qquad (3.72)$$

Any electrochemical cell will have a measured emf which is the algebraic sum of the half-cell electrode potentials,

$$E_{cell} = E_+ - E_- \qquad (3.73)$$

Many oxidations and reductions are used analytically, and the electro-chemical measurements of ion concentrations are also very important. The oxidation of iron(II) salts by cerium(IV) salts may be followed by using a titration, adding small amounts until an oxidation indicator changes colour, or by using electrodes and noting when a rapid change in electrode potential occurs, because the last trace of reactant is removed.

$$Ce^{4+} + Fe^{2+} = Ce^{3+} + Fe^{3+}$$

From Tables, in the acidic medium used, $\Delta G^{\ominus} = -70.45$ kJ mol^{-1}. This means the reaction will be favoured to proceed to the products. The equilibrium constant will be about 2.2×10^{12}. Strong oxidants have high positive $E^{\ominus}$ values, strong reducing agents less positive or negative values. Thus $E^{\ominus}$ for the Ce(IV)/Ce(III) electrode is 1.41 V and for the Fe(III)/Fe(II) it is 0.68 V. The end point will be mid-way between the two, at about 1.05 V. When measured electrochemically, the cell would be

$$Pt|Fe^{3+}, Fe^{2+}|Ce^{4+}, Ce^{3+}|Pt \qquad (3.74)$$

and under standard conditions:

$$E^{\ominus}_{cell} = E^{\ominus}(Ce^{4+}, Ce^{3+}) - E^{\ominus}(Fe^{3+}, Fe^{2+}) = 1.41 - 0.68 = 0.73 \text{ V} \quad (3.75)$$

The electrode potentials will depend on the oxidised and reduced species, and for the iron(II) to iron(III) reaction at 25°C:

$$E = E^{\ominus} + (RT/F) \ln(a_{Fe^{3+}}/a_{Fe^{2+}}) \qquad (3.76)$$

$$E = 0.68 + 0.0257 \times \ln(a_{Fe^{3+}}/a_{Fe^{2+}})$$

so that at the endpoint, the ratio of oxidised to reduced ions is about a million to one!

3.8.7 Complexation

The bonding of species, particularly transition metals, into complexes is another example of analytical and environmental equilibria. Although there are two distinct ways of presenting the equilibrium constants let us consider the stepwise formation of a simple complex:

$$M + L = ML, \qquad K_f = \frac{a_{ML}}{a_M \cdot a_L} \qquad (3.77)$$

In a similar way to the other equilibria given above, large values of the complexation equilibrium constant will show a tendency to give a stable complex ML, whereas small values will indicate poor stability.

One extra problem which may occur here is that the complexing species itself may be involved in several, simultaneous equilibria! For example, the powerful complexing ligand EDTA, (ethylenediaminetetracetic acid) is a tetrabasic acid, H_4Y and can dissociate completely to Y^{4-} or partially into HY^{3-}, H_2Y^{2-}, or H_3Y^- or remain as H_4Y! The concentrations of each of these species will depend on pH, and thus the complexing power will vary with the acidity.

3.9 Reaction kinetics

The Law of Mass Action proposed that the rate of reaction depended on the 'active masses' of the reactants. Besides the thermodynamic driving force for reactions, another chemical effect must be considered. In order to react, species must come together. Therefore their rate of reaction will depend on how often, and how successfully they meet. This is affected by three factors:

1. the collision or encounter frequency of the species.
2. the orientation or geometric factors involved.
3. the activation energy of the reaction.

The collision or encounter frequency will depend on the concentration of the reactive species. The dependence may be very complex, since the species may have to be formed by an initial reaction before it can start a second one. There will also be a *small* dependence on temperature for this factor.

The orientation of species relates the arrangements of the reactants in space relative to each other. The geometry and shape of the species, and the effects of other species or surfaces present may have an effect too.

Most importantly, the species must overcome an energy barrier, called the *activation energy* E_a for the reaction to succeed. Figure 3.16 shows a typical situation. The reactants A and B have an initial energy when far apart of E_i. As they approach, their energy increases, but unless they have enough energy to reach the peak E_a, they will return without reacting. This energy is generally obtained by thermal collisions. After reacting, the energy drops back to a final value E_f, and the difference $(E_f - E_i)$ is the value of ΔH for this reaction.

In practical terms, the rate of reaction, in moles reacting per second (or moles reacting per unit volume per second) is given by

$$\text{Rate} = A \exp(-E_a/RT) \times \text{function (concentrations)} \qquad (3.78)$$

or

$$\text{Rate} = k \times \text{function (concentrations)} \qquad (3.79)$$

where

$$k = A \exp(-E_a/RT) \qquad (3.80)$$

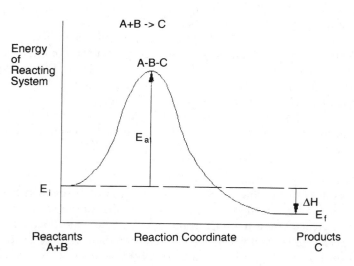

Figure 3.16 Energy and reaction coordinates.

A is called the 'pre-exponential factor' and incorporates collision and orientation parameters. *k* is the rate constant of the reaction at the temperature of interest. Note that it will increase greatly with increase in temperature, chiefly due to the activation energy effect.

For some simple solution and gas reactions, the rate depends directly on the concentrations, so we may simplify to:

$$\text{Rate} = k \times (\text{concentration})^n \tag{3.81}$$

n is an experimental parameter chosen to obtain the best agreement and is called the order of reaction. It may take values from 0 to about 3 and need not be an integer.

3.10.1 Radioactive reactions

3.10 Examples of reaction kinetics

As noted above, certain elements above lead in the period table naturally decay with the emission of α, β or γ particles. Because these reactions are not affected by any outside factors, even temperature, they follow the simplest rate law of a first-order equation. If the radioactive species is *R* and its concentration c_R, then

$$\text{Rate} = - \, dc_R/dt = k \cdot c_R \tag{3.82}$$

This equation may be rearranged and integrated into the form

$$\ln(c_{R,0}/c_R) = kt \tag{3.83}$$

where $c_{R,0}$ is the concentration at $t = 0$. When c_R has reached half the initial value, the left of this equation always equals $\ln(2)$, no matter what the concentrations involved. Thus the half-life $t_{1/2}$ is characteristic of all first-order and radioactive reactions. For example, the long-lived isotope ^{90}Sr ('strontium-90') has a half-life of 28.1 years, and decays by β emission to yttrium-90. Expressing *k* in years^{-1},

$$\text{Rate} = 0.0247 \cdot c_{Sr-90}$$

3.10.2 Ionic reactions

Because of the electrical nature of ions and their mobility in aqueous systems, reactions between ions are generally very rapid. For this reason, they find great application in analytical determinations such as titrations, precipitations and colour reactions. As noted above, the rate may depend on the encounters between ionic species, moving by diffusion through the solvent, and also on the dissociation of molecular species into ions. Organic reactions may also involve ionic species in a multi-step process such as the $S_N 1$ reactions of alkyl halides:

$$RCl + NaOH = ROH + NaCl \qquad (3.84)$$

where R is an alkyl group such as C_4H_9. This may occur in two stages:

$$R - Cl = R^+ + Cl^- \qquad (3.85a)$$

$$R^+ + OH^- = ROH \qquad (3.85b)$$

The first stage is slow, because it involves a large activation energy to break the C–Cl bond, but the second is fast, since it is ionic.

3.10.3   Solid-state reactions

The ability of species to move through solids is very low, and so many solid-state reactions occur extremely slowly, for example on a geological time-scale of millions of years. There are, however, reactions involving solids which go very rapidly indeed, for example explosions and pyrotechnic reactions. This is often due to the fact that the reaction is very exothermic, and, if the heat produced is not dissipated, it can heat up the reactants, raise the reaction rate, produce more heat, and lead eventually to thermal runaway and explosive reaction. Another effect of exothermic reaction is to melt or vaporise the reactants, and so produce a more fluid medium in which reactions can occur more easily.

When reactions involving solids need to be studied, we cannot use the normal concentration methods. Often the fraction *reacted*, α is used instead, and the rate equation would be of the form

$$\text{Rate} = d\alpha/dt = k \cdot (1 - \alpha)^n \qquad (3.86)$$

3.10.4   Photochemical reactions

High energy radiation such as UV or γ-rays when absorbed by molecules may dissociate them into free radicals, that is species with unpaired electrons, or into ions. Unless the radiation is absorbed, no change can take place. In the upper atmosphere reactions are initiated by high-energy radiation, represented by $h\nu$:

$$N_2 + h\nu = 2N$$

$$O_2 + h\nu = O_2^+ + e$$

$$O_3 + h\nu = O_2 + O$$

$$O + h\nu = O^+ + e$$

Subsequent reactions may take place between these active species:

$$O^+ + N_2 = NO^+ + N$$

In organic reactions, the halogenation of hydrocarbons, RH may proceed by a complex sequence of reactions, initiated by the absorption of light producing free radicals:

$$Br_2 + h\nu = 2Br$$

$$Br + RH = R + HBr$$

$$R + Br_2 = RBr + Br$$

$$Br + Br + M = Br_2 + M$$

giving a complex rate law involving the light intensity, I.

$$dc_{RBr}/dt = k \cdot I^{1/2} \cdot c_{RH}/(c_M^{1/2}) \qquad (3.87)$$

3.11 Summary

Throughout all nature, the ways in which atoms, ions and molecules interact depends upon their bonding, structure and their thermodynamic and kinetic properties. Unless we recognise that the chemical behaviour is determined by these fundamental quantities, efforts to analyse materials or to devise better environmental procedures may be doomed from the start!

Two short cautionary tales may illustrate this.

A student chemist was told to prepare a new polymer. He found out what chemicals to use and what apparatus would be suitable for the reaction, and what the possible health hazards were. He carefully followed all the instructions, but ended up with a black, tarry mess! What he had forgotten was that the reaction was highly exothermic and that the chemicals decomposed when heated too strongly!

A second experimenter was determined to get the best possible yield of product from his reaction. He carefully calculated the quantities needed from the equations for the reaction. He even looked up tables to see whether the reaction was exothermic or not. He was most disappointed when he obtained a very small amount, until he realised that this particular reaction had a small equilibrium constant under the conditions he had used!

Further reading

Brescia, F., Arents, J., Meislich, H. and Turk, A. (1988) *General Chemistry*, 5th edition, Harcourt Brace Jovanovich, New York.
Atkins, P. W. (1977) *Physical Chemistry*, Oxford University Press, Oxford.
O'Neill P. (1993) *Environmental Chemistry*, 2nd edition, Chapman & Hall, London.

4 Titrimetry and gravimetry

F.W. Fifield

4.1 Introduction

Both *titrimetry* and *gravimetry* represent techniques which have been employed in analysis for very many years, and for this reason are sometimes referred to as 'classical techniques'. Over the last 50 years, with the growth of instrumentally based methods and new types of demand on analytical chemistry, their dominant position has been eroded. Nevertheless the techniques, especially titrimetry, remain in use and, for the analysis of environmental samples especially, continue to fulfil important roles. Their success has been based upon three key characteristics:

1. good selectivity;
2. good precision and accuracy;
3. simplicity of operation.

The first two are obvious desirable attributes of any method, whilst the last means that few demands are made on complex and expensive laboratory instrumentation. Titrimetry for example is possible, and often used, under quite primitive field conditions. On the debit side it must be recognised firstly that procedures may be lengthy and time consuming. Secondly, a high degree of manipulative skill is required of operators if good quality results are to be obtained. Methods based on titrimetry or gravimetry are readily applicable to the determination of many major and minor components of samples, but with certain exceptions are difficult to apply at lower levels of analyte concentration. A common feature of the two is their reliance upon and exploitation of chemical reactions in solution, which are discussed in general terms in chapter 3. Expanded discussion on more specific points relevant to particular analytical methods will be found below.

4.2 Titrimetry

4.2.1 Introduction

Titrimetry makes use of solution reactions between standard reagents and the analyte. The reactions must be quantitative and reproducible and preferably, although not essentially, rapid. In its simplest form a solution of the reagent of known composition, the *titrant*, is added slowly in measured volume to a solution of the analyte, the *titrand*, until all of the analyte has reacted. This point is known as the *equivalence point* or *endpoint* and must be detected accurately and precisely by use of an *indicator*. The overall process is called a *titration*. In some cases where the analytical reaction between reagent and analyte proceeds slowly, the titration may be expedited by adding a measured excess of the reagent. The unused portion of the titrant is then determined by titration with a second reagent in a back titration. The second reagent is selected as one undergoing a rapid reaction. Volume-based titrations require only simple equipment in most cases, such as pipettes, burettes, standard flasks for volume measurement, auxiliary glassware and an analytical standard balance. In limited circumstances where appropriate reagents are insufficiently stable to be stored in a solution they can be generated *in situ* by passing an electrical current to bring about the appropriate electrochemical reaction. Greater overall sensitivity may also be attainable in this way because of the greater sensitivity of electrical measurements relative to those of volume which are most frequently employed. Titrations based upon this electrochemical generation of the reagent are known as coulometric titrations (see chapter 10). A number of different types of chemical reaction are used in titrimetry. Table 4.1 provides a summary. Of these acid–base complexation and redox are the most important.

Table 4.1 Types of chemical reaction used in titrimetry

Acid–base
Complexometric
Redox
Precipitation
Electrochemical

Precise and accurate detection of the endpoint is obviously a cornerstone of any titrimetric method. Indicators used respond either to the reduction of the analyte concentration as the equivalence point is approached or the excess of the reagent after it is passed. In either case one of two different endpoint detection techniques is employed. Colour change indicators undergo a visible colour change at the endpoint as a result of chemical reactions with titrant of titrand. The colour change may be observed by eye or instrumentally in a photometric titration. An instrumental technique is the alternative to a colour change one. Electrochemical devices are used and may use the phenomena of conductivity, potentiometry or amperometry (see chapter 10). In the following sections titrimetric techniques appropriate to environmental analysis are discussed. Characteristics of relevant endpoint detection procedures are included as part of the overall discussions.

4.2.2   Acid–base titrations

Principles. The general concepts of acids, bases, and pH have been developed in chapter 3, section 3.8.4, together with an indication of the importance of water as a solvent. These general ideas are presented in several different theories. For the purposes of this chapter, the Lowry–Brønsted Theory is the most appropriate to use. In this, an acid is defined as a proton (hydrogen ion) donor, and a base as a proton acceptor. Thus, in an acid–base reaction a proton is exchanged between the acid and the base, as expressed in equation (4.1).

$$AH + B \; = \; A^- \; + \; BH^+ \tag{4.1}$$

$\underset{acid}{\quad} \underset{base}{\quad} \underset{\substack{conjugate \\ base}}{\quad} \underset{\substack{conjugate \\ acid}}{\quad}$

It is immediately clear that in this equilibrium that the reverse reaction must also be acid–base in nature, hence the terms conjugate acid and conjugate base are used to describe the species on the right-hand side of the equation. If the reactions between a range of acids and a single base are considered, the strength of the acid will be reflected in the degree to which the equilibrium is displaced to the right, and measured quantitatively by the associated equilibrium constant (page 58). Because of its widescale function as a solvent involved in acid–base equilibria, water is the most appropriate compound to select for the role of the base in this comparison. When a compound containing an ionisable proton dissolves in water, it will undergo a degree of dissociation as shown in equation (4.2).

$$AH + H_2O \; \rightleftharpoons \; A^- \; + \; H_3O^+ \tag{4.2}$$

$\underset{acid}{\quad} \underset{base}{\quad} \underset{\substack{conjugate \\ base}}{\quad} \underset{\substack{conjugate \\ acid}}{\quad}$

Written in concentration terms the equilibrium constant for the reaction, K_a is given by

$$K_a = \frac{[A^-][H_3O^+]}{[AH][H_2O]} \tag{4.3}$$

and as water is present in large excess $[H_2O] \sim 1$, and the equation usually appears in the form of equation (4.4) below.

$$K_a = \frac{[A^-][H_3O^+]}{[AH]} \tag{4.4}$$

K_a is known as the acid dissociation constant. K_a values vary over very large ranges and it is again convenient to use the 'p-notation' introduced earlier for hydrogen ion concentrations, pH (chapter 3, section 3.8.4). It is therefore usual to find pK_a values quoted for most purposes. The stronger the acid the larger will be the value of K_a. It is customary to use the terms strong and weak acids. The former refers typically to mineral acids such as HCl and HNO_3 which are almost completely dissociated in aqueous solution and have very large K_a values. The second group is of acids for which a significant amount of the original compound remains undissociated. Carbonic acid (H_2CO_3) and undissociated compounds containing –COOH or –OH groups fall into the latter category. It is important to remember this when dealing with environmental samples where the acidity often derives from dissolved carbon dioxide or human substances. A representative list of K_a values for weak acids is shown in Table 4.2.

Table 4.2 pK_a values for some Lowry–Brønsted acids

Acid	pK_1	pK_2	pK_3	pK_4
Acetic CH_3COOH	4.76			
Ammonium NH_4^+	9.25			
Chloracetic $CH_2ClCOOH$	2.86			
Diethanolammonium $NH_2(CH_2CH_2OH)_2^+$	9.00			
Ethanolammonium $NH_3(CH_2CH_2OH)^+$	9.49			
Ethylenediaminetetraacetic acid (EDTA) $(HOOCCH_2)_2N-CH_2-CH_2-N(CH_2COOH)_2$	2.0	2.67	6.16	10.27
Formic $HCOOH$	3.75			
Hydrocyanic HCN	9.22			
Nitric HNO_3	-1.4			
Nitrilotriacetic (NTA) $N(CH_2COOH)_3$	1.66	2.96	10.28	
Nitrous HNO_2	3.29			
Phenol C_6H_6OH	9.98			
Phosphoric H_3PO_4	2.17	7.21	12.36	
Sulphuric H_2SO_4	-1.96	1.96		
Sulphurous H_2SO_3	1.76	7.21		

In the terms of the Lowry-Brønsted theory, acids and bases are defined only within a particular reaction as proton donor and acceptor respectively (equation (4.1)). A compound with a very low tendency to donate protons (very small K_a) reacts with a stronger proton donor then it behaves as a base accepting an additional proton. Water is a compound that readily acts as an acid or base, see equation (4.2) above and equation (4.5) below.

$$AH + H_2O = BH + OH^-$$
$$\text{\small acid \quad base \quad conjugate \quad base}$$
$$\text{\small acid}$$
(4.5)

It is common to think of compounds which donate protons to water, generating H_3O^+ ions, as acids, and those which accept protons from water, generating OH^- ions as bases. This is exemplified by the behaviour of ammonia in water.

$$NH_3 + H_2O \rightleftharpoons NH_4^- + OH^-$$
$$\text{\small base \quad acid \quad conjugate \quad conjugate}$$
$$\text{\small base \qquad acid}$$
(4.6)

These concepts are of practical as well as theoretical importance. For a titration to be carried out satisfactorily the reaction must be 'quantitative', i.e. the equilibrium must be heavily displaced to one side or the other. Hence

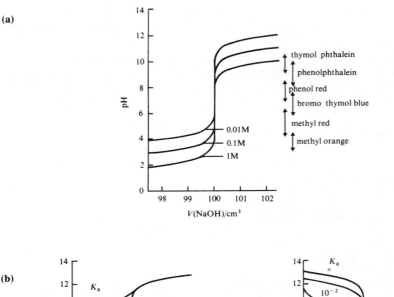

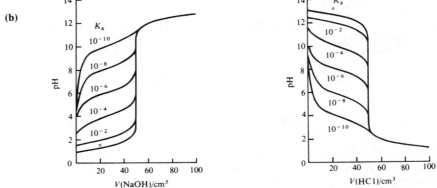

Figure 4.1 Titration curves for (a) strong acid–strong base showing effects of concentration; (b) weak acid–strong base and weak base–strong acid showing different K_a values. Source: Fifield, F.W. and Kealey, D. (1995) *Principles and Practice of Analytical Chemistry*, 4th edition, Blackie Academic & Professional, Glasgow.

in general terms, to titrate a weak base a strong acid is needed and conversely a strong base is needed for a weak acid. Furthermore, only in the case of reactions between strong acids and strong bases will the equivalence point be at pH 7. For the detection of an endpoint, this is an important point to bear in mind. The overall picture is best represented in a titration curve in which the pH is recorded with sequential additions of the titrant (Figures 4.1a and b). In the region of the endpoint the pH will change relatively rapidly and the endpoint is indicated by the point of inflection in the titration curve. The magnitude and prominence of this region of pH change on the curve will reflect the strengths and concentrations of the acids and bases reacting. The weaker and the more dilute they are the less obvious the 'endpoint break' will become.

Endpoint detection. Reliable and reproducible detection of endpoints is essential for satisfactory titrations. The pH of the titration solution may be followed using a glass electrode (chapter 10, p. 206). For manual titrations this is rather cumbersome, but if the endpoint break is not well defined it may be essential. In laboratories where large numbers of acid–base titrations are routinely carried out, automatic instruments using glass electrodes are of great value. For manual titrations, however, colour change indicators are more convenient to use. These indicators act as secondary acids or bases in the titration and show distinct colour changes as they gain or lose protons.

$$\underset{colour\ 1}{In}\ +\ H^+ \rightleftharpoons \underset{colour\ 2}{InH^+} \tag{4.7}$$

Provided that they are selected to have similar K_a values to the titrand, they will change colour over a narrow pH range close to the endpoint of the titration. Table 4.3 gives some details of commonly used indicators.

Table 4.3 A range of visual indicators for acid–base titrations

Indicator	pK_{In}	Low pH colour	High pH colour	Experimental colour change range/pH
Cresol red	ca.1	Red	Yellow	0.2–1.8
Thymol blue	1.7	Red	Yellow	1.2–2.8
Bromo-phenol blue	4.0	Yellow	Blue	2.8–4.6
Methyl orange	3.7	Red	Yellow	3.1–4.4
Methyl red	5.1	Red	Yellow	4.2–6.3
Bromo-thymol blue	7.0	Yellow	Blue	6.0–7.6
Phenol red	7.9	Yellow	Red	6.8–8.4
Phenolphthalein	9.6	Colourless	Red	8.3–10.0
Alizarin yellow R	ca.11	Yellow	Orange	10.1–12.0
Nitramine	ca.12	Colourless	Orange	10.8–13.0

Environmental applications. The determination of acidity and alkalinity as discussed in chapter 18 are important examples of the use of acid–base titrations. A further example is the Kjeldahl determination of total nitrogen, in which all the nitrogen is reduced to ammonia by a catalyst, distilled into a measured excess of hydrochloric acid, with the unreacted acid being titrated with sodium hydroxide.

4.2.3 Complexometric titrations

Principles. The concept of the metal–ligand bond has been introduced in chapter 3, section 3.2.4. It is a covalent chemical bond in which the electrons shared are donated from one electron-rich centre only. The atom or group supplying the electrons is known as a *ligand,* and the reaction product as a *complex.* Once formed, the dative bonds have properties similar to other covalent bonds. Typically, the acceptor atom in the complex is a metal and complexometric reactions can be used in the titrimetric determination of metals. There are parallels between complexometric titrations and acid–base titrations, in that the overall execution of the titrations is similar, with colour change or electrochemical endpoint detection being available. Complexometric reactions do not, however, proceed as rapidly as the ionic reactions on which acid–base titrations are based. Accordingly a very gradual approach to the endpoint may be required in some cases, and back titrations are more frequently encountered.

Analytical uses of complex forming reactions are generally made in aqueous solution. In equilibrium terms this presents a complex system, so that any rigorous approach to complex formation rapidly becomes algebraically complicated. Fortunately, the general principles of complexometric titrations, and, indeed, of most other analytical uses of complexes, can be appreciated by a simplified approach and recognition of a few salient points. A useful model for the formation of a complex ML_n from metal ion M, and ligand groups L, is the stepwise model. In this the ligand groups are considered to attach to the metal ion in sequence as shown by Table 4.4, with each reaction having its own formation constant (K_f). In Table 4.4 charges have been omitted for simplicity. However, their existence must be remembered, and one aspect of their significance figures in the discussion of K_f values immediately following. The successive values of K_f will be expected to decrease for three reasons:

1. As successive ligands are attached the number of available sites is reduced and reaction is statistically less likely.
2. The positive charge is reduced and ultimately a negative charge is present. Coulombic attraction decreases and is ultimately converted into repulsion.
3. Bulky ligand groups already attached will sterically hinder the approach of subsequent groups.

Table 4.4 Stepwise formation constants ($mol^{-1} dm^3$)

M + L =	ML	$K_{f_1} = \dfrac{[ML]}{[M][L]}$
ML + L =	ML_2	$K_{f_2} = \dfrac{[ML_2]}{[ML][L]}$
ML_2 + L =	ML_3	$K_{f_3} = \dfrac{[ML_3]}{[ML_2][L]}$
ML_{n-1} + L =	ML_n	$K_{f_n} = \dfrac{[ML_n]}{[ML_{n-1}][L]}$

Note: Charges have been omitted for simplicity.

Table 4.5 contains data on the Cd^{2+}/CN^- system as an example. The successive decreases in K_f values reflect complexes of decreasing stability. From the practical point of view this means that in the titration solution all complexes will exist together, each predominating in turn until the equivalence point is reached or exceeded. An alternative method of expressing the equilibria is in terms of the *overall formation constant*, β, shown in Table 4.6.

Table 4.5 Formation constants for Cd/CN^- complex showing the decrease as successive ligands are attached.

ML_n	$K_f/mol^{-1}dm^3$
$Cd(CN)^+$	5.0×10^5
$Cd(CN)_2$	1.3×10^5
$Cd(CN)_3^-$	4.3×10^4
$Cd(CN)_4^{2-}$	3.5×10^3

Table 4.6 Overall formation constants ($mol^{-n} dm^{3n}$)

$$M + L = ML \qquad\qquad \beta_1 = \frac{[ML]}{[M][L]}$$

$$M + 2L = ML_2 \qquad\qquad \beta_2 = \frac{[ML_2]}{[M][L]^2}$$

$$M + 3L = ML_3 \qquad\qquad \beta_3 = \frac{[ML_3]}{[M][L]^3}$$

$$M + nL = ML_n \qquad\qquad \beta_n = \frac{[ML_n]}{[M][L]^n}$$

The relations between values of K_f and values of β are simple and easily seen. For example,, β_3 may be multiplied by $[ML]/[ML]$ and $[ML_2]/[ML_2]$ to give $\beta_3 = ([ML_3]/[M][L]^3)([ML]/[ML])([ML_2]/[ML_2])$. *Note.* Charges have been omitted for simplicity.

A further aspect of complex formation and stability is the chelate effect, which is stability enhancing and derives from entropy considerations. Chelate complexes involve the formation of two or more dative bonds between a single acceptor species and a single ligand species having two or more donating centres. Ethylene diamine, which has two electron-rich amino groups, is a simple example, Figure 4.2, and illustrates the resultant

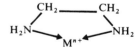

Figure 4.2 Representation of the formation of a chelate by ethylene diamine. Source: Fifield, F.W. and Kealey, D. (1995) *Principles and Practice of Analytical Chemistry*, 4th edition, Blackie Academic & Professional, Glasgow.

ring formation in the complex. In chapter 3, section 3.5, the importance of entropy changes in chemical reactions has been explained. Where an entropy increase occurs, the reaction is favoured and an increase in K_f follows. If entropy is regarded as the degree of disorder in a reaction system it may be related to the grouping of a fixed number of atoms into molecules, i.e. the more molecules involved the greater the disorder. The second key point in explaining the chelate effect requires a slightly more detailed view of the metal–ligand reaction. Metal ions in aqueous solution do not exist as free ions, but are coordinated with up to six water molecules (usually 4 or 6) via the lone pairs of electrons on the oxygen atoms of the water. Any reaction with another ligand requires the displacement of water molecules. Equations (4.8) and (4.9) show in parallel the reaction of nickel ions with two nitrogen-donating centres, firstly in separate ammonia molecules and secondly in one ethylene diamine molecule.

$$[Ni(H_2O)_6]^{2+} + 2NH_3(aq) \rightleftharpoons [Ni(H_2O)_4(NH_3)_2]^{2+} + 2H_2O \qquad (4.8)$$

$$[Ni(H_2O)_6]^{2+} + en \rightleftharpoons [Ni(H_2O)_4en]^{2+} + 2H_2O \qquad (4.9)$$

(en = ethylene diamine). Inspection and comparison of these two equations shows that for reaction with ammonia the total number of molecular species remains at three whilst for ethylene diamine it increases from two to three. The increased entropy indicates that the ethylene diamine complex will be stronger than the ammonia one. This is borne out in practice.

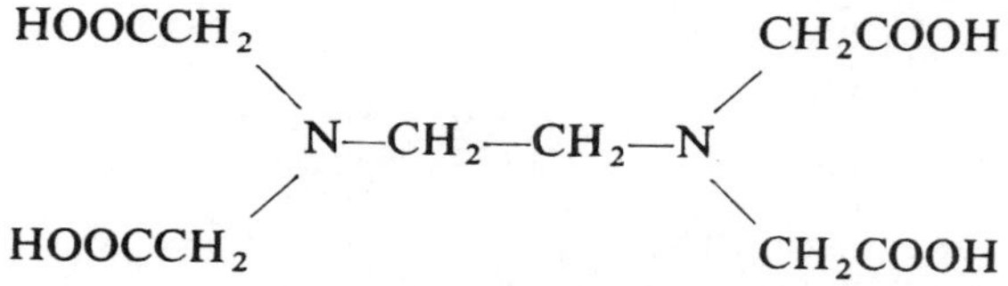

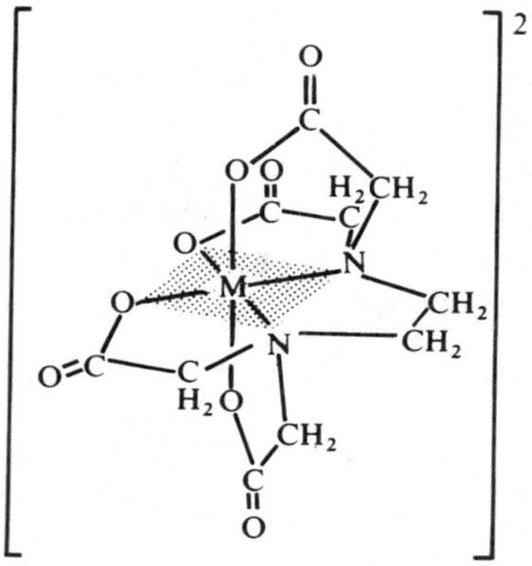

Figure 4.3 A proposed structure for a metal–EDTA chelate showing its octahedral geometry. Source: Fifield, F.W. and Kealey, D. (1995) *Principles and Practice of Analytical Chemistry*, 4th edition, Blackie Academic & Professional, Glasgow.

Complexometric reagents. Ethylenediaminetetraacetic acid (EDTA) is by far the most widely used complexometric reagent in titrimetric analysis. Its molecular structure is shown in Figure 4.3. It is a multidentate ligand and can potentially form six bonds and five chelate rings with a single metal ion, although commonly only four bonds are formed. Complexes of moderate or high stability can be formed with a substantial range of metal ions. A most important structural feature of the complexes formed is their 1:1 stoichiometry. The 'cage-like' structure produced by the multidentate EDTA ligand imposes steric restrictions preventing the binding of more than one ligand group. The pure reagent has rather a low solubility, and it is customary to use the di-sodium salt instead, which has adequate solubility for the preparation of volumetric solutions in water. From equation (4.10)

$$YH_2^{2-} + M^{n+} \rightleftharpoons YM^{(n-2)+} + 2H^+ \qquad (4.10)$$

it will be seen that hydrogen ions are generated as the complex is formed. Hence by varying the pH of the titrand solution a measure of selectivity may be imported to the titrations. In this way two metals with significantly different K_f values may be distinguished, but prior separation, e.g. by ion-exchange chromatography, is frequently required. EDTA may be used as a reagent for metals like Zn, Ca, Cd, Mg, Ba, Pb, Hg. In recent years its use has been eroded by atomic spectroscopic techniques. It is routinely applicable for the determination of calcium and magnesium as 'water-hardness'. This is discussed in chapter 18, section 18.6. Some other reagents of the aminocarboxylic acid type are also available. Interesting to note is that one of these, nitrilotriacetic acid (NTA) is also used as an industrial cleaning agent and environmental analysis may be needed to assess its discharge as an industrial pollutant. Finally, some inorganic ligands such as halide ions and the pseudo halogens CN^- and SCN^- have been used at times for the titration of metals.

Endpoint detection. The detection of endpoints in complexometric titrations can be effected by electrochemical means or by colour-change indicators. In the former case an electrochemical sensor is used to monitor the concentration of the metal ion and produce a titration curve. In modern analytical chemistry this technique is rarely used. Much more likely, as in the example of the determination of water hardness, a colour-change indicator will be employed. The indicator will consist of metal complex with a formation constant higher than that for the analyte–EDTA complex. Hence, when all the analyte is complexed, excess EDTA will displace the indicator equilibrium, equation (4.11), and bring about the colour change.

$$InM = In^{n-} + M^{n+} \qquad (4.11)$$
$$\text{\textit{colour 1}} \qquad \text{\textit{colour 2}}$$

4.2.4   Redox titrations

Principles. Where an analyte can be quantitatively oxidised or reduced by a reagent in solution the basis for a redox titration exists. These titrations may be used for the determination of the species concerned. In the context of environmental science the oxidation state of a metal, e.g. Fe(III)/Fe(II) or Mn(II)/Mn(IV) may well be of importance. Redox titration can be further employed to measure the balance of oxidation states present. Oxidation–reduction equilibria have been introduced together with the concept of the reduction potential as a measure of the oxidising and reducing characteristics of a species. Remembering that a species is considered to be reduced when it gains electrons and oxidised when it loses them, a generalised redox equilibrium is described by equation (4.12).

$$\underset{\text{Form}}{\text{Oxidised}} + \text{Electrons} = \underset{\text{Form}}{\text{Reduced}} \qquad (4.12)$$

The electrons may be supplied or removed by electrochemical means, and techniques based on this principle are discussed in chapter 10. In the present chapter considerations are restricted to chemical reactions in which oxidising agents obstruct electrons and reducing agents supply them. In these circumstances oxidising and reducing agents need always to be considered in complementary pairs in equilibria and can only be specified as oxidising or reducing, in relation to each other i.e.

$$A_{ox} + B_{red} = A_{red} + B_{ox} \qquad (4.13)$$

where 'ox' and 'red' represent oxidised and reduced forms, respectively. Generally where the $E^{\ominus}$ values differ by 0.1 V or more quantitative oxidation and reduction will occur, with the species having the higher value being the oxidant.

Redox reagents. Many possibilities for redox reagents exist but those with current practical applications are small in number. These are summarised in Table 4.7. Sometimes back titrations are used, i.e. by adding excess Fe(II) or Fe(III) which can then be readily titrated, with the analyte concentration being calculated indirectly. In other circumstances iodine may be used, being liberated in solution by the oxidation of iodide ions and then titrated as a measure of the amount of analyte. Chapter 18 contains some examples of redox processes used in the analysis of water.

Endpoint detection. As in previously discussed types of titrimetry, endpoints can be detected by instrumental means or by colour change indicators. An inert electrode, typically platinum or gold will respond to the potential of the titrand solution, and in conjunction with a suitable reference electrode

Table 4.7 Some representative redox reagents

Reagent and formal valency	Half reaction	$E^{\ominus}$ (volts)	Conditions
Oxidising agents			
Manganese(VII)	$MnO_4^- + 8H^+ + 5e^- = Mn^2 + 4H_2O$	1.51	Strong acid
$KMnO_4$	$MnO_4^- + 4H^+ + 3e^- = MnO_2 + 2H_2O$	1.69	Weak acid/ neutral
	$MnO_4^- + e^- = MnO_4^{2-}$	0.56	Strong base
Cerium(IV) $Ce(SO_4)_2$	$Ce^{4+} + e^- = Ce^{3+}$	1.44	Sulphuric acid solution
Chromium(VI) $K_2Cr_2O_7$	$Cr_2O_7^{2-} + 14H^+ + 6e^- = 2Cr^3 + 7H_2O$	1.33	Strong acid
Iodine(V) KIO_3	$IO_3^- + 2Cl^- + 6H^+ + 4e^- = ICl_2^- + 3H_2O$	1.23	Strong hydrochloric acid
Bromine(V) $KBrO_3(+KBr)$	$BrO_3^- + 5Br^- + 6H^+ = 3Br_2 + 3H_2O$ $Br_2(aq) + 2e^- = 2Br^-$	1.05 1.09	Dilute acid Dilute acid
Reducing Agents			
Iron(II) $FeSO_4$	$Fe^{3+} + e^- = Fe^{2+}$	0.771	Sulphuric acid
Arsenic(III) H_3AsO_3	$H_3AsO_4 + 2H^+ + 2e^- = H_3AsO_3 + H_2O$	0.559	Acid
Titanium(III) $TiCl_3$	$TiO^{2+} + 2H^+ + e^- = Ti^{3+} + H_2O$	0.10	Acid
Sulphur(II) $Na_2S_2O_3$	$S_4O_6^{2-} + 2e^- = 2S_2O_3^{2-}$	0.08	Neutral or dilute acid
Carbon(III) $H_2C_2O_4$	$2CO_2(g) + 2e^- + 2H^+ = H_2C_2O_4$	0.49	Dilute sulphuric acid

(1) The division into oxidising and reducing reagents has been made only on the basis of the common ways in which the reagents are employed. In principle it is possible to define oxidising and reducing agents only for a specified reaction.

(2) Most half reactions involve large numbers of hydrogen ions and are therefore pH dependent.

(3) Reduction potentials are thermodynamic quantities and cannot be used to predict the rate at which a redox reaction will occur.

can be used to plot a titration curve which will have a similar form to acid–base titration curves. Colour change indicators are selected to undergo colour changes as they are oxidised or reduced. Their $E^{\ominus}$ values must be matched to $E^{\ominus}$ values for the desired endpoint. Potassium permanganate, $KMnO_4$, is a commonly used oxidant and is unusual in that it is self-indicating. It has an intense purple colour which is removed when it is reduced. Iodine is also detected in an exceptional way. Starch forms a compound with it that is intensely dark blue, and can be used as an indicator in titrations.

4.3 Gravimetry

4.3.1 Principles

Gravimetry or gravimetric analysis comprises those methods of analysis in which the ultimate measurement of the analyte is by weight. Either the analyte itself or a stoichiometric compound of it is weighed, usually with a precision in the region of ± 0.1 mg. Obviously the other stages of the analytical procedure must be carried out with matching levels of precision. The procedures used can be relatively complex, involving a number of stages. Analytical methods can thus be lengthy, time consuming and demanding of high levels of manipulative skill. Broadly speaking, the methods need good quality laboratory facilities and are unsuited to measurements in the field. At one time widely used for the determination of many elements and chemical species, currently the technique has very limited use. One version that has undergone rapid development in the last 20 years, and continues to increase in use is thermogravimetry, in which weight changes are monitored under controlled heating. There are various applications of this in environmental analysis which are considered in chapter 11.

4.3.2   Gravimetric procedures

Some gravimetric procedures are very simple and straight forward. This is exemplified by one of the commonest gravimetric procedures, the determination of moisture in a sample. Typically samples are weighed before and after heating at a moderate temperature (60–80°C). It is important to minimise the possibility of thermal decomposition by using the lowest acceptable temperatures. A second example would involve fiercer heating in a flame in order to determine combustible organic material. A problem here concerns the likelihood, in many cases, of accompanying changes in the chemical composition of the non-combustible material. Other gravimetric procedures are often more complex and involve the chemical processing of the sample in order to separate the analyte in a suitably pure form for weighing. Selective dissolution may sometimes be possible in which the analyte is dissolved leaving the matrix undissolved, or the reverse. In many cases the whole sample is initially taken into solution, sometimes by the use of aggressive chemical reagents such as strong acids, alkalis and oxidising agents. The analyte is then selectively precipitated from solution and separated by filtration. If the analyte is to be weighed in the chemical form precipitated, filtration using a sintered glass crucible followed by oven drying is appropriate. In the frequent cases where the precipitate needs to be ignited in order to produce the final stoichiometric compound, filter funnels and filter papers are more appropriate. This final ignition step is required because some compounds which can be selectively and quantitatively precipitated from solution are not in themselves stoichiometric. Subsequent ignition e.g. to produce refractory oxides, is then needed. Good examples are the precipitation of the hydrous oxides of iron and aluminium which can be ignited to give the stoichiometric oxides Fe_2O_3 and Al_2O_3.

Precipitation is brought about by changing the solution conditions such that a compound of the analyte, having low solubility, is produced. Usually a precipitating agent (precipitant) is added to do this. For example, barium maybe precipitated as $BaSO_4$ by the addition of sulphuric acid, chloride ions as AgCl by the addition of silver nitrate, and various metal ions by use of organic complexing agents such as dimethylglyoxime or oxine (Table 4.8). In practice care must be exercised to ensure that an excess of the precipitant is added and that quantitative precipitation is achieved. Precipitate purity is also important and can be compromised by occlusion of impurities in the precipitate if it is formed from other than dilute solutions. Additionally impurities can be incorporated by sorption onto the precipitate surface and

Table 4.8 Some organic preciptation reagents

Reagent	Uses
8-Hydroxyquinoline (oxine)	A non-specific reagent complexing with over 20 metals. pH control can be exploited to aid selectivity. Precipitates may be brominated to provide a volumetric finish
Dimethylglyoxime	A highly specific reagent complexing with Ni(II) alone in alkaline media and Pd(II) alone in acid
Sodium tetraphenylboron $(C_6H_5)_4 B^- Na^+$	Forms 'salt like' precipitates with K^+ and NH_4^+. Hg^{2+}, Rb^+, Ca^+ interfere
Benzidene	Forms a salt with SO_4^{2-} in acid solution ($C_{12}H_{12}N_2 \cdot H_2SO_4$). The precipitate can be weighed, or titrated with NaOH or $KMnO_4$
Cupferron	Precipitates a large number of heavy metals from dilute acid solution and some, e.g. Fe, Ti, Zr, V, U, Sn, Nb, Ta, from solutions as concentrated as 10% (v/v) HCl or H_2SO_4
α-Benzoinoxime (Cupron)	Used as a specific precipitant for Cu from ammonical solution (Tartrate ions keep Fe, Al in solution), and for Mo from dilute acid. (Ni, Nb, Ta interfere)
Tetraphenylarsonium chloride $(C_6H_5)_4 AsCl$	Gives 'salt like' precipitates with a number of anionic species, e.g. ReO_4^-, MoO_4^-, WO_4^-, $HgCl_4^{2-}$, $SnCl_6^{2-}$, $CdCl_4^{2-}$, $ZnCl_4^{2-}$

can be a particular problem with flocculent precipitates which have high surface areas. Digestion of the particles by heating them in contact with the solution for an extended period produces enhanced granulation, reduces the surface area and hence sorption. For established methods the interferences are generally well understood and practical procedures have been developed to overcome them. When following a recorded method strict adherence to the experimental instructions is essential if good quality results are to be obtained. Examples of precipitation methods used for other than gravimetric analysis is discussed in chapter 5, section 5.4.5.

Further reading

Fifield, F.W. and Kealey, D. (1995) *Principles and Practice of Analytical* Chemistry, 4th edition, Blackie Academic & Professional, Glasgow.

Jefferey, G.H., Basett, J., Mendham, J. and Denney R.C. (1984) *Vogel's Textbook of Quantitative Chemical Analysis*, 5th edition, Longman Scientific and Technical, London.

Separation techniques 5
F.W. Fifield

In chemical analysis, separation of the analyte from the sample matrix may be required for two reasons. Firstly, as already reviewed in chapter 1, section 1.3.3, it may be necessary to avoid an interfering substance in the matrix. Given the complex matrices of most environmental samples this need often arises. It has also been noted in chapter 1, section 1.1.2 that the environmental levels at which elements and compounds are significant can frequently be very low e.g. mg dm^{-3} or µg dm^{-3}. In order to bring such analytes into the operating ranges of some techniques preconcentration may well be needed. All the techniques which are reviewed and discussed in this chapter have the potential for use in either role. Separations may be relatively straight forward in that the analyte is concentrated from the matrix as a whole or separated from a particular interfering substance. At the other extreme, various techniques of chromatography can be used to separate and determine a large number of components in a complex mixture. Modern chemical separation techniques involve the selective transfer of the analyte from one chemical phase to another. The most important techniques used in environmental analysis are summarised in Table 5.1 and discussed in sequence in the subsequent parts of this chapter.

5.1 Introduction

5.2.1 Introduction

Solvent extraction is widely used both for the separation of an analyte from a sample matrix and for the separation of two or more analytes. In the former, subsequent evaporation of the solvent to low volume facilitates its use as an important concentration method. In the latter, various components of a mixture can be separated prior to determination, although chromatographic methods are also now widely used for this purpose.

5.2 Solvent extraction

5.2.2 Solvent extraction of analytes from environmental samples

Organic analytes present in environmental samples from natural or anthropogenic sources are often conveniently extracted with organic solvents. The

Table 5.1 Summary of separation techniques and phase systems.

Technique	Phase system
Solvent extraction	Liquid–solid Liquid–liquid
Liquid (column) chramatography	Liquid–solid Liquid–liquid
Gas chromatography	Gas–solid Gas–liquid
Supercritical fluid extraction and chromatography	Liquid–solid
Gel permeation chromatography	Liquid–solid
Electrophoresis	Liquid–solid
Thin layer chromatography	Liquid–solid Liquid–liquid
Paper chromatography	Liquid–liquid
Ion-exchange chromatography	Liquid–solid
Precipitation	Liquid–solid

solvent must be selected as appropriate for the solution of the analyte. For solid matrices the sample must be finely divided, so that maximum contact between sample and solvent is made (see chapter 19). The mixture is then shaken or stirred vigorously, before separating off the liquid by filtration or centrifugation. A sequence of extractions is needed, typically three, in order to achieve maximum total extraction. This point is discussed more fully in section 5.2.3. Heating the mixture may also improve the dissolution of the analyte. Alternatively, it may be necessary to use a form of continuous extraction as exemplified by Soxhlet extraction (Figure 5.1). In this device the solvent is continuously cycled through the sample, which is held in a porous thimble, by distilling the solvent to a condenser centred over the thimble. The extraction of chlorinated pesticides from a soil is a good example of the use of these extraction techniques. For liquid samples, the extracting solvent must also exhibit a low miscibility towards the sample. Sequential extraction will again be needed. Solid-phase extraction (section 5.3.3) is an increasingly popular alternative for the separation and concentration of traces of organic compounds from water.

5.2.3 Separation of mixtures by solvent extraction

Where the various substances in a sample have sufficiently different solubility characteristics, it is possible, by using multiple extractions, to separate quite complex mixtures. In the past, some important routine analytical procedures were based upon solvent extraction. The analysis of amino acid mixtures and mixtures of metal ions are two examples. In the former case

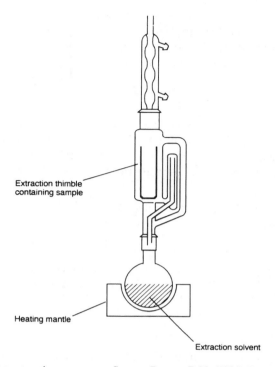

Extraction thimble
containing sample

Heating mantle

Extraction solvent

Figure 5.1 Soxhlet extraction apparatus. Source: Reeves, R.N. (1994) *Environmental Analysis*, Wiley, New York.

chromatography has developed to replace the solvent extraction technique and in the latter, atomic spectrometry or ion-chromatography may not be used without prior separation. However, the practical simplicity of solvent extraction means that it is still employed, and consideration of its fundamental principles is useful.

In most cases solvent extraction employs a phase system of water and an organic solvent the latter selected, in part, to have low miscibility. The solvent–solute interaction is also of central importance in determining suitable organic solvents. When the two solvents are agitated together, the solute is distributed between them in proportion to its solubility in the respective phases. Provided solution conditions and temperature remain constant this ratio is independent of the total amount of solute present. The Nernst Partition (Distribution) Law expresses this as a Distribution Coefficient, K_D.

$$K_D = \frac{[A]_0}{[A]_{Aq}} \qquad (5.1)$$

where $[A]_0$ and $[A]_{Aq}$ represent the concentrations of solute A, in the organic and aqueous phases, respectively.

The Nernst Law only applies strictly where the solute has the same chemical form in each solvent. The very nature of the extraction process often involves the exploitation of chemical differences so it is more valuable in practice to define a Distribution Ratio, D

$$D = \frac{[C_A]_0}{[C_A]_{Aq}} \tag{5.2}$$

where $[C_A]$ represents the total concentration of the solute irrespective of its chemical form. From this more practical viewpoint it is obviously important to be able to evaluate the efficiency (E) of an extraction. This will also be influenced by the relative volumes of the two solvents as well as D. The value of E is given by

$$E = \frac{100\,D}{\left[D + \left(\dfrac{V_{Aq}}{V_0}\right)\right]} \tag{5.3}$$

where V_{Aq} and V_0 are the respective volumes of aqueous and organic phases. If the volumes are equal, as is frequently the case, equation (5.3) is simplified to

$$E = \frac{100\,D}{(D + 1)} \tag{5.4}$$

where $D > 10^2$ a single extraction will give virtually quantitative solute transfer but as D decreases so will the efficiency. The effectiveness of a series of extractions may be assessed by calculating the amount of solute remaining in the aqueous phase after n extractions, $[C_{Aq}]_n$ where

$$[C_{Aq}]_n = C_{Aq}\left[\frac{V_{Aq}}{DV_0 + V_{Aq}}\right]^n \tag{5.5}$$

C_{Aq} being the original amount of solute present in the aqueous phase.

When separating two solutes (A and B) the ratio of the distribution ratios, β, will indicate the effectiveness of separation. β is known as the separation factor.

$$\beta = \frac{D_A}{D_B} \quad \text{where } D_A > D_B \tag{5.6}$$

This point is further illustrated in Table 5.2

Table 5.2 Separation of two solutes with one extraction, assuming equal volumes of each phase

D_A	D_B	β	% A extracted	% B extracted
	10	10	99.0	90.9
	1	10^2	99.0	50.0
10^2	10^{-1}	10^3	99.0	9.1
	10^{-2}	10^4	99.0	1.0
	10^{-3}	10^5	99.0	0.1

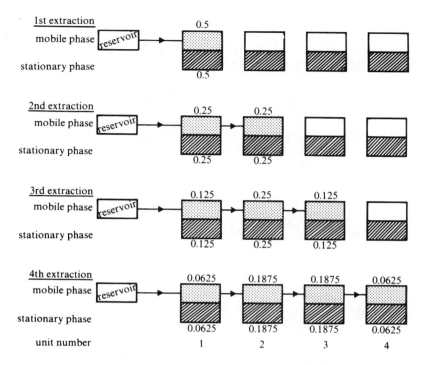

Figure 5.2 Extraction scheme for a single solute by Craig countercurrent distribution. (Figures represent the proportions in each phase for $D = 1$ and equal volumes. Only the first four extractions are shown.) Source: Fifield, F.W. and Kealey, D. (1995) *Principles and Practice of Analytical Chemistry*, 4th edition, Blackie Academic & Professional, Glasgow.

Where components of a mixture have values of D which are relatively close together, separation may still be possible by use of a large number of sequential extractions. The Craig countercurrent principle represents an elegant approach to this problem. If aliquots of one phase are placed in a series of stationary containers and a sequence of mobile phase aliquots are equilibrated with the stationary phase aliquots in turn as shown in Figure 5.2, the various components will move through the system at different rates dependent upon their different distribution ratios. Figure 5.3 illustrates some typical profiles. The peaks will have Gaussian characteristics.

The differing affinities which various solutes show for the liquid phases may arise from a whole range of chemical sources. In some cases e.g. the extraction of organic compounds from water, the basis is simply one of preferential solubility of the compounds themselves, but in others, especially where metal ions are involved, preliminary treatment to change their chemical form is often employed. Metal ions can for instance be complexed with reagents such as Dithizone for lead and dimethylglyoxime for nickel, to form coloured compounds which can be extracted into an organic solvent

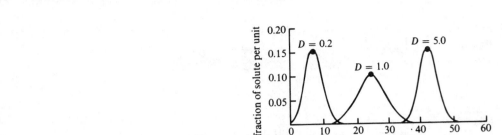

Figure 5.3 Distribution profiles for solutes having different values of *D*. Source: Fifield, F.W. and Kealey, D. (1995) *Principles and Practice of Analytical Chemistry*, 4th edition, Blackie Academic & Professional, Glasgow.

and subsequently used for both qualitative and quantitative analysis. The simplicity of operation means that fieldwork kits for 'on-site' water testing often contain such methods.

5.3 Chromatography   *5.3.1 Introduction*

In modern analytical chemistry, *chromatographic techniques* play major roles. The detail of the various techniques varies enormously but all are based on the same basic principles. A separation of substances is achieved on the basis of their differing rates of migration across or through a stationary phase, under the influence of a mobile phase. This concept has been introduced in the preceding section with the discussion of the Craig countercurrent principle. If the transfer and equilibration processes are continuous between the two phases, rather than compartmentalised as in the Craig system, a model for chromatography is produced.

Sorption mechanisms. The mechanisms by which the stationary phase attracts the solute from the mobile phase vary but it is the strength of this attraction which determines the relative rate of movement of the solutes through the system. The four sorption mechanisms summarised below are used.

1. *Surface adsorption*: Separation is based upon polarity differences between the mobile phase and a solid stationary phase.
2. *Partition*: If a liquid is coated upon an inert solid support, a liquid stationary phase can be produced. Partition between the phases is then dependent upon differences in solubility, for a liquid mobile phase, or in volatility for a gaseous mobile phase.
3. *Ion exchange*: Polymeric materials can be chemically modified to contain large numbers of functional ionic groups. Mobile counter ions

attached to these groups can interchange with solute ions in the mobile phase. Separation is effected on the basis of differing affinities between ions and functional groups.

4. *Size exclusion*: Large molecules having 'cage-like' structures can be used to influence the rate of migration of solute species. Small molecules can diffuse into the cage structure which larger molecules are unable to do. Thus the smaller the species, the more it will be retarded by the stationary phase. Separation is thus effected on the basis of the size of the solute species. This is not strictly a sorption process but for the purposes of chromatography may be classed as such.

It would be wrong to assume that in a system only one sorption process exists. However, with certain exceptions, one mode of sorption will predominate with others playing minor roles.

5.3.2   *Characteristics of chromatograms*

Operationally, chromatography is used in two main ways. Either the stationary phase is packed into a tube, known as a *column*, or spread as a *thin layer* on a plate support. *Paper* chromatography may be regarded as a special version of the latter. In column chromatography the elute is monitored for the components as it leaves the column, with a detector generating a signal proportional to the amount of component present. This produces a series of peaks in the signal which are characterised by the volume of mobile phase required for their elution, whilst the peak area represents the amount of analyte present. The peak shape develops as a result of *mass transfer* and *diffusion processes* which are discussed below. In thin layer or paper chromatography the analytes if not already coloured are visualised on the plate by the use of colour reagents and characterised by the distance that they have migrated. Mass transfer and diffusion effects are also relevant to the appearance of these chromatograms. As the solute moves through the particles of the stationary phase, some molecules will travel by slightly longer paths than others. This multiple path effect also contributes to band broadening. The effect of these factors on the efficiency of a separation is often expressed in the van Deemter equation (equation (5.7))

$$H = A + B\bar{u} + C\bar{u} \tag{5.7}$$

A is the multiple path term, $B\bar{u}$ the molecular diffusion term, and $C\bar{u}$ the mass transfer term. H is a measure of efficiency known as the plate height, defined in equation (5.10), and $\bar{u}$ the mean velocity of the mobile phase. Thus the value of H is dependent on the mobile phase velocity.

Column chromatography.   In column chromatography the sample mixture is injected onto the beginning of the column in a very small volume of liquid

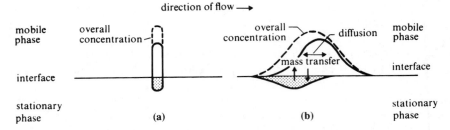

Figure 5.4 Effects of diffusion and mass transfer on peak width. (a) Concentration profiles of a solute at the beginning of a separation. (b) Concentration profiles of a solute after passing some distance through the system. Source: Fifield, F.W. and Kealey, D. (1995) *Principles and Practice of Analytical Chemistry*, 4th edition, Blackie Academic & Professional, Glasgow.

(ca. 1 µl) or possibly vapour, and initially forms a narrow band. As a result of the mass transfer between mobile and stationary phases band broadening occurs, because, whilst those species in the mobile phase move with it, those in the stationary phase remain immobile until transferred back to the mobile phase. At the same time diffusion processes taking place, largely in the mobile phase, will contribute to the band broadening. These concepts are represented diagrammatically in Figure 5.4 In the ideal, the peaks are symmetrical with Gaussian shapes although in practice some deviation from ideality can be expected. The narrowness or sharpness of the peaks has a direct bearing on the ease with which adjacent peaks can be resolved. Hence it is useful to have a numerical representation of the sharpness or *efficiency* as it is termed. This measures the ability of a system to resolve peaks which are eluted in close sequence after the passage of similar volumes of the mobile phase. This volume for a particular component is known as its *retention volume*, V_R. The formulae used are derived from theoretical principles which will not be reviewed here. Either the efficiency can be calculated from

$$N = 16 \left(\frac{V_R}{W_B} \right)^2 \qquad (5.8)$$

or

$$N = 5.54 \left(\frac{V_R}{W_{h/2}} \right)^2 \qquad (5.9)$$

where N is known as the plate number, W_B the base width of the peak, and $W_{h/2}$ its width at half height. In practice, because chromatographic columns are usually run at constant flow rates, retention time V_T and peak width in time units, are more readily measured, and may be substituted for volumes. Direct comparison of values of N calculated by the different formulae is not valid as they do not necessarily yield the same values. It is important therefore, to ensure that the same formula is used throughout. The latter (5.8) containing $W_{h/2}$ is the easiest to measure practically. A further term that may be met is the plate height (H) or HETP (Height Equivalent to a Theoretical Plate), which is defined by

$$H = \frac{L}{N} \qquad (5.10)$$

where L is the column length. Chromatographic theory was originally developed by analogy to distillation theory and some of the terminology of the latter has persisted in use. Values of N can be large, i.e. many thousands with corresponding values of H less than 1 mm.

The factors above and their influence on efficiency of separations are expressed in the van Deemter relationship (equation (5.7) and represented diagrammatically in Figure 5.5.

The ability of a chromatographic method to distinguish between two adjacent peaks in a chromatogram is known as its resolution. R_S. In practical terms this is measured by the comparison of the peak widths to the distance between the peaks, and is defined by

$$R_S = \frac{2\Delta V_R}{(W_1 + W_2)} \qquad (5.11)$$

where ΔV_R is the distance between the peak maxima, with W_1 and W_2 being the respective peak widths. As with efficiency measurements, provided the

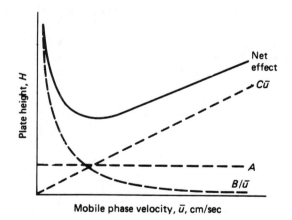

Figure 5.5 Efficiency as a function of mobile phase velocity and the effect of each term in equation (5.11). Source: Fifield, F.W. and Kealey, D. (1995) *Principles and Practice of Analytical Chemistry*, 4th edition, Blackie Academic & Professional, Glasgow.

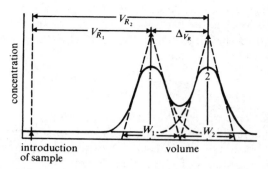

Figure 5.6 Resolution of adjacent peaks. Source: Fifield, F.W. and Kealey, D. (1995) *Principles and Practice of Analytical Chemistry*, 4th edition, Blackie Academic & Professional, Glasgow.

flow rate is constant, time measurements may be substituted for volume. Figure 5.6 shows those ideas diagrammatically.

It is important to note that both efficiency and resolution are highly temperature dependent. Careful temperature control is needed especially in GLC where separations are routinely carried out at elevated temperatures. Temperature effects may be used to advantage as will be explained in later comment on GLC.

Thin-layer and paper chromatography.   Thin-layer and paper chromatography are much less sophisticated in operation than column techniques. The sample is placed on the plate close to one edge, in the form of a small spot or occasionally as a streak. The plate then develops when the edge is placed in contact with the mobile phase. As the mobile phase moves across the plate, the process of mass transfer and diffusion operate just as for column chromatography. The net result is that the spot becomes larger and elliptical as it moves across the plate. Characterisation of the solutes is in terms of R_f value only, where

$$R_f = \frac{\text{distance travelled by solute spot}}{\text{distance travelled by solvent front}} \qquad (5.12)$$

5.3.3   High-performance liquid chromatography

High-performance liquid chromatography or HPLC, as it is almost universally known, is one of the major techniques of modern analytical chemistry. It provides high efficiency and high resolution for a wide range of organic compounds, and is extensively used in environmental analysis. Separations are based upon stationary phases of small particulate size and consequently high surface areas, packed into short columns of 20 cm length or less. With

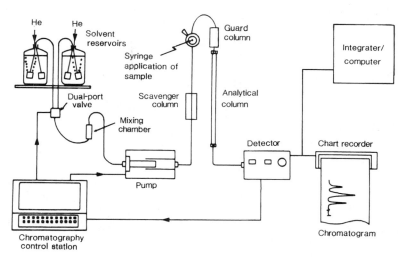

Figure 5.7 Schematic diagram of a binary (two-solvent) HPLC system. Source: Fifield, F.W. and Kealey, D. (1995) *Principles and Practice of Analytical Chemistry*, 4th edition, Blackie Academic & Professional, Glasgow.

the mobile phase pumped through the system at high pressures (200 bar, 3000 psi) complex separations may be achieved in only a few minutes at flow rates of 1–3 cm^3 min^{-1}. Figure 5.7 illustrates a typical chromatograph.

Prior to being pumped through the column, the mobile phase must have dissolved gases removed to prevent bubble formation when under high pressure in the chromatograph. This is done by sweeping the gases out with a current of helium which is passed through for about a minute. Some separations involve changes in mobile phase composition during the separation and a switching and mixing system is incorporated to allow for this.

Pumps of varying design may be found in use but ultimately they must all conform to similar specifications. They must operate at the high pressures needed to force the mobile phase through the tightly packed column and must do this with a steady flow which is as near pulseless as possible.

The sample solution is introduced into the mobile phase between the pump and the column by an injection loop. This loop is of fixed volume (5–20 μl) and is fitted so that the mobile phase can be directed through it. This is done by a switching valve once the sample solution has been injected into the loop. The stationary phase is packed into a column, usually made of stainless steel, typically 10–20 cm in length and 5 mm internal diameter, although longer or narrower columns may be encountered. Sometimes a pre-column is used with the double function of removing an interfering solute and filtering out particulate material which would otherwise collect in the analytical column and shorten its life.

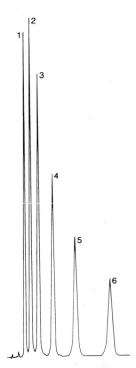

Figure 5.8 An illustrative HLPC chromatogram. Analysis by Microbore LC. Column, 130 × 2 mm Spheri-5ODS; flow, 500 μl min⁻¹. Peak identity: 1, benzene; 2, toluene; 3, ethyl benzene; 4, propyl benzene; 5, butyl benzene; 6, amyl benzene. Source: Pye Unicam Ltd., Cambridge.

After leaving the column the eluate passes through the detector which responds to the presence of analytes and develops an electrical signal with a magnitude determined by the amount of analyte passing through the detector. This signal is plotted against solvent flow or time to give a chromatogram. A typical example is shown in Figure 5.8. For most purposes a detector is required which will respond to as comprehensive a range of solutes as possible. In this respect a detector based upon the difference in refractive index between the solute and the mobile phase has the widest applicability. However, *refractive index detectors* lack sensitivity relative to others, and tend to be reserved for use when other techniques cannot be used. The most commonly encountered detector type is based upon the absorption of UV radiation by the solute. As discussed in chapter 8, many organic compounds absorb UV radiation with considerable intensity. Hence the detectors have high sensitivity. Furthermore, because compounds have characteristic absorption spectra variable wavelength detectors may be used to monitor for a selected analyte or multiwavelength ones for a whole range of compounds simultaneously. The latter type are often referred to as *diode array detectors* by reference to the technology employed in their construction. These detectors can also be used to scan a peak for impurities by looking for absorption characteristics foreign to that particular component. Other types of detector which may be met include fluorimetric and amperometric (polarographic). Interfacing with mass spectrometry (chapter 8) is also a possibility and this can enable unknown solutes to be identified. It is, however, complex, being difficult and expensive to achieve, and with an applicability rather more limited than the GLC-MS interface.

Mobile and stationary phases.   Various combinations of mobile and stationary phases are employed. Probably the most common, uses a stationary phase consisting of porous silica particles with a surface onto which long chain hydrocarbon groups, R, are chemically bonded by siloxane (Si–O–Si–R) bonds. *Reverse phase* is a term sometimes used to distinguish them from particles with chemically unmodified surfaces which are described as *normal phase*. A widely used stationary phase has an octadecyl group (C_{18}) often being referred to as ODS. Other bonded groups include octyl, hexyl and more ionic groups for use as ion exchangers (see section 5.3.4). Partition of solutes between the nonpolar stationary phases and more ionic mobile phases is commonly the basis for separation with the components eluting in decreasing order of polarity. The composition of the mobile phase may be adjusted so as to optimise its polarity for the separation of particular analytes. Examples of mobile phase mixtures are:

1. Hexane (low polarity), methanol (high polarity)
2. Hexane (low polarity), acetonitrile (high polarity)
3. Methanol (high polarity), aqueous phosphate (very high polarity)

Applications.   In environmental analysis many organic compounds are determined by HPLC. Pesticides, herbicides and poly-nuclear aromatic hydrocarbons are particularly noteworthy examples. More information on these can be found in chapter 19. Short columns or cartridges (ca. 0.5 cm) containing HPLC-type stationary phases can be very useful in some field-work. Compounds such as those above may be concentrated on to the cartridges by passing relatively large quantities of water through them. The cartridges can then be returned to the laboratory where the compounds can be desorbed and determined using sophisticated equipment. *Solid phase extraction* is a term often applied to this technique and is more fully discussed in chapter 19.

5.3.4 Ion-exchange chromatography

Ion-exchange chromatography (IC) is a long established technique for the separation of charged species in solution, and their concentration prior to analysis. In its early form, large columns (5–50 cm long, 0.5–2 cm i.d.), operating under gravitational flow or low pressure were used. Such systems still have a useful role to play. In more recent years *ion chromatographs* have been developed which are broadly similar to HPLC systems (Figure 5.7) The key factors in this development are the availability of stationary phases which can resist high pressures used and ultrasensitive techniques for detecting and measuring ions in solution.

In both cases the principle of separation is the selective exchange of ions between a mobile phase and a stationary phase which has a higher and varying affinity for the ions to be separated. The ion-exchange process for a cationic system may be represented by the generalised equation

$$nR^-H^+ + M^{n+} = (R^-)_n M^{n+} + nH^+ \tag{5.13}$$

where R represents the stationary phase matrix. The equilibrium constant, K, for this reaction is known as the *selectivity coefficient* and is a measure of the affinity between the ion and the stationary phase. From the equation

$$K = \frac{[M^{n+}]_R [H^+]^n}{[M^{n+}][H^+]_R} \tag{5.14}$$

where $[M^{n+}]_R$ and $[H^+]_R$ are the respective concentrations of M^{n+} and H^+ in the stationary phase, it can be seen that K increases as the affinity of an ion for the stationary phase increases relative to hydrogen. K will also be dependent upon the charge carried by an ion. Typical selectivity ratios for common cations and anions vary over the range $K = 0.1$ to $K = 10$. The resolution and efficiency of ion-exchange separations also depend upon the factors discussed in section 5.3.1. The solution chemistry involved in ion-exchange separations is often complex. For example, cations and anions may be separated on the basis of differing affinities for stationary phases. Some metallic cations may, however, be complexed in solution with anionic species involving subsequent chromatography on anion-exchange stationary phases. Zinc is one such cation, forming $ZnCl_4^{2-}$ species in solutions which contain chloride ions. Uranium and the transuranic elements are also separated on the basis of anion-exchange chromatography (see chapter 16). Generally included in discussions of ion-exchange systems are chelating stationary phases for the separation of traces of metal ions. Although not strictly ion exchangers they are employed in very similar ways.

Instrumentation.   Where column chromatography is being used under gravitational or low pressure flow simple glass columns, funnels and connectors are used. For high performance use, instruments are analagous to HPLC chromatographs (Figures 5.7 and 5.8). Indeed, some equipment is constructed with dual use in mind. Special care is needed to avoid construction materials which release ions into the mobile phase but otherwise the significant differences lie in the stationary phase and the detectors used.

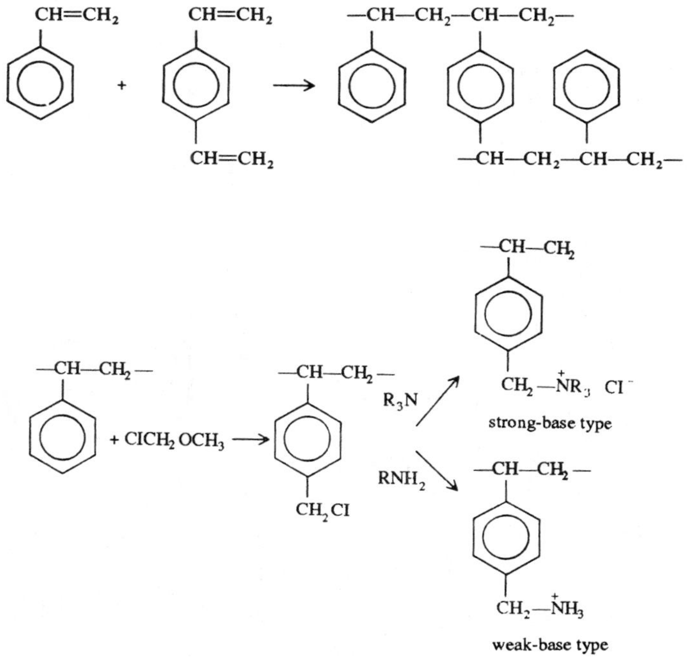

Figure 5.9 (a) The formation of a divinyl benzene polymer. (b) The introduction of functional groups for ion exchange.

Some ion chromatographs use a dual column system enabling anions and cations to be determined in sequence.

Stationary phases. Polymeric, resinous materials containing large numbers of functional groups provide the basis for ion-exchange stationary phases. They are exemplified by styrene–divinyl benzene copolymers as illustrated in Figure 5.9. For low pressure chromatography beads of pure resin ca. 1 mm in diameter are suitable. For high-performance ion-exchange chromatography smaller beads with much greater rigidity to withstand the high pressures are required.

Detectors. The most commonly used detector for ion-exchange chromatography is based on conductivity, which is a particularly sensitive technique for detecting ions in solution but is non-selective in its response. Thus the analyte must be separated from other electrolytes before measurement. Although the chromatography will have separated an analyte from other components in the sample mixture the mobile phase itself has a high ionic strength and will interfere in the conductivity measurements. A suppressor

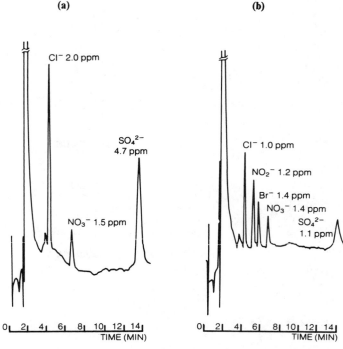

(a) **(b)**

CI⁻ 2.0 ppm

SO_4^{2-}
4.7 ppm

CI⁻ 1.0 ppm

NO_2^- 1.2 ppm

Br^- 1.4 ppm
NO_3^- 1.4 ppm
SO_4^{2-}
1.1 ppm

NO_3^- 1.5 ppm

0 2 4 6 8 10 12 14
TIME (MIN)

0 2 4 6 8 10 12 14
TIME (MIN)

Figure 5.10 Separation of ions in environmental water: (a) anions; (b) cations. Source: Pye Unicam Ltd., Cambridge.

unit, placed between the column and the detector, is used to remove the interfering species.

Applications.   IC clearly has wide potential applications. In environmental analysis it is routinely used for the analysis of water samples. Figure 5.10 shows a chromatogram for the determination of anions in water.

5.3.5   Thin-layer and paper chromatography

Thin-layer chromatography, TLC, and paper chromatography, PC, are essentially different forms of the same technique and can be considered as such. A liquid mobile phase is employed with a stationary phase spread out in a thin layer or sheet. For TLC this layer is spread on an aluminium, glass or plastic plate. In PC the structure of the cellulose itself comprises the sheet. Operationally the sample solution is loaded by pipetting a small volume onto the plate in the form of a spot at the origin, close to (1–2 cm) one edge of the plate. The plate is then allowed to develop upwards by capillary action. This process is relatively fast and most chromatograms

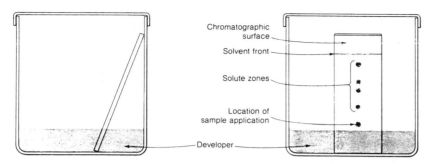

Figure 5.11 End view and side view of plates for thin layer chromatography. Source: Manahan, S.E. (1986) *Quantitative Chemical Analysis*, Brooks Cole.

can be developed in less than an hour and many within a few minutes. Plate development is carried out in a closed tank (Figure 5.11) so that the enclosed atmosphere becomes saturated with the vapour of the mobile phase. In this way evaporation from the surface of the plate is restricted.

A similar range of stationary phases to that used in other forms of chromatography is available. Hence many different types of sample separations can be achieved. The separation of organic compounds by surface adsorption on silica-based stationary phases is especially important. Sometimes complex mixtures can be more effectively resolved if a second solvent is used for successive plate development at 90°C to the initial one. PC is a little unusual in that it functions on the basis of liquid–liquid partition with the liquid stationary phase provided by the water molecules which are bonded to the cellulose structure of paper.

The components of the mixture will be recognised as separate spots which have migrated different distances across the plate with the developing mobile phase (Figure 5.12). They are characterised qualitatively by their R_f values (section 5.3.2). In truth R_f values are rather crude qualitative identifiers because many compounds will exhibit similar or identical values. Further analysis will often be needed for reliable identification. Quantitative analysis may involve use of a densitometer directly on the plate to measure the amount of light absorbed or perhaps fluoresced. It has been previously noted that many compounds absorb UV radiation so UV light may well be used. Removal of the spot from the plate by scraping or cutting out prior to measurements is also a possibility. Where components do not interact appropriately with radiation themselves derivitisation either before or after removal from the plate is a possibility.

TLC and PC provide simple and rapid techniques for scanning a sample for particular analytes or assessing the purity of compounds, and may be encountered and used in these ways.

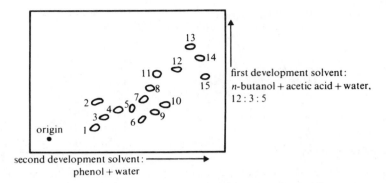

Figure 5.12 Two-dimensional chromatogram; standard map for amino acids. 1, Taurine; 2, glutamic acid; 3, serine; 4, glycine; 5, threonine; 6, lysine; 7, alanine; 8, α-amino-*n*-butyric acid; 9, arginine; 10, hydoxyproline; 11, tyrosine; 12, valine; 13, leucine; 14, β-phenylalanine; 15, proline. Source: Fifield, F.W. and Kealey, D. (1995) *Principles and Practice of Analytical Chemistry*, 4th edition, Blackie Academic & Professional, Glasgow.

5.3.6   Gas chromatography.

Compounds which are gaseous or low boiling liquids (b.p. up to 300°C) may be separated and determined by gas chromatography, GC. Both gas–solid and gas–liquid systems are used but it is the latter gas–liquid chromatography, GLC, which is most extensively employed. Separation is achieved by partitioning the analytes between the gaseous mobile phase and a stationary phase consisting of a high boiling liquid coated onto an inert solid support, which is packed into a narrow column. Permanent gases and other low boiling analytes can be separated on the basis of the different strengths of their absorption on polar surfaces.

Instrumentation.   Figure 5.13 shows the principal parts of a typical gas chromatograph.

Gas supply.   The mobile phase or carrier gas is supplied at a rate between 2 and 50 cm^3 min^{-1} depending upon the nature of the column. Pressures up to 40 psi (3 bar) may be needed to achieve this. A flow controller is used to ensure that the flow rate does not vary significantly even when the column temperature is programmed to rise during some separations. Commonly used carrier gases are nitrogen for packed columns and hydrogen or helium for capillary columns.

Sample injection.   Samples are usually dissolved in a readily volatile solvent and flash vaporised by injection into a pre-column chamber at ca. 50°C above the column temperature. Occasionally thermally unstable compounds are injected directly onto the top of the column in order to

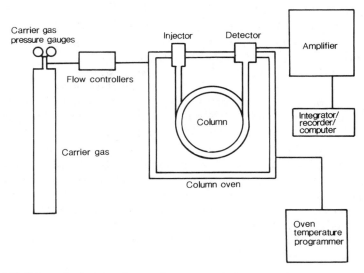

Figure 5.13 Schematic diagram of a gas chromatograph. Source: Smith, R.M. (1988) *Gas and Liquid Chromatography in Analytical Chemistry*, Wiley, New York.

minimise decomposition. Injection is through a self-sealing silicone rubber septum from a microsyringe (1–10 µl capacity). This process is often automated. By doing so, better precision may be attained together with more efficient instrument usage. Split and splitless modes are used. In the latter the whole injection passes straight to the column whereas in the former the sample is split in ratios of 50:1 and up to 500:1 with a large part being vented to waste. This avoids overloading the column and is an especially important facility where capillary columns are used.

Thermal desorption cartridges are a valuable adjunct to sample collection and injection. They can be used to collect samples during fieldwork and to improve sensitivity by concentrating volatile compounds from large volumes of air. The cartridges consist of a small tube containing a strong sorbent which will collect a wide range of compounds. Air is pumped through the cartridge, which strips out, and concentrates pollutants (such as hydrocarbons and petroleum combustion products). In a specially adapted gas chromatograph, the cartridge is heated to desorb these compounds which are then carried onto the column for separation and determination. Examples may be found in chapter 14. This technique is analagous to that described in section 5.3.3 for HPLC.

Columns. The columns used in GC are of two broad types. So called packed columns, are generally made of stainless steel or glass, up to 2 m in length and 1 or 2 mm diameter. The stationary phase in particulate form is tightly packed into the column. Characterised by a tolerance of high sample loading volumes relative to capillary columns, they exhibit poor resolution and efficiencies of a few thousand theoretical plates only. This is due partly to the turbulent flow as the mobile phase moves through the interstices of the stationary phase and partly to peak broadening brought about by the multiple path effect (section 5.3.2). With capillary columns, most commonly made of fused silica, and up to 50 m long, and with an internal diameter of 0.1 mm or less, plate numbers of 150 000 are attainable. The stationary phase is attached to the internal surface of the tube, presenting a relatively smooth profile to the mobile phase. Hence turbulence and multiple path effects are dramatically reduced. An alternative name used is *wall coated open tubular* (WCOT) column. For low boiling analytes porous layer open tubular (plot) columns may be preferred, with the separation based on surface adsorption characteristics. Where a higher column loading is essential, megabore columns with internal diameters in the range 0.5–0.6 mm, can be valuable. The performance of packed and capillary columns is compared in Figure 5.14.

Oven and temperature control.   It has been established (section 5.3.2) that chromatographic performance is sensitive to temperature change. The oven must maintain a well-controlled temperature, ±0.1°C within the range of ambient to 400°C which encompasses most GC separations. With many mixtures where analytes have widely divergent boiling points it is an advantage to be able to raise the temperature after the lower boiling components have eluted, to expedite the elution of the higher boiling ones. Temperature programmers allowing for rise at programmed and reproducible rates are routinely available.

Detectors   used in GC vary in nature depending upon the characteristics of the analyte and the circumstances of its determination. The one with the most comprehensive response is based on the change in thermal conductivity in the mobile phase as a component is eluted. Principally, this results from the mass difference between the molecules eluted and those of the carrier gas. Thermal conductivity detectors (TCDs) have only modest sensitivity and the usually preferred option for general purpose GC use, is the flame ionisation detector (FID). The eluate stream is mixed with air and hydrogen allowing combustion of most components. Ionic species produced in the combustion facilitate the passage of an electronic current between two electrodes in contact with the flame. The magnitude of the current reflects the number of ions and thus the amount of the eluted compound present. For some purposes more selective detectors have advantages, some with particular value in the analysis of environmental samples. One such, is the electron capture detector (ECD). High energy electrons (β-particles) from a radioactive source such as ^{63}Ni or ^{3}H, are used to ionize the carrier gas

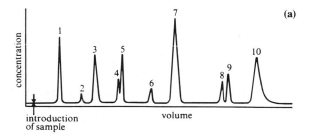

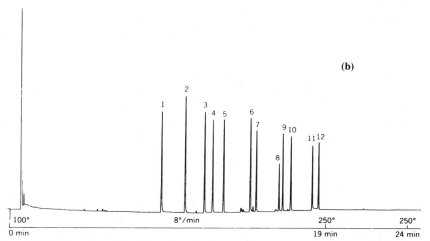

Figure 5.14 Comparison of peak resolution by packed column and capillary column GC. (a) A typical elution profile; the separation of aliphatic esters by gas–liquid chromatography on a packed column. 1, Methyl formate; 2, methyl acetate; 3, ethyl formate; 4, ethyl acetate; 5, *n*-propyl formate; 6, iso-propyl acetate; 7, *n*-butyl formate; 8, *sec*-butyl acetate; 9, iso-butyl acetate; 10, *n*-butyl acetate. (b) Separation of some pesticides on a capillary column. 1, HCB; 2, α-HCH; 3, Lindane; 4, Heptachlor; 5, Aldrin; 6, Δ-HCH; 7, Heptachlorepoxide; 8, *p,p*-DDE; 9, Dieldrin; 10, Endrin; 11, *p,p*-DDD; 12, *p,p*-DDT. Sources: (a) Fifield, F.W. and Kealey, D. (1995) *Principles and Practice of Analytical Chemistry*, 4th edition, Blackie Academic & Professional, Glasgow. (b) Nordian Instruments Oy Ltd., P.O. Box 1, SF-00371, Helsinki, Finland.

stream. As a result, in the absence of an eluted component, a steady electronic current can flow between the electrodes. An eluting component which is electronegative will 'capture' electrons and reduce the current flowing, leading to a peak in the chromatogram.

ECDs have proved to be very sensitive detectors and have been extensively employed in the determination of chlorinated pesticides where the electronegative chlorine makes their application particularly appropriate. Good sensitivity for sulphur containing compounds is also of value in the analysis of environmental samples. It is appropriate also to mention the flame photometric detector (FPD) which selectively monitors radiation emitted by species such as S_2 and HPO produced in the combustion of sulphur or phosphorus containing compounds.

In recent years, the interfacing of gas chromatography with dedicated mass spectrometers has become routine. The mass spectrometer can function as an aid to peak identification. Additionally it may be adjusted for a selective response to a single type of ion originating in one component of the mixture (chapter 8, section 8.5.5).

Applications.   GC and GLC are widely employed for the determination of volatile analytes from different matrices. Many variants of the technique may be met. For environmental analysis the determination of trace amounts of organic compounds represents the most important area of application. GLC has been particularly effective in the study of chlorinated pesticides.

5.4 Other separation techniques

The techniques discussed in the preceding sections of chapter 5 are those which are most widely employed in environmental analysis. Other techniques, not included above, can also be important in rather more limited ways. Some of those which may be encountered from time to time are summarised below.

5.4.1   Supercritical fluid chromatography (SFC)

Supercritical fluids are produced when volatile materials are subjected to temperatures and pressure in excess of critical values. The fluids produced have properties intermediate between liquids and gases. Carbon dioxide with a critical temperature of 31.1°C and critical pressure 72.9 bar has proved most successful. SFC based on CO_2 is currently attracting attention as a technique for preliminary sample treatment or 'clean up'.

5.4.2   Gel permeation chromatography (GPC)

When a mobile phase containing sample molecules passes over and through a stationary phase with a 'cage-like' structure the rate of movement of solute molecules will be affected by the degree to which they penetrate into the interstices of the stationary phase. Large molecules which cannot penetrate at all move across the surface and are eluted first. The smaller the molecule, the more easily it passes into the stationary phase and the more it will be retarded. Hence molecules are eluted in direct order of molecular size. Gels are used for the stationary phase and the technique is known as gel permeation chromatography (GPC). It is also, sometimes, appropriately known as size exclusion chromatography. GPC is useful for the separation of large molecules, of biological origin, and can operate for relative molecular masses of a few hundred up to 1 million or more. The separation of proteins represents a good example of its application.

5.4.3   Electrophoresis

Charged species can be caused to migrate across a surface or through a column under the influence of a potential gradient. For a fixed stationary phase species will migrate at differential rates dependent upon their molecular size, shape and charge. This technique is known as electrophoresis, and has been applied for some years in the semiquantitative separation of biological molecules. More recently, especially in the drug industry, it has attracted much attention in a form known as capillary electrophoresis (CE). Column separations for some types of compounds are rapid and can have resolution and efficiency superior to comparable HPLC separations. Its use can be expected to become more widespread in future. Figure 5.15 illustrates CE-separations currently being attained.

5.4.4   Distillation and volatilisation

Distillation has long been a technique used for the separation of volatile compounds, but even in its most highly developed form of fractional

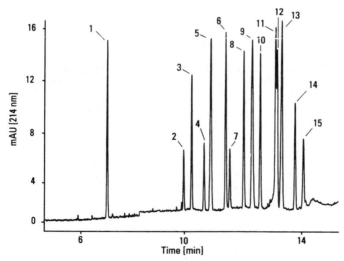

Figure 5.15 An illustration of the separation of some pesticides by capillary electrophoresis. 1, Desethylatrizine; 2, Hexazinone; 3, Metoxuron; 4, Monolinuron; 5, Simazine; 6, Cyanazine; 7, Metabromuron; 8, Chlortoluron; 9, Isoproturon and Atrizine; 11, Linorun; 12, Methabenz-thiazuron; 13, Sebutylazine; 14, Terbutylazine; 15, Metolachlor. Source: Hewlett Packard (1994) *Applications of the HP3D Capillary Electrophoresis System*, Vol. 1, Hewlett Packard Co.

distillation it has very poor resolution compared to modern separation techniques such as chromatography. In the analysis of environmental samples for readily volatile components, it is helpful sometimes to separate these volatiles as a group, prior to analysis by some other technique. One particular application of note is *headspace analysis* in which the volatilisation is brought about in a closed container. The vapour is then sampled using a gas syringe and injected into a gas chromatograph for analysis. Alternatively the volatiles may be concentrated by condensation in a cold trap, prior to analysis, or concentrated onto a thermal desorption cartridge.

5.4.5   Precipitation

Precipitation from solution as a technique of analysis has been discussed in chapter 4. The tendency of some precipitates to sorb other materials as impurities has also been noted. This latter process occurs naturally e.g. in sediments where trace metals may have been concentrated by the hydrous oxides of Fe, Al and Mn. The same processes are also used in water treatment to remove undesirable impurities. Parallel applications may be found in the 'clean up' of samples in the laboratory or for the concentration of trace analyte species from large volumes of solution. For instance, ultra-trace amounts of chloride ions can be concentrated onto the mixed phosphate precipitate of Pb and Mg, or mixed radionuclides from water, using a precipitate of hydrated iron oxide or magnesium hydroxide.

Further reading

Anderson, R. (1987) *Sample Pre-treatment and Separation*. Wiley, Chichester.
Fifield, F. W. and Kealey, D. (1995) *Principles and Practice of Analytical Chemistry*. 4th edition, Blackie Academic & Professional, Glasgow.
Sewell, P. and Clark, B. (1987) *Chromotographic Separations*. Wiley, Chichester.

General principles of spectrometry **6**

P.J. Haines

For any analysis, we need to find the qualitative nature of the sample and often the quantity or concentration present. For this reason, we must make precise measurements of the parameters for any analytical technique used.

6.1 Introduction

The older methods, where the spectrum was observed and which could be described as 'spectroscopy', are now improved so that measurements of amounts, energies, and characteristic interactions can be made. These give the many methods of spectrometry.

As discussed in chapter 3, the structure of atoms and molecules involves the distribution of the species amongst definite sets of energy levels, which may be described by their quantum numbers and geometry.

6.2 Energy levels

Figure 6.1 shows a summary of the electronic energy levels of a hydrogen-like atom, showing the typical transitions which may occur.

Figure 6.2 shows the electronic, vibrational and rotational levels for a typical diatomic molecule and typical transitions which may occur.

For all these sets of energy level, the probability of a transition depends on the population of the levels involved as given by the Boltzmann equation (chapter 3, equation (3.7)) and upon the selection rules which apply for the transition. The absorption of energy, from whatever source, be it thermal, electrical, collision or absorption of radiation, will change the distribution of the species amongst the energy levels, and may even change their chemical nature by ionising them or dissociating them.

In general, there are four types of transition, as shown in Figure 6.3a–d. If the species undergoes an absorption of energy, it is raised from the lower energy, E_L, to the upper energy level, E_U. This is possible because of the absorption of radiation or by thermal activation. The energy absorbed is $(E_U - E_L)$ as shown in Fig 6.3a.

6.3 Types of transition

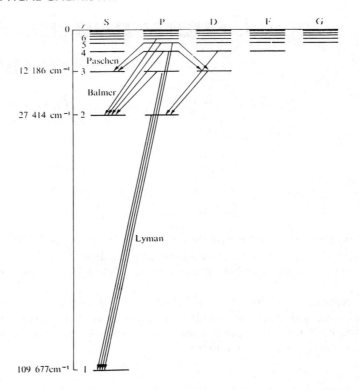

Figure 6.1 Energy levels of the hydrogen atom showing allowed transitions. Source: Atkins
P.W. (1978) *Physical Chemistry*, 1st edition, Oxford University Press, Oxford.

If the species has been activated already into the upper energy level, E_U,
it can fall back to the lower level, with the emission of energy $(E_U - E_L)$ as
shown in Fig. 6.3b.

6.3.1   Lasers

A special case of energy emission occurs when many atoms have been
promoted by 'pumping' (with light or electrical energy) to a higher energy
state, but have not returned spontaneously. A single photon emitted can
cause other excited atoms to return to a lower energy state by the stimulated
emission of radiation. Because of the way in which this occurs, the waves
are all in phase or coherent, and the light is intense, monochromatic and
polarised. This gives Light Amplification by the Stimulated Emission of
Radiation, or LASER action.

If the species absorbs more energy, it may undergo more than one
transition. For example, if it absorbs enough energy it can be raised to the
level labelled E_U, but then it falls back to the level E_3 before dropping back

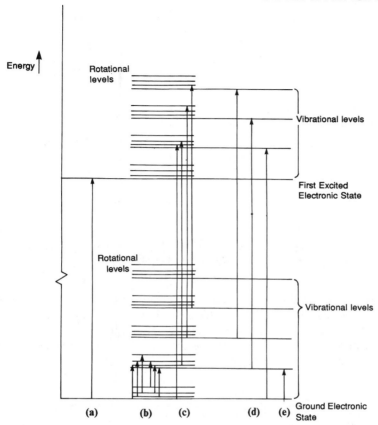

Figure 6.2 Electronic, vibrational and rotational energy levels of a molecule, with typical transitions (not to scale): (a) electronic; (b) rotation + vibration; (c) rotation + vibration + electronic; (d) vibration + electronic; (e) vibration. (Note: vibrational levels equally spaced; rotational levels at levels 1:3:6:10).

to the lowest level E_L. This is known as fluorescence, and the energy emitted may be different from the energy absorbed, as shown in Figure 6.3c. If the excited species fall back directly from the uppermost level E_U to the ground state, as shown by the dashed line, the emission is resonance fluorescence.

If the absorption of energy is very large, and is followed by emission of a slightly different wavelength, then the Raman effect is observed, as depicted in Figure 6.3d. On the left-hand side of Figure 6.3d the molecule in the lowest state E_L is activated, by high energy, high frequency source, and then re-emits a slightly lower frequency to return to a slightly higher state E_U. The difference between the exciting and emitted frequencies corresponds to the difference in the levels $(E_U - E_L)$. This is called the Stokes–Raman Radiation. On the right-hand side is shown the opposite. A molecule in the upper energy level is excited by the same exciting radiation, and emits a

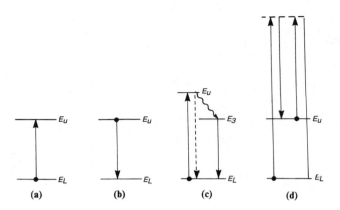

Figure 6.3 Types of transition between energy levels: (a) absorption; (b) emission; (c) fluorescence; (d) Raman.

slightly higher frequency to fall back to the ground state. This is the Anti-Stokes Radiation. If the exciting radiation is re-emitted unchanged, due only to scattering, this is the Rayleigh Radiation.

If the transition is such that an electron is lost completely, the remaining species is then charged, and we have an ionisation process.

6.4 Molecular dissociation

The energies holding molecules together are often large, but not infinite! If enough energy is supplied, bonds may be broken, or molecules be converted back to atoms. As described in chapter 3, the average value for many different molecules is called the bond energy, while for specific bonds (A–B) it is the bond dissociation energy $D(A-B)$.

For example, if we subject molecules of HCl to sufficient energy, perhaps by bombardment with high-energy electrons in a mass spectrometer, they may break down into gaseous atoms of H and Cl, which may become ionised.

$$\text{HCl(gas)} = \text{H(gas)} + \text{Cl(gas)}   \Delta H = D(\text{HCl}) = 428 \text{ kJ mol}^{-1}   (6.1)$$

Absorption of ultraviolet radiation in the upper atmosphere can break down or ionise the molecules of nitrogen and oxygen.

$$\text{N}_2\text{(gas)} + h\nu = 2\text{N(gas)}   \Delta H = D(\text{N–N}) = 942 \text{ kJ mol}^{-1}   (6.2)$$

The breaking of specific bonds in molecules will depend on their particular environment and strength. Weak bonds such as the O–O bond in peroxides $(D(\text{O–O}) = 213 \text{ kJ mol}^{-1})$ may be broken easily, and peroxides are a source of free radicals for polymerisation and other reactions. Stronger bonds, such as the C–C bonds in the benzene molecule are much more difficult to break and the aromatic ring structure will persist while substituents are broken off or react.

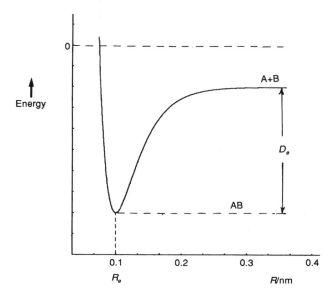

Figure 6.4 Schematic of the dissociation of a molecule AB with equilibrium dissociation energy, D_e.

The fragmentation pattern produced by high-energy ionisation techniques such as electron impact (EI) in mass spectrometry depends on the bonding and structure of the parent molecule and also on the stability of the daughter fragments produced. The soft-ionisation techniques, such as chemical ionisation, often produce less fragmentation of the parent molecules.

Dissociation, or cleavage, of molecules is favoured by the introduction of branching into straight chain molecules and by the presence of a 'heteroatom' such as O, N or S (Figures 6.4 and 6.5).

6.5 Electromagnetic radiation

Electromagnetic radiation is a type of energy which may be transmitted through empty space. It may be described in terms of waves of particular energy and length, or as a stream of particles of particular energy. For many analytical purposes the wave description is very useful.

Electromagnetic waves consist of an alternating electric field and an associated alternating magnetic field. Figure 6.6 shows a single wave travelling in the x-direction.

The wave has:

1. a definite velocity in space, the velocity of light, $c = 2.998 \times 10^8$ m s^{-1};
2. a definite wavelength λ between the maxima; and
3. a definite energy.

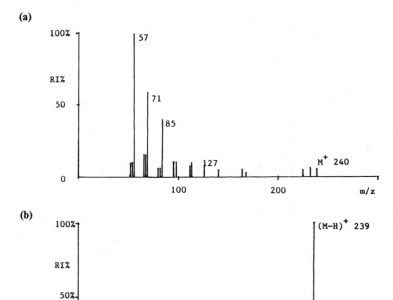

Figure 6.5 Mass spectra of the hydrocarbon $C_{17}H_{36}$ using (a) electron impact; (b) chemical ionisation with isobutane.

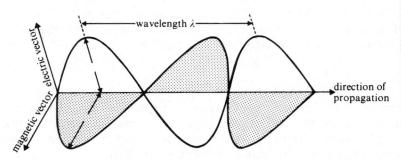

Figure 6.6 Wave representation of electromagnetic radiation. Source: Fifield, F.W. and Kealey, D. (1990) *Principles and Practice of Analytical Chemistry*, 3rd edition, Blackie Academic & Professional, Glasgow.

These are related by the equation

$$E = hc/\lambda \tag{6.3}$$

where h is Planck's constant, $h = 6.626 \times 10^{-34}$ J s

The waves travelling past per second give the frequency,

$$\nu = c/\lambda \tag{6.4}$$

and so we may write.

$$E = h\nu = hc\tilde{\nu} \tag{6.5}$$

where $\tilde{\nu}$ is the wavenumber $= 1/\lambda = \nu/c$. This is usually expressed in cm^{-1} These are most useful equations, because the measured quantities of c, λ and ν allow the energy to be calculated. Higher frequencies relate to higher energies.

In vacuum all electromagnetic radiation travels at the speed of light, c. The range of energies, frequencies or wavelengths possible is very large, and gives the electromagnetic spectrum shown in Figure 6.7 and summarised in Table 6.1.

6.6 The electromagnetic spectrum

Table 6.1 Analytical spectrometric techniques

Name of technique	Principle	Major applications
Arc/spark spectrometry or spectrography Plasma emission spectrometry Flame photometry	Atomic emission	Qualitative and quantitative determination of metals, largely as minor or trace constituents; quantitative determination of metals as minor or trace constituents
X-Ray fluorescence spectrometry	Atomic fluorescence emission	Qualitative and quantitative determination of elements heavier than nitrogen as trace to major constituents
Atomic fluorescence spectrometry		Quantitative determination of metals as minor or trace constituents
Atomic absorption spectrometry	Atomic absorption	Quantitative determination of metals as minor or trace constituents
γ-Spectrometry	Nuclear emission	Qualitative and quantitative determination of elements at trace levels
Ultraviolet spectrometry	Molecular absorption	Quantitative determination of elements and compounds, mainly at trace levels
Visible spectrometry		Quantitative determination of elements and compounds, mainly as trace and minor constituents
Infrared spectrometry		Identification and structural analysis of organic compounds
Nuclear magnetic resonance spectrometry	Nuclear absorption	Identification and structural analysis of organic compounds
Mass spectrometry	Structural fragmentation	Identification and structural analysis of organic compounds

Source: Fifield, F.W. and Kealey, D. (1990) *Principles and Practice of Analytical Chemistry*, 3rd edition, Blackie Academic & Professional, Glasgow.

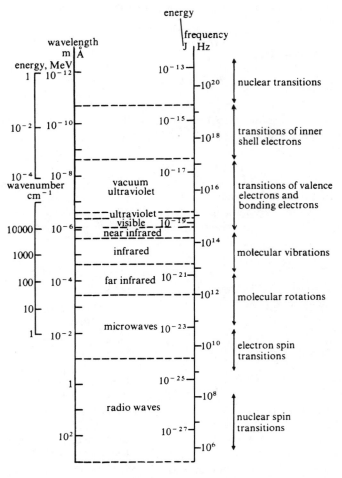

Figure 6.7 The electromagnetic spectrum (γ-ray to radiowave). Source: Fifield, F.W. and Kealey, D. (1990) *Principles and Practice of Analytical Chemistry*, 3rd edition, Blackie Academic & Professional, Glasgow.

We have seen above that there is a progression of energies from nuclear, through electronic to vibrational and rotational. If the electromagnetic radiation is absorbed, or emitted in a particular region, it corresponds to particular changes in the atomic or molecular energy levels of the species which absorb. Thus, we may write:

$$\Delta E = (E_U - E_L) = h\nu = hc/\lambda$$

Consider the absorption of energy in the far infrared region where the wavelength is about 1×10^{-6} m or 100 μm. This means that the frequency is around 3×20^{12} s^{-1}, and the energy about 2×10^{-21} J/molecule. This is only sufficient to cause the molecule to rotate, that is for transition from a low rotational level to a higher one.

Now what about a sodium atom, excited electrically to its upper electronic energy level? When this energy is emitted, the change in electronic energy is about 203 kJ mol^{-1}, or 3.4×10^{-19} J/atom, corresponding to visible light of wavelength 589.3×10^{-9} m, or 589.3 nm.

We should also note the extremes of the range. At very high energies, the X-rays and γ-rays may be associated with inner electronic energy levels or nuclear reactions. At the low energy end, the tiny energies are only sufficient to change the spin of electrons or nuclei. Also, these spin changes involve the magnetic component of the electromagnetic radiation.

In order for species to interact with electromagnetic radiation certain conditions must be fulfilled.

6.7 Interaction of species with electromagnetic radiation

1. The radiation must be absorbed. Radiation which passes straight through the material unchanged, or is at the wrong energy or frequency to match the energy levels will have no effect.
2. The absorption of radiation will depend on the populations of the initial and final energy levels. As an example, consider the rotational energy levels of HCl, as shown in Figure 6.8. If we consider the degeneracy (see section 3.1.4), these have the relative populations shown. The absorption spectrum involves transitions from one level to the next higher one, and the intensities clearly relate to the populations, as shown in Figure 6.9.
3. The absorption will depend on the change in electric (or magnetic) dipole which the transition causes. Consider three examples.

    (a) The HCl molecule has an electric dipole, while the N_2 molecule does not. The rotation of the HCl changes the direction of the dipole, and the stretching vibration of the molecule changes its magnitude. Therefore these can interact with the electric field of the electromagnetic radiation and the rotation and vibration transitions will absorb in the infrared region. Nitrogen has no dipole, and does not get one during rotation or vibration, so cannot interact in the IR.
    (b) The change in the electronic distribution which occurs as a result of any transition between electronic energy levels will alter the charge distribution and change the dipole. Most electronic transitions absorb radiation in the visible and ultraviolet regions.
    (c) The magnetic moment μ is related to the nuclear spin quantum number, I. While certain nuclei such as ^1H and ^{13}C have a magnetic spin quantum of 1/2, and others have higher values, e.g. for ^{11}B, $I = 3/2$, there are other nuclei like ^{12}C and ^{16}O which have $I = 0$. While the nuclear spin energy levels of ^1H interact with the magnetic field, ^{12}C has no effect.

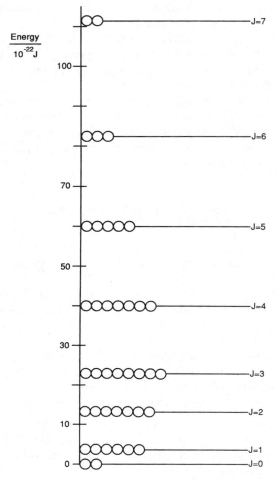

Figure 6.8 Relative populations of the lower rotational levels of the HCl molecule at 25°C.

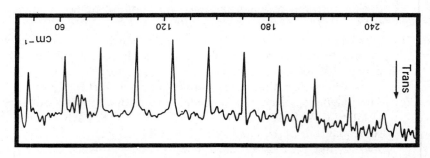

Figure 6.9 The pure rotational spectrum of HCl in the far-infrared. Source: Compton, D.A., George, W.O. and Haines, P.J. (1976) *Educ. Chem.* **13**, 18.

4. The transition probability depends on the wave functions of the initial and final states. It is often summarised by selection rules, where some transitions are 'allowed', and some 'forbidden'. For example, we saw above in Figure 6.7 that only rotational transitions where the rotational quantum number J changes by ± 1 are allowed for the HCl molecule.

$$\Delta J = \pm 1$$

Similarly, vibrational transitions are favoured when

$$\Delta v = \pm 1$$

although 'hot bands' can occur when $\Delta v = 2$ or 3.

When radiation such as UV, visible or IR passes through matter, and interacts with it, there are laws governing the behaviour. Suppose monochromatic light of initial intensity I_0 strikes a sample of concentration c and thickness l. as shown in Figure 6.10. The amount of light absorbed will depend on I and c across the whole of the sample, so that when we integrate we find that we get the Beer–Lambert Absorption Law:

6.8 Absorption laws

$$A = \log_{10}(I_0/I_t) = \varepsilon cl \qquad (6.6)$$

where A is the absorbance; I_t the transmitted intensity; ε the molar absorptivity of the sample at this wavelength; c the concentration; and l the path length through the sample.

If we use SI units, ε will be expressed in $m^2\ mol^{-1}$, which is very logical since it represents the effective interacting area of the molecules of the sample.

In older units, c was expressed in mol dm^{-3}, (molarity) and l in cm, so that the older values of ε are one-tenth of the SI values.

As an example, if 1 cm of a solution with 3.75 g of a compound of RMM = 126 per 1000 cm^3 in aqueous solution lets through 30% of the incident light of wavelength 600 nm, we can calculate:

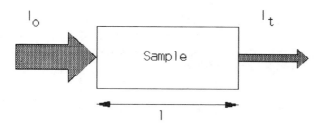

Figure 6.10 The absorption of light: the Beer–Lambert Law.

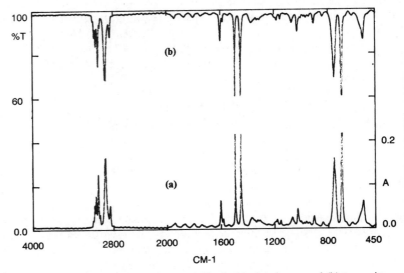

Figure 6.11 Infrared spectra of polystyrene film in (a) absorbance and (b) transmittance.

$$A = \log_{10}(100/30) = 0.523$$
$$c = (3.75/126) = 0.0298 \text{ mol dm}^{-3} = 29.8 \text{ mol m}^{-3}$$
$$l = 1 \text{ cm} = 0.01 \text{ m}$$

and
$$\varepsilon = 0.523/(29.8 \times 0.01) = 1.76 \text{ m}^2 \text{ mol}^{-1}$$

or
$$\varepsilon = 0.523/(0.0298 \times 1) = 17.6 \text{ dm}^3 \text{ mol}^{-1} \text{ cm}^{-1}$$

If we think about the amount of light that is transmitted, this is less directly related to the concentration, but is often plotted, particularly in IR spectroscopy.

$$T = 100 \times I_t/I_0 = 30\% \text{ in the above example} \tag{6.7}$$

T is called the percent transmission.

Figure 6.11 shows the same infrared spectrum expressed as transmission and as absorption plotted against wavenumber.

6.9 Spectrometric instrumentation

The general ideas of spectrometry are similar for most optical spectrometric methods from X-ray to radiofrequency wavelengths. The instrumentation for mass spectrometry is distinct, but the components perform the same functions, although in very different ways!

6.9.1 Single-beam spectrometer

The simplest device is a single-beam spectrometer as shown in Figure 6.12.

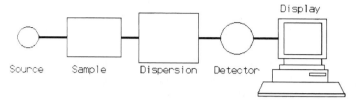

Figure 6.12 Schematic diagram of a single-beam spectrophotometer.

The source may be a tungsten light bulb, giving a continuous spectrum in the visible, a heated rod for infrared, or a discharge tube for UV and atomic absorption. In NMR spectra the source is a radiofrequency transmitter. The sample may also be incorporated into the source when we study atomic emission spectra or mass spectra.

The sample is the material under investigation. Sometimes, when we wish to disturb it as little as possible, the instrumentation is designed to make measurements *in situ*, for example by reflecting radiation from the surface. More often, a small amount is dissolved in a suitable solvent and place in a container in the beam from the source. In flame atomic absorption the 'container' is the flame itself!

Spectral resolution. The term 'spectrum' refers to the separation of the different components of radiation, and by extension, different species in mass spectrometry. In every type of analytical spectrometry we shall need to separate or disperse the components. How this is done will depend on the technique, the energy and the resolution required.

In general, the resolution or resolving power, R, may be defined for a general signal, S, as

$$R = S/\Delta S \qquad (6.8)$$

where ΔS is the smallest difference which may be detected; and S the average value of the two components.

For example, in infrared, a very good spectrometer should have be able to resolve the doublets in the HCl gas spectrum which are about 2 cm^{-1} apart.

$$\Delta \tilde{v} = 2 \text{ cm}^{-1}$$

The average value of the wavenumber for these bands is around 3000 cm^{-1}. Therefore:

$$R = 3000/2 = 1500.$$

The same principle applies in mass spectrometry, where the signal is the mass/charge ratio. The criterion for resolving the signals as separate is that there shall be a valley of 10% between the peaks. If this is to be the case for the peaks of nitrogen (RMM = 28.0062) and of carbon monoxide, (RMM = 27.9949) then the resolution would have to be:

$$R = 28/(0.0113) = 2478$$

The dispersion unit may be a prism of glass or other material, a diffraction grating, or scanning electromagnetic field for mass spectrometry. Sometimes the source is scanned, rather than separating afterwards. The time taken to scan the spectrum by mechanically changing the position of a prism or grating or by altering the electrical imput to a controller will often be quite long, especially if we wish to obtain the highest resolution, or scan over an extended range.

The detector senses the signal transmitted through the instrument from the source and must therefore match the characteristics of the source. A thermal sensor is appropriate for the detection of 'heat' signals in the infrared. A radiofrequency receiver must be used to detect the NMR signal. Some detectors complement the dispersion element by scanning the spectrum. The older type used a photographic film to record the whole spectrum as a function of distance across the film. A photoelectric detector may be traversed across a spectrum, or a series of photoelectric detectors placed at appropriate points to detect specific radiation. The modern diode array detectors consist of many pairs of photodiodes of small area, sensitive to the whole range of radiation required. By monitoring each diode in turn through the electronic circuitry, the whole spectrum can be measured very rapidly.

The display frequently uses a recorder, or a computer screen to process the results and often to store them for future use or calculations.

Single-beam spectrometers are particularly useful when complicated optics are required, or when the primary purpose is quantitative measurements at a single wavelength. However, they do suffer if there are fluctuations of energy in the source.

6.9.2 Double-beam spectrometer

The same source is used for both beams, and is equally divided, usually by means of mirrors to pass through the sample and the reference (Figure 6.13). The reference can be a cell with the pure solvent used for the sample solution, it can be an empty cell, or even air.

By using a rotating sector mirror, or other device to measure the sample and reference beams alternately, we may pass both beams through the same set of optics. If the frequency of alternation is fixed, the detector amplifier may be 'tuned' to this frequency and some noise eliminated. Any fluctuation in the source, detector or amplifier, plus any absorption by the solvent is compensated by the measurement of both beams.

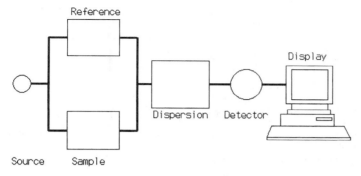

Figure 6.13 Schematic diagram of a double-beam spectrophotometer.

6.9.3 Fourier-transform instruments

If we combine a spectrum of waves of different wavelengths, and different amplitudes, we obtain a complex periodic function which may not resemble any of the original waves. J.B. Fourier showed mathematically how this wave may be obtained, and it can also be shown mathematically how the complex function may be analysed back to discover the wavelengths and amplitudes of the original spectrum. The mathematical technique is called a Fourier-Transform, and thus we have Fourier-Transform (or FT) spectrometry (Figure 6.14).

If a whole range of wavelengths is passed through the sample simultaneously, the number of elements, M, measured per unit time is increased, and hence we can get either a signal/noise (S/N) improvement by a factor of $M^{1/2}$, or, for an equivalent S/N, we can reduce the scan time by $1/M$. This is called Fellgett's Advantage or multiplex advantage.

The optical throughput which can be tolerated to get a good spectrum is greater for FT spectrometry and so we can measure spectra for both very low signals, and also, by subtraction for very high signals. This is the Jacquinot Advantage.

Since we need a computer to carry out the Fourier-Transform, there are often other advantages gained with FT instruments. They are useful for

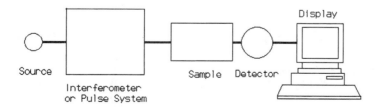

Figure 6.14 Shematic diagram of a Fourier-transform spectrophotometer.

samples which require many scans to augment the signal, or where the sample is highly absorbing, or requires a specialist device, such as a reflection unit to obtain the spectrum. Calibration may be carried out by an in-built He/Ne laser system to give very high accuracy, and the spectrum will show no discontinuities due to filter or grating changes.

Further reading   Fifield, F. W. and Kealey, D. (1990) *Analytical Chemistry*, 3rd edition, Blackie Academic & Professional, Glasgow.

Silverstein, R. M. Bassler, G.C. and Morill, T.C. (1991) *Spectrometric Identification of Organic Compounds*, 5th edition, John Wiley, New York.

Straughan, B.P. and Walker, S. (Eds.) (1976) *Spectroscopy*, Vols. 1–3, Chapman & Hall, London.

Whitfield, R.C. (1970) *Spectroscopy in Chemistry*, Longman, London.

Atomic spectrometry 7
F.W. Fifield

In chapter 3 the origin of the different energy states in which the atoms of an element can exist has been discussed. Atoms may exist in the *ground state* or *excited states*. The latter include the complete loss of one or more electrons to give positive ions. *Excitation* is produced when an atom absorbs energy and moves to a higher energy state. The excited states have a very short lifetime and excess energy is emitted, largely in the form of electromagnetic radiation, as they *relax* to the ground state. Both absorption and emission processes can be used as the basis for analytical techniques. The wavelength of electromagnetic radiation absorbed or emitted is characteristic of the element involved and can be used as the basis for qualitative analysis. At the same time the intensity of emission or absorption reflects the number of atoms undergoing the process and can thus be used for quantitative analysis. Techniques based upon the absorption or emission of electromagnetic radiation by atoms are known as *atomic absorption* and *atomic emission*, respectively. Where the initial excitation is brought about by the absorption of electromagnetic radiation, but with the analytical measurement being made on the emitted radiation accompanying relaxation, the term *atomic fluorescence* is used. The wavelengths of the radiation used in established techniques of atomic spectrometry are in the visible, ultraviolet and X-ray regions of the spectrum.

Figure 7.1 shows a typical energy level diagram for an atom together with appropriate energy transitions. The number of transitions is limited by *selection rules* which have their origin in quantum mechanics, and actual spectra are simplified accordingly, containing less lines than might otherwise be expected. Nevertheless, a full atomic spectrum can be very complex with many lines. The instrumentation employed in atomic spectrometry must be capable of resolving and isolating the single line which is being used for an analysis.

A related, but somewhat different technique is *atomic mass spectrometry*, in which, following excitation to produce atomic ions, the principles of mass spectrometry (chapter 8) are used to identify and determine elements and isotopes. Apart from better levels of detection for many elements, the ability to determine isotopic ratios gives this technique an added dimension which has considerable implications in environmental analysis.

7.1 Introduction

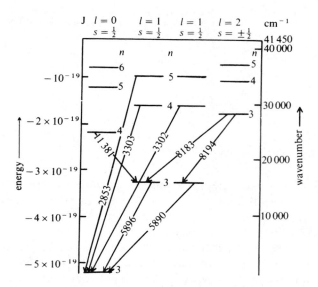

Figure 7.1 Energy level diagram of the sodium atom. The energy levels are denoted by the values for the principle quantum number n, the orbital quantum number l, and the spin quantum number s. Levels with $l = 0$ are not split; for $l = 0$ two separate levels are drawn ($s = -\frac{1}{2}$, $s = -\frac{1}{2}$); for $l > 1$ the splitting is too small to be shown in the figure. Wavelengths of a few spectral transitions are given in nanometers. Source: Fifield, F.W. and Kealey, D. (1995) *Principles and Practice of Analytical Chemistry*, 4th edition, Blackie Academic & Professional, Glasgow.

Methods based on atomic spectrometry are extremely important in the determination of the elements present in environmental samples. Techniques are available for analysis at major constituent, minor and trace levels. Table 7.1 summarises the important atomic spectrometric techniques currently in use.

Table 7.1 Important atomic spectrometric techniques used in environmental analysis

Name of technique	Method of sample excitation	Analytical parameters measured
Flame emission spectrometry (Flame photometry) (FES)	Chemical flame	Emitted visible/UV radiation
Inductively coupled plasma-atomic emission spectrometry (ICP-AES)	Gas plasma	Emitted visible/UV radiation
Inductively coupled plasma-mass spectrometry (ICP-MS)	Gas plasma	Mass and number of atomic ions
Atomic absorption spectrometry (AAS)	Absorption of electromagnetic radiation	Electromagnetic radiation absorbed
Electron probe microanalysis	Energetic electron beam	Primary X-rays emitted
X-ray fluorescence spectrometry (XRFS)	Irradiation with X-rays	Fluoresced X-rays

An obvious, and long established technique for bringing about atomic excitation is the use of a chemical flame. Flames may be used which employ normal mains gas supplies as fuel with air as the support gas or special fuels such as acetylene. The technique of *Flame Emission Spectrometry* (FES) is based on flame of this type. *Flame photometry* is a synonymous term.

7.2 Flame emission spectrometry

7.2.1 The chemical flame

In order to use flame based techniques effectively it is important to have a general idea of the structure of a flame. In nearly all cases where a flame is used, the fuel and support gases are mixed in the appropriate proportions and then, using a device known as a '*nebuliser*' the sample solution is incorporated into the gas stream as an aerosol. Nebuliser design is an important topic and the nature of the one used will depend upon a number of factors. For example, if the sample solution has a high dissolved solids content a special design may be needed. More specialist texts should be consulted if a more detailed account is needed. The flame is required to carry out three functions with regard to the sample:

1. evaporate the solvent;
2. as far as possible breakdown chemical associations so that signals come from the free atoms or ions of the element;
3. bring about excitation.

A practical point to note, concerns the rate at which the sample solution passes through the nebuliser. This rate clearly has an influence on the signal and can be affected by the viscosity of the solution. It needs to be standardised between sample and reference solutions and care must be exercised when using different solvents or, indeed, changing solution conditions. For example, small amounts of sulphuric acid can bring about substantial viscosity increases with the accompanying decrease in flow rate. One answer is to pump the solution to the nebuliser using a peristaltic pump with a set flowrate.

Figure 7.2 shows a typical construction for a flame emission spectrometer. The flame is not homogeneous and the very best performance for a particular analyte-matrix combination requires optimisation of the point of optical observation. However, in most cases a compromise has already been established in the construction of the instrument.

A further flame characteristic which has a direct bearing on the use of FES is the flame background. At the temperatures reached in typical flames, e.g. 2000–3000°C, many molecular species can still exist, emitting and absorbing radiation in the UV/visible range and with a broad band profile. Such species may originate in the sample solution itself or in the combustion products of the flame. In practice this means that FES signals from atomic transitions often have to be observed against a significant broad band

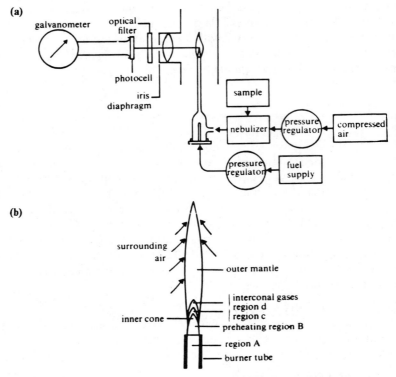

Figure 7.2 (a) Schematic diagram of a flame photometer. (b) Structure of a flame. Source: Fifield, F.W. and Kealey, D. (1995) *Principles and Practice of Analytical Chemistry*, 4th edition, Blackie Academic & Professional, Glasgow.

background, thus limiting the detection limits of the technique. It should also be noted that at the relatively low temperatures of the flame only easily ionised elements undergo significant excitation. Thus in normal use FES is limited to the determination of the alkali and alkaline earth elements. Because of the limited excitation sophisticated monochromators or polychromators are not needed and simple optical filters can be used to isolate the analytical spectral lines. The simplicity of operation and relative cheapness of FES make it an important technique attracting substantial use for the analysis of alkali metals and alkaline earth metals.

7.3 Plasma spectrometry

7.3.1 Introduction

Where analysis is based upon the atomic emission principle there are distinct advantages in using a high temperature. By doing so, chemical associations can be broken down (section 7.2.1) and their attendant problems overcome. At the same time a high degree of excitation and ionisation can be achieved.

In this way good selectivity and sensitivity for elemental analysis may be attained. Over the last 25–30 years spectroscopic sources based upon the use of high temperature gas plasmas have been developed to fulfil this role. The presence of ions and electrons in the gas enable energy to be transferred to the plasma by electromagnetic induction processes.

7.3.2 *Inductively coupled plasma-atomic omission spectrometry*

The most successful plasma source, and most widely used in *inductively coupled plasma torch* (ICP) (Figure 7.3) uses radiofrequency induction in argon and can generate temperatures in excess of 8000°C. This source is extensively used for atomic emission spectrometry in the technique which has become known as ICP-AES. Occasionally the synonym ICP-OES is used where OES stands for optical emission spectrometry. Although ICP-AES largely eliminates chemical interferences, the high temperature leads to a multiplicity of intense emission lines in the spectrum. It follows that the possibility of spectral interferences resulting from line overlap is greatly increased relative to FES. The problem can be usually, although not always, overcome by use of sophisticated computer software in order to make corrections for interferences. Switching to an alternative elemental line for the measurement can often avoid any outstanding problems. However, in a minority of cases, some degree of chemical separation e.g. by ion-exchange chromatography may be necessary. A particularly important case in the

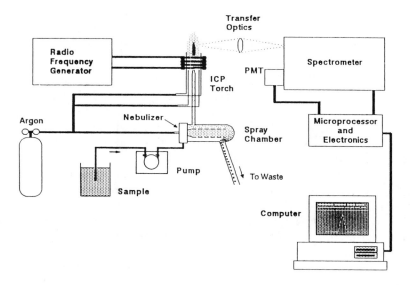

Figure 7.3 Schematic layout of an ICP-AES spectrometer. Source: Perkin-Elmer Instruments.

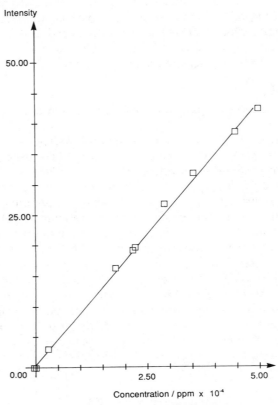

Figure 7.4 ICP-AES calibration for the determination of potassium in a rock sample, showing the long linear range of operation. Courtesy of Dr. S. de Mars, Kingston University.

determination of rare earth elements which are used as geochemical indicators. In such circumstances a switch to a related technique, ICP-*mass spectrometry* may be the best solution. A distinct practical advantage of ICP-AES derives from the near elimination of chemical effects in the plasma. As a result, the emission intensity retains a linear relationship to the analyte concentration over a wide range, e.g. 10^{-3}–10^{2} mg dm^{-3}. Apart from the convenience of simple calibration procedures this means that major, minor and even trace constituents may be determined in the same solution without re-calibration. Figure 7.4 shows a typical calibration curve.

Figure 7.3 shows a schematic representation of the ICP torch, which consists of three concentric silica tubes. The plasma is established by the flow of auxiliary argon through the outer tube with the argon being initially 'seeded' with electrons and ions by a silent electric discharge. The radio-frequency induction heater then produces the high-temperature fireball at the mouth of the torch. Leakage argon is introduced through the intermediate

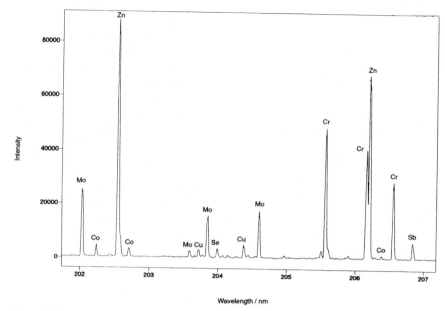

Figure 7.5 ICP-AES scan showing the separation of peaks from a number of elements in the same sample. Source: Instruments SA (UK) Ltd., with permission.

tube to ensure that the fireball is kept clear of the tube ends. The sample, usually as an aerosol of an aqueous solution, is then injected through the centre. In limited circumstances a laser may be used to vaporise a solid sample or an electric spark to produce finely divided particles of the sample. These give the sample introduction techniques of *laser microprobe* and *spark erosion*, respectively. An example of an ICP-AES spectrum is shown in Figure 7.5 For ICP-AES the tail of the plasma is viewed at a point just above the fireball where the spectrum is richest in lines and has the highest emission intensity. Two types of instrument exist with slightly different operating principles, both being used in environmental analysis. The element being determined is characterised by the wavelength of a line and its concentration by the line intensity. *Sequential instruments* are programmed to make measurements on a series of lines in turn with the experimental parameters for best performance changing under computer control. Typical modern instruments can give results on a portfolio of about ten elements in a running time of a few minutes. This approach has the advantage of flexibility in that the portfolio of elements may readily be redefined and the instrument reprogrammed accordingly. Furthermore, optimum instrument settings may be used for each element.

The alternative is *simultaneous operation* in which measurements are made on a number of lines simultaneously using a series of preset optical channels built into the instrument. In principle the full range of elements (70) which is accessible to ICP-AES determination could be measured in less than a minute. However, physical limitations with the fitting of the optical channels will normally restrict this to 30–35. In this way, results for a large number of elements may be obtained in a minute or less. It must be recognised that compromise settings for instrumental parameters have to be used. Changing the portfolio of elements in a simultaneous programme is both time consuming and costly, so there is a degree of inflexibility involved in this mode of operation. Where a large throughput of samples is required on a fixed programme, there are clear advantages. An environmental analysis example is the statutory measurement of over 30 elements in water supplies. At the time of writing a new type of instrument is being introduced, which, if successful, will materially change the above pattern. Using a modified optical detection system multichannel, simultaneous analysis is being combined with the flexibility of the rapid programming which is a feature of simultaneous operation. The development is based on the use of the diode array detection mentioned in chapter 6.

7.3.3   Inductively coupled plasma-mass spectrometry

In the preceding section the efficiency of the ICP as a device providing a rich source of elemental ions has been described. These ions may also be characterised according to their mass/charge ratio, in a mass spectrometer, the general principles of which are discussed in chapter 8, section 8.5. Interfacing a mass spectrometer to an ICP torch gives the technique of *Inductively coupled plasma-mass spectrometry* (ICP-MS). Figure 7.6 shows in outline how this is done. The mass spectrometer used is similar to the quadrupole type also described in chapter 8, section 8.5.1. With the

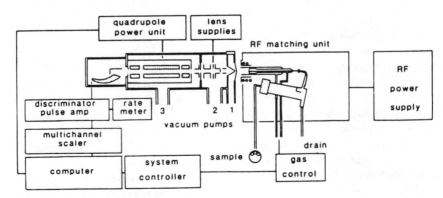

Figure 7.6 Schematic diagram of a typical ICP-MS system. Source: Fifield, F.W. and Kealey, D. (1995) *Principles and Practice of Analytical Chemistry*, 4th edition, Blackie Academic & Professional, Glasgow.

analytical signals now dependent only upon the mass/charge ratio and the number of ions with a particular ratio a comprehensive technique for all elements is produced. This does not of course mean that ICP-MS may be applied to all elements with equal success, as some interferences can be anticipated. The most troublesome are caused by molecular ions such as Ar_2O^+, ArN^+ and ArO^+ which can obscure some isotopic peaks. The laser microprobe mentioned in section 7.3.2 has found considerable favour for rapid preliminary analysis, as it may be used with a minimum of sample preparation. ICP-MS can also be used to measure the isotopic ratios within elements. This characteristic is most useful in analysis for geochemical indicators where isotopic ratios such as $^{87}Rb:^{87}Sr$, $^{147}Sm:^{143}Nd$ can indicate the date of crystallisation of a rock. The ratio, fixed at the time of crystallisation, changes as radioactive decay of one to the other proceeds with a knowledge of the half life the original date of crystallisation can be calculated.

In pollution studies the tracing of the origin of lead pollution has been of particular interest. This is based on the fact that as a result of its radiogenic origin, lead isotope ratios $^{206}Pb:^{204}Pb$ and $^{207}Pb:^{204}Pb$ vary from deposit to deposit in the world with particular sources, a topic more fully explained in chapter 15, section 15.6.1. An ICP-MS spectrum is illustrated in Figure 7.7. The relation of the analytical signal directly to the mass number of the isotope means that in most cases it is unambiguous even at ultra-low concentrations (e.g. 0.01 ng kg^{-1} or ppb). Consequently ICP-MS is one of the most sensitive techniques available for elemental analysis.

ICP-MS has a wide potential use for non-radioactive tracer studies in chemistry and biochemistry but these are largely beyond the scope of this text.

7.4 X-ray emission techniques

7.4.1   Introduction

When atomic transitions occur in which the electron movements involve the inner orbitals (principal quantum numbers 1, 2 and 3) the movements are between energy levels which are widely spaced (in terms of energy) (Figure 7.8). Hence the electromagnetic radiations which are absorbed or emitted have high quantum energies and short wavelengths, falling in the X-ray region of the electromagnetic spectrum. The general principles which govern such behaviour are similar to those for atomic spectrometry in general but the high energy of the radiation means that the instrumentation operates in rather different ways although carrying out the same basic functions of measuring wavelengths and line intensities. Two important X-ray emission techniques are in use. One in which X-ray emission is stimulated by the impact of an energetic electron beam is known as *electron probe micro-analysis* and the other which uses X-rays from an X-ray generator in a fluorescence process is known accordingly as *X-ray fluorescence*

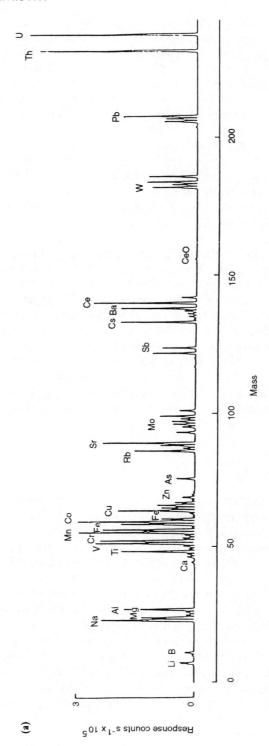

Figure 7.7 Illustrative ICP-MS spectra: (a) a set of 26 standards consists of sharp smooth symmetrical peaks; the blocking-in that has occured is due to the plotter.

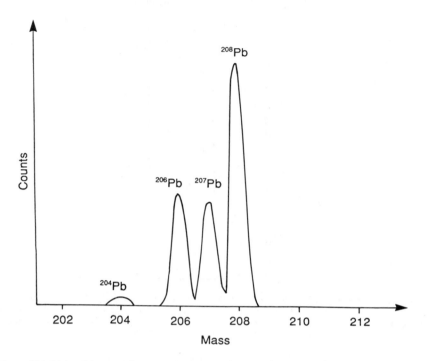

Figure 7.7 (b) head isotopes in vegatation from the vicinity of the M25 motorway in the UK. Courtesy of Dr. N.I. Ward, University of Surrey.

spectrometry (XRFS). Because X-rays can penetrate through solid material the possibility exists of direct analysis of solid specimens, with a minimum of sample preparation. X-rays are detected and measured, by a *wavelength dispersive* (WDX) system or an *energy dispersive* (EDX) one. The terms WDAX and EDAX are also used for these detection techniques. Of these, the former intrinsically has the best resolution and is to be preferred where resolution is paramount. In the former the X-rays are dispersed by an analysing crystal and the angle of diffraction characterises the X-ray. In the latter a semiconductor detector, based on silicon, generates an electrical pulse which is related in size to the energy of the X-ray. A spectrum is then generated electronically.

Three key processes should be noted which directly determine the nature and quality of X-ray analytical signals.

1. The production of primary X-rays by particle impact (usually energetic electrons)
2. Absorption of X-rays by a target
3. Fluorescent emission of X-rays

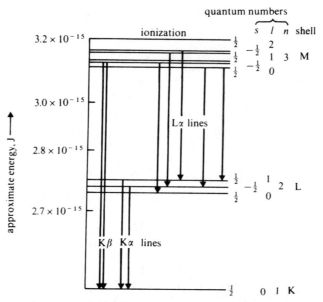

Figure 7.8 X-Ray transitions in the molybdenum atom. Lines are designated by the letter referring to the lower quantum group and α where $\Delta n = -1$ and β where $\Delta \dot{n} = -2$. Source: Fifield, F.W. and Kealey, D. (1995) *Principles and Practice of Analytical Chemistry*, 4th edition, Blackie Academic & Professional, Glasgow.

The first is clearly important in electron probe analysis and the others in XRFS, as well as in the general assessment of background effects and interferences.

The wavelengths of the emitted X-rays are related to the atomic numbers of the elements by the Moseley relationship (equation (7.1)), where a is a constant and b has a specific value for all the lines in one series.

$$\lambda^{-1/2} = a(Z - b) \tag{7.1}$$

Absorption of X-radiation in a target is expressed by equation (7.2),

$$I = I_0 \exp(-\mu\rho b) \tag{7.2}$$

in which I_0 is the intensity of a monochromatic beam of X-rays, I is the intensity at a distance b into a target of density ρ, and μ is the *mass absorption coefficient* and depends upon the wavelength of the radiation and the atomic number of the element. Typical values are 5.5 m² kg⁻¹ for carbon and 241 m² kg⁻¹ for lead.

Up to half of the radiation absorbed may be re-emitted in a fluoescence process. Equation (7.3) shows the relationship between the absorbed radiation intensity, I_a and that re-emitted I_F in terms of the *fluorescence yield factor* φ which may be up to 0.5 for heavy elements but falls below 0.01 for elements with atomic numbers below 15.

$$I_F = \varphi I_a \qquad\qquad (7.3)$$

Thus, X-ray techniques are best applied to heavy elements especially where the matrix elements are of low relative atomic mass and exert a minimum of interference.

7.4.2   Electron probe microanalysis

If a specimen is subjected to bombardment by an energetic electron beam (20–50 keV) ionisation and excitation of the electrons in its atomic orbitals will be effected. As the atoms subsequently relax to the ground state and the inner electron orbitals are refilled X-radiation will be emitted. The characteristic X-ray emission spectrum obtained will contain photopeaks whose wavelength is determined by the element concerned (Figure 7.9). Thus the wavelength and intensity of the peaks can be used for qualitative and quantitative analysis, respectively. Because the impact of the electrons excites a wide range of electrons in both the analyte and the matrix, the analytical signal will usually have a high background, which is a factor ultimately limiting the sensitivity of the technique. The electron beam will generally impact upon an area of the specimen surface about 1 µm across

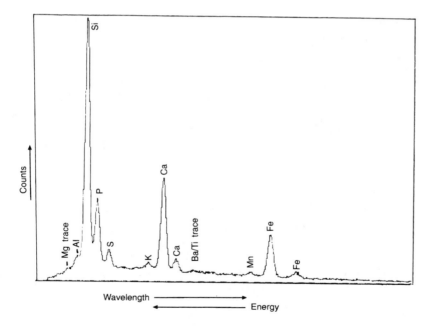

Figure 7.9 Electron probe spectrum obtained using an energy dispersive detector and a scanning electron microscope. It shows elements present in the insoluble residue of a fresh-water mussel from the Lago Presidente Rios in Southern Chile.

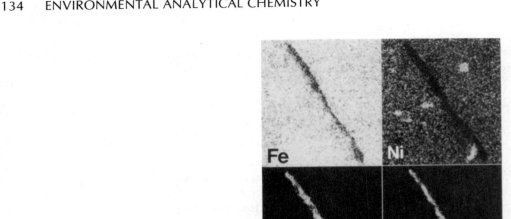

Figure 7.10 Elemental mapping of part of the surface of an iron meteorite, by electron probe analysis. Light areas indicate concentrations of the element mapped. It is obvious that the inclusion contains Cr and S, but not Ni or Fe. Source: *JEOL News*, Vol. 22E, No. 2(46).

and penetrate about 1 μm into the surface layers. Thus the instantaneous signal will be representative of a localised part of the sample. Larger areas can be analysed if the instrument is programmed to scan backwards and forwards across the surface. Modern imaging techniques will allow the display to be tuned to a particular element so that a distribution map can be obtained (Figure 7.10). However, the complexity of the initial interaction of the electron beam with the specimen and the subsequent effects of the matrix on the emitted X-rays means that truly quantitative data are difficult to obtain. Hence electron probe analysis is a powerful technique for

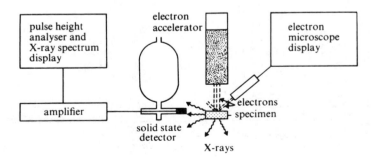

Figure 7.11 Schematic layout of an electron microscope—electron probe analyser. Source: Fifield, F.W. and Kealey, D. (1995) *Principles and Practice of Analytical Chemistry*, 4th edition, Blackie Academic & Professional, Glasgow.

establishing the identity and distribution of elements in solid samples but performs poorly for quantitative analysis. The distribution of heavy metals at cellular level in biological materials or in mineral particles are good environmental examples. Instruments for electron probe analysis can be custom designed for that purpose alone but are most frequently made interfaced with scanning electron microscopes where the electron beam serves the dual purpose of producing the microscope image and the X-ray signal. Figure 7.11 shows a schematic instrumental arrangement using an energy dispersive detector.

7.4.3   X-ray fluorescence spectrometry

Inner electron excitation leading to X-ray emission and relaxation (Figure 7.8) can also be brought about by irradiation of a specimen with X-rays from an X-ray generator. Radiation is thus absorbed and re-emitted in the process, and the technique based upon it is known as *X-ray fluorescence spectrometry* (XRFS). Only radiation which has exactly the right quantum energy will be absorbed and excitation will be rather more selective than in the case of electron beam excitation. As a consequence, the spectrum will have a relatively low background compared to electron probe spectra (Figure 7.12). The intensity of the fluorescent signal will again be affected by interactions with the sample matrix. However, because the general

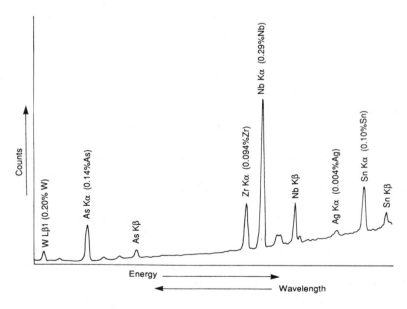

Figure 7.12 An example of an X-ray fluorescence spectrum containing peaks for W, As, Zr, Nb, Ag and Sn.

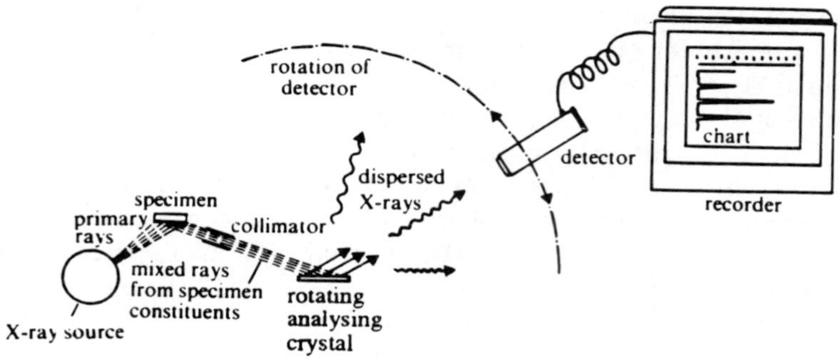

Figure 7.13 Schematic layout of a dispersive X-ray fluorescence spectrometer. Source: Fifield, F.W. and Kealey, D. (1995) *Principles and Practice of Analytical Chemistry*, 4th edition, Blackie Academic & Professional, Glasgow.

processes are rather simpler than in the case of electron probe measurements, and given modern computational facilities, good quality, quantitative data can frequently be attained.

Figure 7.13 schematically shows the construction of a spectrometer for XRFS. The target used in the X-ray generator can alternatively be W or Mo allowing for line overlap between the photopeaks of the incident X-rays and the analyte to be avoided.

A wavelength dispersive analyser is depicted in Figure 7.13 and where this is used it has become common practice to calibrate the energy/wavelength axis in units of 2θ. This reflects the fact that the angle of diffraction θ is related to these by the Bragg Equation.

$$n\lambda = 2d \sin \theta \qquad (7.4)$$

where n is an integer; λ the wavelength; and d, the crystal spacing in the diffracting crystal. For certain types of fieldwork, low resolution, portable instruments have been developed. These are based on the use of radioactive sources to produce excitation and fixed X-ray filters to isolate particular spectroscopic lines.

Figure 7.13 also shows the analysing collimator arranged at an angle of approximately 90° to the incident beam. This helps to minimise the background from scattered radiation passing straight through the specimen from the incident beam.

The irradiating X-rays penetrate deeply into the specimen so that the analytical signal is representative of the sample as a whole rather than a localised part of it. Obviously the X-rays do not have infinite penetration and there is a sample size limit as a consequence. It must be remembered also that the degree of penetration and absorption is dependent upon overall sample density (equation (7.2)). Thus it may be appropriate to prepare a sample in a light matrix for example for dissolution in water or fusion in a borax bead. XRFS is a well established technique particularly suited to the routine, rapid analysis of minerals and related samples for major and minor constituents. Its application does not normally extend to trace elements.

7.5.1 Introduction

If free atoms of an element in the ground state are irradiated with electromagnetic radiation of appropriate wavelength, they will absorb radiation and be raised to an excited state. The wavelength of absorption and the amount of radiation absorbed provide qualitative and quantitative parameters for the determination of the element present. Where the irradiating radiation is sharp-line (very narrow bandwidth) a linear relationship exists similar to that described by the Beer–Lambert Law (Chapter 6, section 6.8). The technique established upon these principles in known as *atomic absorption spectrometry* (AAS). The atomic vapour required is produced in most cases by *flame or electrical heating* of a solution of the sample. In a small number of cases a volatile hydride may be produced and in the singular case of Hg a *cold vapour* technique is used.

7.5.2 Sharp-line radiation

A key feature of the technique of AAS is the sharpness of the absorption line when an atom is raised from its ground state to the *first excited state*. These lines are very intense and as a consequence have been called *resonance lines*, and it follows that AAS has both excellent resolution and good sensitivity. The absorption line width is ca. 10^{-2} nm so the irradiating radiation line must be of the same order or narrower if appropriate sensitivity and linearity of response are to be attained (see Figure 7.14). A device known as *the hollow cathode lamp* (HCL) (Figure 7.15) has been developed to produce radiation with the necessary characteristics. The HCL consists of a glass envelope containing anode and cathode electrodes. The cathode is either

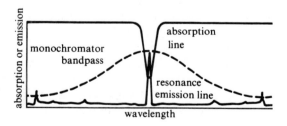

Figure 7.14 Profiles of an absorption line, an emission line from a sharp line source and the bandpass of a monochromator. Source: Fifield, F.W. and Kealey, D. (1995) *Principles and Practice of Analytical Chemistry*, 4th edition, Blackie Academic & Professional, Glasgow.

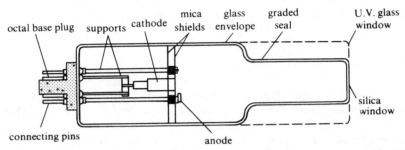

Figure 7.15 Diagram of a hollow cathode lamp. Source: Fifield, F.W. and Kealey, D. (1995) *Principles and Practice of Analytical Chemistry*, 4th edition, Blackie Academic & Professional, Glasgow.

fabricated from the analyte or packed with a non-volatile form of it. After evacuation the lamp is filled with argon at low pressure. A silent electric discharge produces positive argon ions which are attracted to the cathode. On striking the cathode in a process known as *sputtering*, atoms of the cathodic material are ionised and volatilised. On relaxation, radiation with the appropriate sharpline characteristics is produced. With individual lamps being used for each element determined spectral interference from other lines is minimal. In a small number of cases for non-metallic elements suitable HCLs cannot be constructed. In these circumstances an electrode-less discharge lamp (EDL), based on microwave heating, is used.

7.5.3   AAS measurements

The layout of a typical spectrometer is shown in Figure 7.16. The key feature is the production of free, ground state atoms from the sample which pass through the light beam from the HCL, absorbing radiation and reducing the electric current from the detector. For optimum performance the number of free, ground state atoms, needs to be maximised.

Chemical interference with the production of free atoms can result if analyte atoms are in the form of compounds such as oxides, phosphates or sulphates which have considerable thermal stability. Steps need to be taken to minimise their formation. At the same time the temperature must be optimised to ensure maximum compound breakdown, whilst ensuring minimum thermal excitation of the free atoms. *Physical interference* arises when the volatilisation is inhibited because they are trapped inside large particles or in a solid solution with a large amount of another element, such as iron, in the matrix. *Spectral interference* from line overlap is rare, but interference occurs as a result of the emission of radiation at the analytical wavelength by excited atoms relaxing to their ground state. In addition, molecular species can exhibit broad band absorption and emission.

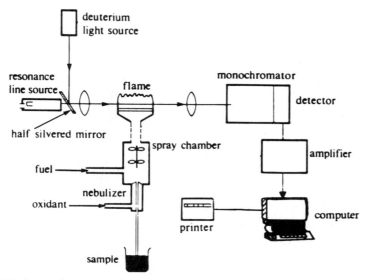

Figure 7.16 Practical system for flame atomic absorption spectrometry including a deuterium background corrector. Source: Fifield, F.W. and Kealey, D. (1995) *Principles and Practice of Analytical Chemistry*, 4th edition, Blackie Academic & Professional, Glasgow.

Excited atoms of the analyte emitting radiation at the analytical wavelength as they relax constitute a continuous interfering signal. A distinction from the HCL beam can be made by pulsing the lamp current and hence its output, so that an electronic circuit can separate the two signals. The effect of broad-band emission by molecular species can be overcome at the same time. Broad-band absorption is assessed by irradiation from a background correcting source and then subtracted from the analytical signal. *Deuterium lamp, Zeeman* or *Smith-Hieftje* corrections are used. For explanations a more specialist text should be consulted.

Calibration curves for AAS measurements show straight line relationships only over narrow concentration ranges. Typically 1–20 ppm or 1–50 ppm. At the higher ends of the graphs curvature towards the concentration axis is observed. From the practical point of view, dilution of sample solutions may be needed in order to bring them into the operating range for the method. A typical calibration curve for the determination of chromium by flame AAS is shown in Figure 7.17.

7.5.4 Flame AAS

Where a chemical flame is used as the spectroscopic source the characteristics have to be carefully optimised to meet the requirements set out above. This is done initially by selecting the fuel gas/support gas mixture

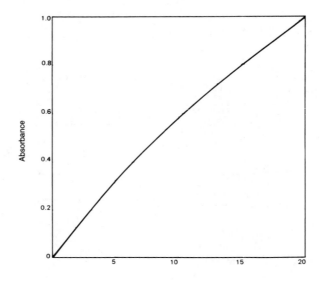

Concentration / μg cm^{-3}(ppm)

Figure 7.17 Calibration curve for the determination of chromium by flame atomic absorption (spectrometry). It shows slight curvature towards the concentration axis. This is a common feature.

and the proportions of mixing. Flow rate and burner height relative to the optical path are also important factors. The gas composition and flow rate, in conjunction with the thermal currents in the flame, determine the *residence time* of the atoms in the light path. Obviously the longer the residence time the greater the chance of an absorption process occurring and the better the sensitivity of the measurements. Typically air–acetylene mixtures are used and nitrous oxide–acetylene where higher temperatures are required. The burner used is commonly of the slot type and 10 cm in length. In practice flame AAS is simple and easy to use for a wide range of elements giving good sensitivity, mostly at below 1 mg dm^{-3} (1 ppm).

7.5.5 Electrothermal AAS

Electrical heating in a carbon tube furnace (Figure 7.18) allows greater control to be exercised over the various stages in the production of free ground state atoms. A typical heating cycle is shown in Figure 7.19. This enables a higher proportion of free atoms to be produced with an accompanying improvement in sensitivity. The sensitivity improvement

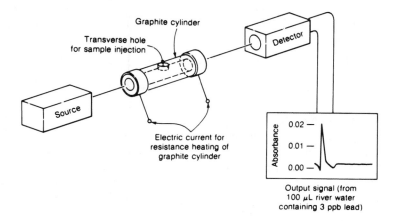

Figure 7.18 Graphite furnace for atomic absorption analysis and typical output signal. Source: Manahan, S.E. (1988) *Quantitative Analytical Chemistry*, Brooks Cole.

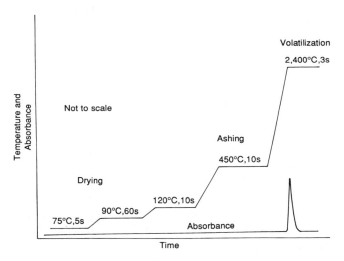

Figure 7.19 A possible electrothermal heating cycle for the determination of nickel, showing drying, ashing and volatilisation stages.

is enhanced by the *'atom trap effect'* of the tube which by holding the atoms in the light path increases the residence time to ca. 1 s. As a consequence, detection limits can be lowered by between 100 and 1000 times.

In operation, a small volume (10–100 µl) is injected into the furnace via a transverse hole onto a platform. The latter ensures a uniform, reproducible heating profile for the sample, and improves the precision of the measurements. Automated injection ensures that the poor precision resulting from manual injection is removed. The signal from this system is transient and is observed as a peak, the area of which represents the quantity of analyte present.

7.5.6   *AAS using mercury vapour or volatile hydrides*

Mercury is a dangerous pollutant which in its elemental form is unique in being a metal which is liquid at ordinary temperatures. It has a high vapour pressure and can readily be volatilised to produce free, ground-state atoms for AAS. Mercury compounds are first reduced to the metal by a strong reducing agent such as sodium borohydride before being volatilised and swept into the light path of the spectrometer, using a special cell. Under similar conditions some elements such as Se, Bi and Sn, are reduced to hydride forms, e.g. AsH_3 which are also readily volatile. In order to produce free atoms for AAS gentle heating of the optical cell is needed to break up the compounds. Sensitivity may be improved by collecting the vapour in a special cell or loop before passing it into the light path and in the case of mercury forming an amalgam with gold or copper which is subsequently heated to release the mercury vapour into the light path.

7.5.7   *Use of atomic spectrometry*

The various techniques of atomic spectrometry, with the exception of ICP-MS, are similar in that they are all based upon the exploitation of electronic changes in atoms. However, the details of instrument construction and operation are very different as are the characteristics of the results of the techniques. In order to select the one most appropriate to a particular analysis or to appreciate fully the limitations of experimental data it is important to have an understanding of the basic principles of their operation and performance. Key features to assess are:

1.   detection limits and range of operation;
2.   concentrations (for detection limits see Table 7.2);
3.   possible interferences;
4.   potential for single element and multi-element operation;
5.   complexity;
6.   ease of operation;
7.   cost of equipment and operation.

Atomic spectrometric methods account for a very high proportion of elemental analyses carried out on environmental samples. Their detection limits are summarised in Table 7.2.

Table 7.2 Atomic spectroscopy detection limits (μg/l) (from Perkin Elmer, *Guide to Techniques and Applications of Atomic Spectroscopy*, 1988)

Element	Flame AA	Hg/hydride	GFAA	ICP emission	ICP-MS
Ag	0.9		0.005	1	0.04
Al	30		0.04	4	0.1
As	100	0.02	0.2	20	0.05
Au	6		0.1	4	0.1
B	700		20	2	0.1
Ba	8		0.1	0.1	0.02
Be	1		0.01	0.06	0.1
Bi	20	0.02	0.1	20	0.04
Br					1
C				50	50
Ca	1		0.05	0.08	5
Cd	0.5		0.003	1	0.02
Ce				10	0.01
Cl					10
Co	6		0.01	2	0.02
Cr	2		0.01	2	0.02
Cs	8		0.05		0.02
Cu	1		0.02	0.9	0.03
Dy	50				0.04
Er	40				0.02
Eu	20				0.02
F					100
Fe	3		0.02	1	1
Ga	50		0.1	10	0.08
Gd	1200				0.04
Ge	200		0.2	10	0.08
Hf	200				0.03
Hg	200	0.008	1	20	0.03
Ho	40				0.01
I					0.02
In	20		0.05	30	0.02
Ir	600		2	20	0.06
K	2		0.02	50	10
La	2000			1	0.01
Li	0.5		0.05	0.9	0.1
Lu	700				0.01
Mg	0.1		0.004	0.08	0.1
Mn	1		0.01	0.4	0.04
Mo	30		0.04	5	0.08
Na	0.2		0.05	4	0.06
Nb	1000			3	0.02
Nd	1000				0.02
Ni	4		0.1	4	0.03
Os	80				0.02
P	50000		30	30	20
Pb	10		0.05	20	0.02
Pd	20		0.25	1	0.06
Pr	5000				0.01
Pt	40		0.5	20	0.08
Rb	2		0.05		0.02
Re	500			20	0.06
Rh	4			20	0.02
Ru	70			4	0.05
S				50	500
Sb	30	0.1	0.2	60	0.02

Table 7.2 *Cont'd.*

Element	Flame AA	Hg/hydride	GFAA	ICP emission	ICP-MS
Sc	20			0.2	0.08
Se	70	0.02	0.2	60	0.5
Si	60		0.4	3	10
Sm	2000				0.04
Sn	100		0.2	40	0.03
Sr	2		0.02	0.05	0.02
Ta	1000			20	0.02
Tb	600				0.01
Te	20	0.02	0.1	50	0.04
Th					0.02
Ti	50		1	0.5	0.06
Tl	9		0.1	40	0.02
Tm	10				0.01
U	10000			10	0.01
V	40		0.2	2	0.03
W	1000			20	0.06
Y	50			0.2	0.02
Yb	5				0.03
Zn	0.8		0.01	1	0.08
Zr	300			0.8	0.03

All detection limits are given in micrograms per litre and were determined using elemental standards in dilute aqueous solution. Atomic absorption (Model 5100) and ICP emission (Plasma II) detection limits are based on a 95% confidence level (2 standard deviations) using instrumental parameters optimised for the individual element. ICP emission detection limits obtained during multielement analyses will typically be within a factor of 2 of the values shown. Cold vapour mercury AA detection limits were determined using an MHS-20 mercury/hydride system with an amalgamation accessory. Furnace AA (Zeeman/5100) detection limits were determined using STPF conditions and are based on 100–μl sample volumes. ICP-MS (ELAN) detection limits were determined using operating parameters optimised for full mass range coverage and a 98% confidence level (3 standard deviations). ICP-MS detection limits using operating conditions optimised for individual elements are frequently better than the values shown. ICP-MS detection limits for fluorine and chlorine were determined using the ELAN's negative ion detection capabilities.

Further reading

Fifield, F.W. and Kealey, D. (1995) *Principles and Practice of Analytical Chemistry*, 4th edition, Blackie Academic & Professional, Glasgow.

Willard H.H. *et al.* (1988) *Instrumental Methods of Analysis*, 6th edition, Wadsworth.

Boumans P.W.J.M. *et al.* (1987) *ICP Emission Spectrometry*, Wiley, New York.

Date, A.R. and Gray, A.L. (1989) *Applications of Inductively Coupled Plasma Mass Spectrometry*, Blackie Academic & Professional, Glasgow.

Molecular spectrometry 8

P.J. Haines

8.1 Introduction

The analysis of samples from almost every source is frequently done by studying their molecular spectra. Techniques are available to study every type of sample, gas, pure liquid or solution or solid, but especially organic chemicals from many sources. In chapter 3, the ideas of molecular energy levels were introduced, and in chapter 6 the general concepts of spectrometry were reviewed. This chapter will give details of the main types of molecular spectrometric apparatus and techniques and examples of their use.

For environmental analysis we are often restricted to very small sample amounts, and therefore special methods must be used to concentrate the material, or often, special instrumental techniques used to deal with small samples, or samples which have low concentration. Long path-length cells, computerised difference and accumulation methods and adsorption onto a surface are all used where the amount or concentration is small.

Table 8.1 shows the chief molecular spectrometric techniques and the molecular processes involved.

Table 8.1 Molecular spectrometric processes

Technique		Process	Samples
Ultraviolet/visible	UV/vis	Electronic transitions	Solutions
Infrared	IR	Molecular vibrations	Gas, liquid or solid
Nuclear magnetic resonance	NMR	Nuclear spin changes	Solutions
Mass spectrometry	MS	Molecular fragmentation	Gas, liquid or volatile solid

8.2.1 Instrumentation

8.2 Ultraviolet and visible spectrophotometry

The same instrument is often used for the whole of the spectral region from the near ultraviolet limit of about 190 nm through to the visible limit of about 800 nm. Different components may have to be used for parts of the spectrum, however.

In the region 190–350 nm, a high-pressure deuterium discharge or a xenon lamp is used as a source, with fused quartz cells containing the sample in solution. Diffraction gratings or quartz prisms are the dispersion components and the detector may be a photomultiplier, or a diode array system.

For the visible region, a tungsten filament lamp may be used as the source, glass or plastic cells for the sample, and the dispersion and detection are similar to the UV instrument. Some instruments change sources and detectors at pre-set wavelengths as the spectrum is scanned.

For UV/vis and IR, absorption spectra are the most frequently studied, and the Beer–Lambert law is applied to many samples.

$$A = \varepsilon c l \qquad (8.1)$$

If more than one species is present, then each species will contribute to the total absorbance at a particular wavelength:

$$A_{mixture} = \varepsilon_A c_A l + \varepsilon_B c_B l + \varepsilon_D c_D l + \dots \qquad (8.2)$$

For a two-component mixture, we may solve for two concentrations c_A and c_B by measuring at two wavelengths for pure solutions of components A and B to find their absorptivities, and then, measuring the mixture at the same two wavelengths, λ_1 and λ_2.

$$A_{mix, \lambda 1} = \varepsilon_{A, \lambda 1} c_A l + \varepsilon_{B, \lambda 1} c_B l \qquad (8.3)$$

$$A_{mix, \lambda 2} = \varepsilon_{A, \lambda 2} c_A l + \varepsilon_{B, \lambda 2} c_B l \qquad (8.4)$$

It should be recognised, however, that there are cases where deviations may occur from this equation.

1. For species which undergo reactions, or equilibria which may change with concentration, such as complexation, hydrogen bonding or dissociation, there will be deviations.
2. If the radiation is not monochromatic, then the values of ε may change with the wavelength and we may get negative deviations.
3. Stray light within the instrument can also cause problems.

Therefore it is most important that the system should be checked for its adherence to a true Beer–Lambert plot.

8.2.2 Band spectra

The electronic bands of organic molecules and complexes are also associated with vibrational and rotational bands, as shown in Figure 6.2. The transitions may occur between many of these levels, and consequently the spectra, while centred in the UV/vis region, incorporates a 'band' of transitions. These are often blurred together, since the solvents used interact with the solute molecules. The most precise measurements are generally found to be at the band maximum.

8.2.3   Polyatomic organic molecules

The structures of almost all organic species are based on σ and π overlap of atomic orbitals. There are therefore, in general, electronic energy levels corresponding to the σ and π bonding plus the σ* and π* antibonding, and also the non-bonding levels occupied by electrons not involved in bonding, such as lone-pairs on oxygen atoms. The relative energies of these levels are shown in Figure 8.1.

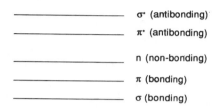

Figure 8.1 Electronic energy levels of polyatomic molecules.

Certain selection rules apply. It is more probable for transitions to occur between σ and σ* levels and between π and π*, so that these will have a large ε. Transitions from n–σ* or from n–π* are less probable, or less favoured, and ε will be smaller.

Molecules with only σ bonding, such as aliphatic hydrocarbons like hexane, can only show the high energy σ–σ* transitions which occur in the far UV, often below 200 nm. Saturated molecules with non-bonding electrons, such as water or methanol, can undergo n–σ* transitions, which occur around 200 nm. Hexane, water and methanol are therefore very suitable as solvents for UV work, since they show little absorption across the majority of the range.

Molecules with π levels, such as unsaturated and aromatic species like butadiene and benzene will show both σ–σ* and π–π* transitions, but the latter will be more noticeable at about 200 to 250 nm.

Carbonyl compounds, like propanone $(CH_3)_2 \cdot C=O$ have σ*–σ*, π–π* and n–π* transitions. The n–π* occur at 250 to 300 nm, and are generally rather weak.

The energy levels may interact, and if there are several π levels, or several double bonds, they may conjugate and give smaller energy gaps, or bands at longer wavelengths.

Groups which show typical absorptions are called 'chromophores', and some are listed in Table 8.2, together with typical solvents for UV work.

Groups which are added or substituted onto the chromophores may change their absorption maxima. These are called 'auxochromes' and rules have been suggested for the calculation of the effects of these groups (Table 8.3).

Table 8.2 Absorption characteristics of some typical chromophores

Chromophore	Example	Transition	λ_{max}/nm	ε/mol^{-1} m^2
$C = C$	Ethylene	$\pi \rightarrow \pi^*$	165	1500
$C = O$	Acetone	$\pi \rightarrow \pi^*$	188	90
		$n \rightarrow \pi^*$	279	1.5
$-N = N-$	Azomethane	$n \rightarrow \pi^*$	347	0.45
$-N = O$	Nitrosobutane	$\pi \rightarrow \pi^*$	300	10
		$n \rightarrow \pi^*$	665	2
Benzene		$\pi \rightarrow \pi^*$	200	800
			255	21.5

Source: Fifield, F. W. and Kealey, D. (1990) *Analytical Chemistry*, 3rd edition, Blackie Academic & Professional, Glasgow.

Table 8.3 Rules for diene absorption

	Wavelength (nm)
Base value for heteroannular diene	214
Base value for homoannular diene	253
Increments added for:	
Double bond extending conjugation	30
Alkyl substituent or ring residue	5
Exocyclic double bond	5
Polar groups OAc	0
OAlk	6
SAlk	30
Cl, Br	5
N(Alk)$_2$	60
Solvent correction	0

Source: As Table 8.2.

8.2.4   Solvent effects

Since the spectrum of the species is measured in solution, we must consider any possible effects of the solvent on the transitions.

1. If we increase the polarity of a solvent say from hexane to water, the more polar n level is stabilised to lower energy, and the π^* is slightly lowered. The n–π^* transition increases in energy, or decreases in wavelength. This is a 'blue' or hypsochromic shift.
2. If we increase the polarity of a solvent, the more polar π^* is slightly stabilised while the π level is hardly affected. Thus the π–π^* transitions show a 'red' or bathochromic shift.
3. If we alter the pH of an aqueous solvent, we may change the nature of acidic or basic species. For example, in acidic solutions, phenols exist as ROH species, which absorb at about 270 nm, whereas in alkaline solution they exist as RO$^-$ which absorbs at about 210 nm.

8.2.5 Metal complexes

The d levels of transition metals are split by complexation. We can measure the concentration by measurement of the d–d transition, or occasionally the transitions of the ligand groups, or charge transfer transitions involving electron transfer between two orbitals, one predominantly associated with the metal, the other with the ligand.

These may all be used quantitatively, although the d–d transitions are sometimes of low intensity.

8.2.6 Applications

Some applications of quantitative visible and ultraviolet spectrometry are listed in Table 8.4. Some representative UV and visible spectra are shown in Figure 8.2.

Table 8.4 Applications of quantitative visible and ultraviolet spectrometry

Element or compound determined	Reagent	Example of application
Fe	o-Phenanthroline	Natural waters, petroleum products
Cu	Neocuproine	Minerals, alloys
Mn	Oxidation to MnO_4^-	Steels
Cr	Diphenylcarbazide	Alloys, minerals
Hg, Pb	Dithizone	Food products, fish
P, PO_4^{3-}	Reduction to molybdenum blue	Fertiliser residues, soils
F⁻	Lanthanum alizarin complexone	Drinking water, analgesic preparations
Aspirin	—	
Cholesterol	Liebermann–Burchard reaction	Body fluids
Vitamin A	Glycerol trichlorohydrin	Foodstuffs
Sulphonamides	Diazo derivatives	Drug preparation
Proteins	Biuret reaction	Tissue, body fluids
DDT	Nitrated derivative	Soils, fish

Source: As Table 8.2.

8.2.7 UV fluorescence methods

The absorption of radiation at one wavelength can sometimes be re-emitted at a longer wavelength, due to the processes of fluorescence (see Figure 6.3).

Instruments for measuring the fluorescent emission generally have a monochromator to control the exciting wavelength, and a second monochromator and optics at right angles to the exciting beam, to measure the wavelength and intensity of the fluorescent beam. An example of the apparatus is shown in Figure 8.3, and a fluorescent spectrum in Figure 8.4.

Vitamin E occurs in nature as several oil-soluble compounds. After extracting with cyclohexane/ethanol, exciting at 295 nm gave fluorescent emission at 330 nm.

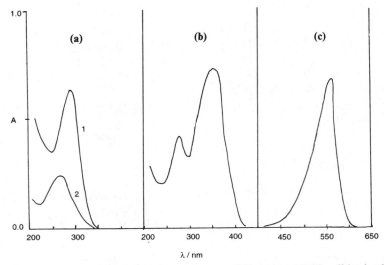

Figure 8.2 Typical UV and visible spectra. (a) Spectra of (1) aspirin and (2) caffeine in alkaline solution. Mixtures of them may be analysed by UV spectrometry; (b) vitamin A_2 in aqueous solution. (c) Cobalt complex $Co(H_2O)_6^{2+}$ in aqueous solution.

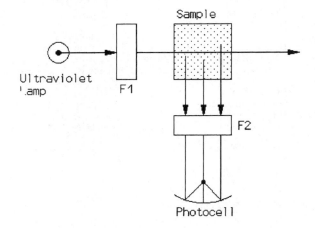

Figure 8.3 Schematic of a fluorescence spectrophotometer. F1 and F2 may be filters or monochromators.

The laws relating to fluorescent intensity depend on the amount of incident radiation absorbed. Provided the absorbance is not too high ($A < 0.05$) then the fluorescent power F is given by

$$F = Kc \tag{8.5}$$

and a plot of F versus concentration c is linear at low concentrations.

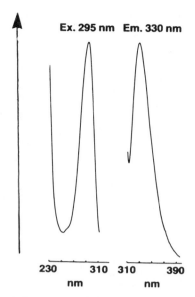

Figure 8.4 Excitation and emission spectra of vitamin E. Source: Rhys Williams, A.T. (1985) *An Introduction to Fluorescence in Biological Analysis*. Perkin-Elmer, Beaconfield.

Metal species may be measured by forming fluorescent complexes, for example, aluminium will complex with Alizarin garnet R to give a complex excited at 470 nm, fluorescing at 500 nm.

Sensitivity. The high sensitivity of some fluorescent measurements make then very suitable for measuring some pharmaceutical substances. For example, carminomycin may be measured down to concentrations of 2 ng/ml.

8.2.8 Combined separation and UV techniques

The use of UV detectors with liquid chromatographic separations is most important. Choice of a suitable wavelength, for example around 250 nm for species having aromatic groupings also give some structural information. The complete spectrum of each component may be obtained using diode array detectors (Figure 8.5) (see chapter 6).

In order to identify and quantify components separated by thin-layer chromatography (TLC) we may extract them into suitable solvents and then measure their UV spectrum. Separation of amino acids by liquid of thin-layer chromatography allows the simultaneous determination of all components with fluorescent detection.

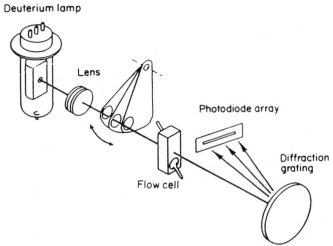

Deuterium lamp

Lens

Photodiode array

Diffraction grating

Flow cell

Figure 8.5 Optical path in a UV photodiode array detector. Source: Fifield, F.W. and Kealey, D. (1990) *Principles and Practice of Analytical Chemistry*, 3rd edition, Blackie Academic & Professional, Glasgow.

The measurement of the UV or visible absorbance of a spot on a TLC plate may also be accomplished directly by reflectance methods.

8.3 Infrared spectrometry

The absorption of radiation in the infrared region, between 50 and 2 μm or 200 cm^{-1} and 5000 cm^{-1} is due to the excitation of molecular vibrations, and sometimes of rotations as well.

The source used in most infrared systems is a glowing metal ribbon or a ceramic rod, heated electrically. Since many materials absorb infrared, the optics are most frequently of front-reflective mirrors and reflection gratings, although occasionally prisms of ionic solids (e.g. NaCl, TlBr/TlI) are used. Continuous wave dispersive instruments most often have a double-beam arrangement, although Fourier-Transform instruments use a single beam. The detector is sensitive to the heating effect of infrared and can be a small thermocouple, or a multiple thermocouple ('bolometer'). For faster response, a pyroelectric detector based on deuterated triglycine sulphate (DTGS) which changes its electrical characteristics when illuminated with IR, or a cooled semiconductor detector such as the mercury–cadmium telluride (MCT) unit may be used (Figure 8.6).

8.3.1 Sampling

Solid samples are often studied making them into a thin paste or 'mull', by grinding them with paraffin oil ('Nujol') and then placing a thin layer of mull between NaCl plates.

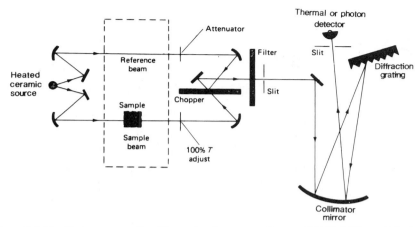

Figure 8.6 Double-beam recording IR spectrophotometer. Source: Fifield, F.W. and Kealey, D. (1990) *Principles and Practice of Analytical Chemistry*, 3rd edition, Blackie Academic & Professional, Glasgow.

Liquid samples, or solids in solution are placed between plates of NaCl, or, if the mixture contains water, between insoluble plates of AgCl or BaF$_2$. Suitable solvents for infrared are tetrachloromethane, CCl$_4$, or cyclohexane, C$_6$H$_{12}$.

Gaseous samples, having a lower concentration at ordinary pressures require a long path-length cell of 0.1 to 20 m, depending on the partial pressure and absorptivity of the gas to be studied.

Very small solid or liquid samples may be examined using infrared microscopic techniques or beam condensation accessories.

Samples where the surface concentration of a species is important can be examined using reflectance methods, such as attenuated total reflectance (ATR) or multiple internal reflectance (MIR). In these, the sample is placed in intimate contact with the surface of a high refractive index prism which is illuminated with IR as shown in Figure 8.7. The IR strikes the surface at

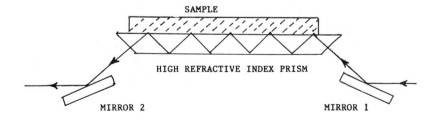

Figure 8.7 Schematic of an attenuated total reflectance (ATR) assembly for infrared spectrometry. Mirrors are shown flat for simplicity.

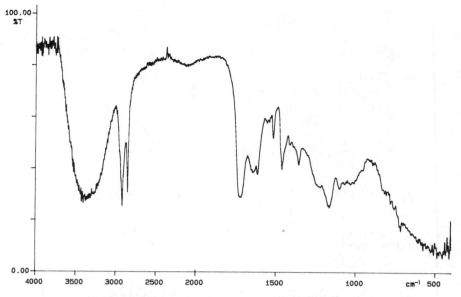

Figure 8.8 ATR spectrum of a leaf treated with insecticide.

an angle greater than the critical, and is reflected. However, at the surface some reduction of intensity occurs because the sample absorbs at specific wavelengths, giving the spectrum shown in Figure 8.8.

8.3.2 Infrared absorption

Considering a molecule with N atoms, we must define $3N$ degrees of freedom in order to be sure of every atom. However, if the molecule moves about in space complete, the coordinates of the centre of mass define three coordinates. If it rotates as a whole, three more are defined (except for a linear molecule, when it is only two). Therefore there can be $3N - 6$ different degrees of freedom for vibration, (or $3N - 5$ if a linear molecule). Thus there are $3N - 6$ fundamental vibrations for a general, non-linear molecule.

In order to interact with the electromagnetic radiation, the motion of the molecule must alter the electric dipole. Only molecules with a permanent dipole will interact because of their rotations. Homonuclear diatomics such as N_2 have no dipole. Vibration of the molecule does not alter that fact, so they do not interact in the IR.

A heteronuclear diatomic molecule such as CO has only one fundamental, as shown in Figure 8.9.

The vibration of diatomic molecule AB may be described, approximately, in terms of Hooke's Law:

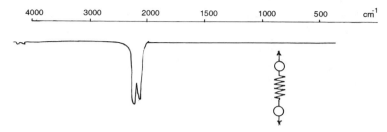

Figure 8.9 Infrared spectrum of carbon monoxide, CO. The fundamental stretching vibration is at 2150 cm^{-1} and the first overtone gives the very weak band near 4200 cm^{-1}. The band shape is due to the rotational fine structure.

$$\nu = (1/2\pi) \times (k/\mu)^{1/2} \qquad (8.6)$$

where ν is the fundamental vibration frequency; μ the reduced mass = $m_A m_B/(m_A + m_B)$; and k the force constant of the vibration.

This formula shows that alteration of the force constant, or the atomic masses will alter the vibrational frequency. Molecules may also show weak 'overtone' vibrations approximately at multiples of the fundamental frequency.

Sulphur dioxide, SO_2, is a bent triatomic molecule and can vibrate in $3 \times 3 - 6 = 3$ different ways, as shown in Figure 8.10.

Note that all the above vibrations change the dipole of the molecule, so each will be active in the infrared.

Carbon dioxide, CO_2, is a linear molecule and can therefore vibrate in $3 \times 3 - 5 = 4$ fundamental ways as shown in Figure 8.11. The symmetric stretching vibration ν_1 does not change the dipole, so is inactive in the infrared. The antisymmetric stretch, ν_3 does change the dipole and so do the two bending vibrations ν_2 (a and b), which occur at the same energy.

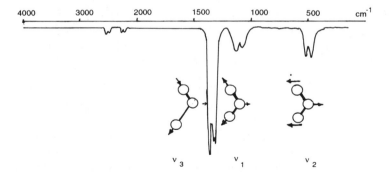

Figure 8.10 Infrared spectrum of sulphur dioxide, SO_2. This is a bent molecule and shows three fundamentals in the IR: ν_1 (symmetric stretching) at 1150 cm^{-1}, ν_2 (bending) at 520 cm^{-1} and ν_3 (antisymmetric stretching) at 1367 cm^{-1}. The very weak peaks near 2400 cm^{-1} are overtones and combination vibrations. The band shapes are due to rotational fine structure.

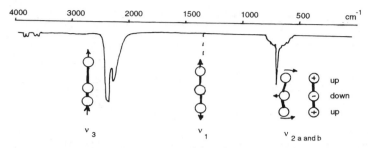

Figure 8.11 Infrared spectrum of carbon dioxide, CO_2. This is a linear molecule and shows only two peaks in the IR (v_1 (symmetric stretching) at 1340 cm^{-1} is absent!): v_2 (a and b) (two bendings) which are degenerate and both occur at 667 cm^{-1} and v_3 (antisymmetric stretch) at 2349 cm^{-1}. The band shapes are due to rotational fine structure.

Note that we may also get combinations of these vibrations, e.g.

$$v_1 + v_3, \ 2v_2 + v_3. \tag{8.7}$$

8.3.3   Polyatomic molecules

To apply the '$3N - 6$' rule to a molecule with 12 atoms, or 30 atoms would be very difficult, especially as we should consider the combination and overtone vibrations too!

It has been found that we may consider some of the groups within molecules as giving approximately constant vibrational 'group frequencies'. This can be ascribed to the effects of the masses of the atoms and force constants of bonds.

1. H–X groups: the hydrogen atom has such a low atomic mass that the groups attached to it move comparatively little. The frequencies of the H–X group vibrations will depend on the force constant of the bond, but the bond energies of H–X bonds are mostly in the region of 400 kJ mol^{-1} (e.g. C–H in benzene $\Delta H(CH) = 469$ kJ mol^{-1}, OH in alcohols $\Delta H(OH) = 425$ kJ mol^{-1}). The H–X stretching vibrations occur within the region 4000–2500 cm^{-1}. The H–X bending vibrations are more scattered, as shown in Figure 8.12.

2. X=Y groups: multiple bonds are considerably stronger than single bonds, and will thus be somewhat less affected by the effects of attaching other groups. This is particularly important for groups such as the C=N and C=O. These have significant dipoles and very characteristic group frequencies. In general, the group frequencies of X–Y bonds increase with the bond order between X and Y, for example:

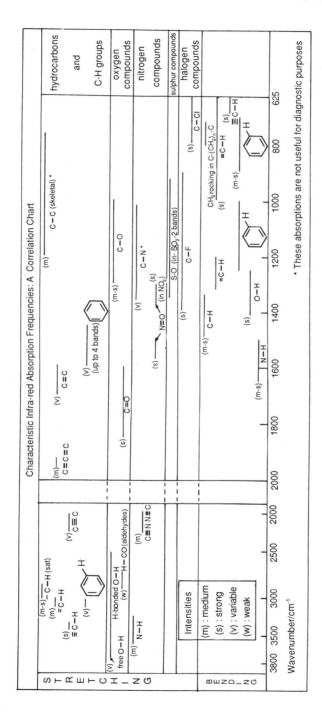

Figure 8.12 Infrared group frequency correlation chart. Source: *S24/S25 DB Chemistry Data Book*, Open University Press, Milton Keynes, 1973.

Bond	Order	Approx $\nu(\text{cm}^{-1})$
$-C-O-$	1	1100
$-C=O$	2	1700
$C\equiv O$	3	2200

3. By contrast, single bond frequencies, especially bonds such as C–C, are extremely variable, and specific to the compound. The region 1500–600 cm^{-1} has many of these bonds and is decribed as the 'fingerprint' region because of its use in identification.
4. The bending vibrations of molecules are very characteristic, and may be described as being 'in plane' or 'out of plane'. This refers to the principal plane of the molecule, for example plane of a benzene ring or of an olefinic bond. Figure 8.12 shows simplified infrared correlation charts for many groups.
5. The interaction of groups within the molecule, and the changes in bonding due to interaction will affect the vibrational bonds.

For a molecule which has a group such as –OH or –NH which may hydrogen bond to other molecules, or to other groups within the molecule, the –O–H bond strength will be altered by hydrogen bonding, and the vibrational band will be broadened. For example, in gaseous acids, a sharp band due to the free –OH stretch is observed around 3600 cm^{-1}, whereas in pure liquid acid, the band is stronger and stretches between 2500 and 3500 cm^{-1} due to the hydrogen bonding in the liquid.

For vibrations of adjacent groups which occur near the same frequency, there may be a strong coupling interaction. For example, the stretching vibration of the carbonyl group alone occurs at around 1700 cm^{-1} and the bending vibration of the –NH group around 1650 cm^{-1} while for the amide gouping –CO–NH– these vibrations couple to give several 'amide bands' over the region 1500–1700 cm^{-1} where the vibrations contain components of both vibrations.

8.3.4   Combinations of infrared and separation techniques

GC–IR. The addition of an infrared spectrometer as a detector for a gas chromatograph has been of great value. The carrier gas for the GC may be He, H$_2$ or N$_2$, none of which absorbs in the infrared region. The gas stream from the GC column is often split unequally 10:1, the smaller amount going to a GC detector, such as a flame ionisation detector, the larger amount through a short, heated transfer line into a heated IR cell. This cell has a path length of about 0.1 m and reflective walls. Better sensitivity and signal are obtained using FTIR especially with a cooled semiconductor detector rather than a pyroelectric detector. Detection limits are not very low, and under the most favourable conditions nanogram amounts of material may be detected.

Pyrolysis-GC-FTIR. The chromatography of solid samples, for example, polymers or pesticides, to determine their breakdown and the possible effects of environmental changes may be carried out using a heated probe to pyrolyse the sample into the injection volume of the gas chromatograph. The sample is placed onto a wire coil or a ribbon which is then heated electrically to temperatures up to 800°C. An alternative method uses wires of a ferromagnetic material. The sample is placed in a loop on this wire which is then heated inductively up to the Curie Point of the metal, for example an iron wire would heat to 760°C.

The pyrolysis generally produces a large number of gaseous products which are separated by the GC and detected by FID and IR. The pattern of the chromatogram is quite characteristic and the products may be identified by IR (Figure 8.13).

Liquid chromatography-FTIR. Samples from a liquid chromatograph may be collected and their infrared spectra determined, especially if the chromatograph is operated in a 'preparative' mode. This is very time consuming, and other techniques have been suggested. The mobile phase stream from the standard LC detector may be passed through a small volume IR liquid -sampling cell and the spectra scanned as with GC-IR. Horizontal flow-through ATR cells have also been used. The collection and concentration of HPLC fractions has been carried out by mixing the eluant with a heated gas stream and spraying through a jet at a moving planar collection disc. The mobile phase evaporates and the solutes are deposited on the collection surface which is then transferred into an FTIR spectrometer.

8.3.5 Applications of infrared spectrometry in environmental analysis

1. *Air pollutants.* The detection of gases which absorb infrared is discussed more fully in chapter 14. The detection limit of hydrocarbons, where the C–H stretching vibration absorbs around 3000 cm^{-1} is of the order of 1 ppm in a 10-m cell. Carbon monoxide with a sharp absorption band at 2250–2050 cm^{-1} is detectable to about 10 ppm, and SO_2 to about 1 ppm using the absorption at 1250–1000 cm^{-1}.

2. *Solid pollutants.* Collection onto filters, cooled surfaces or GC adsorbents such as Tenax may be followed by examination of the surface directly by IR, surface analysis or IR microscopy. Contaminants such as pesticides, organic or inorganic sprays or dust on plant surfaces may be examined.

3. *Degradation.* The effects of environmental exposure on samples may be followed by infrared measurements. The degradation of polymers under the combined effects of exposure to UV radiation, chemical attack by free radicals produced by natural processes and microbial

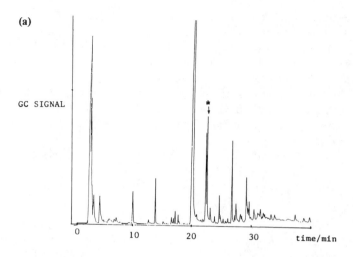

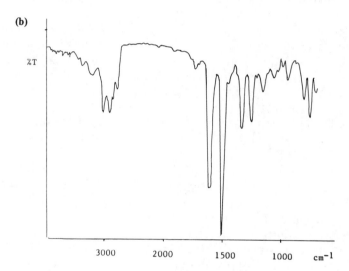

Figure 8.13 Pyrolysis-GC-FTIR of an epoxy resin. (a) The complex gas chromatogram after pyrolysis at 850°C. (b) The FTIR spectrum of the peak at 22.4 min. Comparison with library vapour phase spectra shows this to be *p*-toluidine.

attack can be followed by recording the IR spectrum after successive exposures.

8.4 Nuclear magnetic resonance spectrometry   The nuclei of certain isotopic species of some elements possess a magnetic moment characterised by a nuclear spin quantum number, *I*, as shown in Table 8.5.

Table 8.5 Nuclear spin quantum numbers and magnetic properties of selected nuclei

Nucleus	Nuclear spin quantum number I	Magnetic moment, μ (ampere square metre $\times 10^{27}$)	Resonance frequency in MHz at 1.4092 T	Relative sensitivity at the natural isotopic abundance
^{1}H	1/2	14.09	60.000	1.00
^{2}H	1	4.34	9.211	1.5×10^{-4}
^{12}C	0	–	–	–
^{13}C	1/2	3.53	15.085	1.8×10^{-4}
^{14}N	1	2.02	4.335	1×10^{-3}
^{16}O	0	–	–	–
^{17}O	5/2	−9.55	8.134	1×10^{-5}
^{19}F	1/2	13.28	56.446	0.834
^{31}P	1/2	5.71	24.288	0.066

Source: As Table 8.2.

The nuclei with non-zero values of I may take any position in space in the absence of a magnetic field. However, when a magnetic field, B, is imposed in a particular direction (e.g. the z direction), the component of the magnetic moment in the z direction, μ_z may only take certain values given by:

$$\mu_z = m_I \gamma(h/2\pi) \tag{8.8}$$

where m_I is a magnetic quantum number which may take values $I, I - 1, I - 2, \ldots, 2 - I, 1 - I, - I$, and γ the gyromagnetic ratio characteristic of that isotope.

For ^{1}H, $\gamma = 2.675 \times 10^8$ rad T^{-1} s^{-1}, e.g. for ^{1}H when $I = 1/2$, m_I may be $+1/2$ and $−1/2$ *only*, for ^{2}H when $I = 1$, m_I may be 1, 0 and $−1$ only.

Each allowed orientation of the nucleus will interact with the magnetic field to give defined energy levels:

$$E = - \mu_z B = - m_I \cdot \gamma(h/2\pi) \cdot B \tag{8.9}$$

Transitions between the levels may occur by the absorption or emission of radiation, so that, for ^{1}H, transition from the lower level, where $m_I = +1/2$ to the upper level, where $m_I = - 1/2$ requires an energy change of

$$\Delta E = \gamma(h/2\pi) \cdot B = h\nu \tag{8.10}$$

Using the value of γ given above we find that

$$\nu = \gamma B/2\pi \tag{8.11a}$$

and if $B = 1.409$ T

$$\nu = 60 \text{ MHz} \tag{8.11b}$$

This places the nuclear magnetic resonance effect in the radiofrequency region of the electromagnetic spectrum.

Absorption of radiation at this frequency will raise the ^{1}H nuclei present in this magnetic field from the lower to the upper energy level.

Other nuclei, having different gyromagnetic ratios (or magnetic moments) will resonate at different frequencies.

8.4.1   Instrumentation

The magnetic field may be produced by a permanent magnet for fields up to 2 T or an electromagnet up to 2.4 T, and for the highest fields up to 14 T, superconducting magnets are employed.

The homogeneity of the magnetic field is very important, and the sample, in solution in a thin glass tube is spun between the poles of the magnet to improve apparent homogeneity. The 'sweep coils' can alter the field by 10–20 parts per million. The radiofrequency transmitter may operate at a fixed frequency, for example 60 MHz, or alternatively, the magnetic field can be fixed and the frequency scanned. The radiofrequency receiver detects the absorption of radiation by the sample (Figure 8.14).

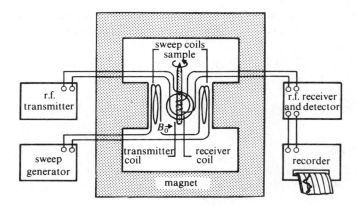

Figure 8.14 Schematic of an NMR spectrometer. Source: Fifield, F.W. and Kealey, D. (1990) *Principles and Practice of Analytical Chemistry*, 3rd edition, Blackie Academic & Professional, Glasgow.

Pulsed Fourier-Transform NMR (FT-NMR) spectrometers operate by supplying a high-intensity pulse of radiofrequency, which may be regarded as a combination of many frequencies. The detected signal following the pulse is a 'free induction decay' curve containing information on each frequency. This may be computer processed to yield a conventional NMR spectrum. FT-NMR has the advantage of being more rapid and allowing the averaging of many spectra with improvement in the signal/noise ratio and the sensitivity. This is particularly important for ^{13}C NMR, where there is only 1.1% of the active isotope present in the natural carbon-containing compounds.

The relaxation of the nuclei back to their equilibrium states, either by transferring energy to the surrounding lattice or solvent (*spin–lattice relaxation*) or by interchange of spin states (*spin–spin relaxation*) affects the line width. With ^{1}H NMR spectra in solution, the relaxation has a small effect, and the integrated area of the peak is proportional to the number of protons in that environment. For solid samples, the peaks are broader and less easy to integrate, while differences in relaxation times of carbon nuclei in different environments mean that integration cannot be used for ^{13}C NMR.

8.4.2 Solvents for NMR work

The commonest solvent for both ^{1}H and ^{13}C NMR work is tetrachloromethane (or carbon tetrachloride), CCl_4. In cases where this is not suitable, for example for ionic compounds, deuterated solvents can be used, for example D_2O, $CDCl_3$, C_6D_6 or $(CD_3)_2SO$, [d_6–DMSO].

Residual signals from the remaining, un-deuterated protons may be observed.

8.4.3 The chemical shift

If all protons had exactly the same energy levels, and hence resonated at exactly the same field and frequency, we could not distinguish between them. Each proton will be in an environment determined by the shielding effects of its surrounding electrons which will circulate in the applied field and create an additional field to alter the applied field. The more dense the electron environment, the greater the shielding it causes, the less the effect of the applied field, B. To bring it to resonance at the required frequency will require a higher applied field.

Each environment will have a characteristic effect on the nucleus. The chemical shift is defined with respect to a reference signal generated by an added compound, most often tetramethyl silane (TMS) $(CH_3)_4Si$, which is added to the sample in solution.

The chemical shift δ in parts per million (ppm) is defined by

$$\delta/\text{ppm} = (\nu(\text{sample}) - \nu(\text{TMS})) \times 10^6/\nu(\text{instrument}) \qquad (8.12)$$

TMS will have a δ value $= 0$.

As an example, using a 60 MHz instrument, if a sample signal occurs at a frequency 156 Hz higher than the reference TMS, then

$$\delta = 156 \times 10^6/60 \times 10^6 = 2.6 \text{ ppm}$$

Table 8.6A shows the chemical shift values for some common environments for ^{1}H nuclei, and Table 8.6B shows the values for ^{13}C nuclei.

Factors affecting the chemical shift. In order that the protons may be exactly equivalent, they must firstly be chemically equivalent, for example the three protons of the same methyl group, or the two protons on a methylene group are equivalent. One useful rule is that, if a proton is replaced by a substituent, X, then equivalent protons will give the same product, while non-equivalent protons give different products. Thus, the protons of the vinyl group $CH_2=CH-X$ are all different and give three NMR signals at about 5.4, 5.6 and 6.2 ppm. Similarly, the o-, m-, and p-protons of a monosubstituted benzene have distinct signals.

Table 8.6 Chemical shifts in NMR spectra

Group	Chemical shift, δ/ppm from TMS	
	(A) Proton ^{1}H	(B) Carbon ^{13}C
RCH$_3$	0.9–2.0	6–30
R$_2$CH$_2$	1.3–2.5	15–55
R$_3$CH	1.5–3.3	22–60
R$_4$C	–^{a}	30–50
C=CH	4.6–5.9	100–160
Aromatic, ArH	6.0–8.5	120–150
CH–Cl	3–4	10–80
CH–OH	3.4–4.0	50–85
CH–CO–R	2.0–3.6	20–50
H–CO Aldehyde	9–10	190–210
OH	1–6	–
CO	–	150–230
COOH	10–12	170–190
NH$_2$	1–5	–

^{a} Dash (–) indicates no signal.

The effects of temperature may be to cause a slower rotation about a bond, or a slower alternation of configurations and alter the interactions and hence the appearance of the NMR spectrum.

The electronegativity of the groups attached. For protons directly attached, or on the same carbon as electronegative atoms, such as halogens, the signals are shifted downfield, to higher δ values. Conversely, electropositive atoms will shift to lower δ values. The effect of changing from $Si(CH_3)_4$ ($\delta = 0.0$) through $CH_3 \cdot CH_3$ ($\delta = 0.9$) to $CH_3 \cdot Cl$ ($\delta = 3.1$), CH_3OH ($\delta = 3.4$) to CH_3F ($\delta = 4.3$) is due to the increasing electronegativity of the attached group. Similar effects are noticed with the ^{13}C carbon atoms.

The anisotropy of chemical bonds. The chemical shift of protons will be affected by the magnetic fields due to the circulation of the electrons in the applied magnetic field. If this field is not the same in all directions, it is said to be diamagnetically anisotropic. A benzene ring will have such an induced field which will increase the effective field at the protons around the ring and thus 'deshield' them so that they resonate at higher δ values around 7.3 ppm (Figure 8.15).

Double bonds behave in a similar fashion and the olefinic protons and carbon atoms have higher δ values and the carbon and proton of the aldehyde group have very high δ values.

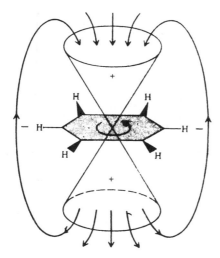

Figure 8.15 Anisotropic shielding around a benzene ring. The shielding zones are indicated by the '+' signs, the deshielding zones by '−' signs. Source: Fifield, F.W. and Kealey, D. (1990) *Principles and Practice of Analytical Chemistry*, 3rd edition, Blackie Academic & Professional, Glasgow.

For ethynic compounds, the electron ciculation is along the axis of the triple bond, so that the ethynic protons are shielded

Hydrogen bonded protons can exchange rapidly between environments, and their chemical shifts may also vary widely, for example from 0.5 to 5 for alcohols and amines. If the sample in solution in CCl_4 is shaken with D_2O, any OH, NH or SH protons will exchange and their signal will be removed from the spectrum.

8.4.4   The peak area

The intensity of the signals generated depends on the number of nuclei resonating at that frequency and their relaxation times. For ^1H, or proton NMR, the lines are generally sharp and relaxation is rapid. The integrated areas of the peaks are in the ratio of the number of protons in each environment. For example, in Figure 8.16, the signals at $\delta = 0.9$, 2.5 and 7 correspond to 9, 2 and 5 protons, respectively.

The integrated trace is presented as well as the original signal, and the step heights indicate the ratio of the protons in each peak.

For ^{13}C NMR, the peak integrals are not in proportion to the number of equivalent carbons, since the relaxation times of the carbon atoms vary widely.

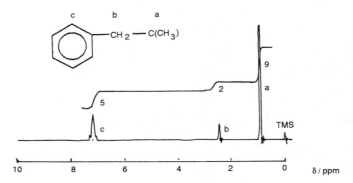

Figure 8.16 Typical proton NMR spectrum. The signals from the three sets of protons of neopentyl benzene occur at different chemical shifts and have different integrated areas.

8.4.5   Spin–spin coupling

If there are magnetically active nuclei on nearby atoms, they may interact to divide the signal of proton or carbon into a multiplet. This is due to the small variations in the field caused by neighbouring nuclei in different spin states. It is transmitted through the bonds joining the atoms. Generally, coupling decreases very greatly with distance, and in saturated compounds is restricted to about three bond lengths. In aromatic and unsaturated compounds, it may extend further.

Coupling in ¹*H NMR.*   If the chemical shifts of two signals for protons on neighbouring atoms are well separated, then the system is described as an AX system, and the splitting as '*first-order*' (Figure 8.17). The proton signal for the 'A' proton is affected by the two allowed spin states of the 'X'

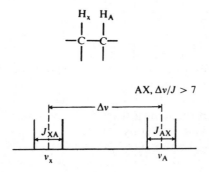

Figure 8.17 First-order coupling between adjacent protons AX. Source: Fifield, F.W. and Kealey, D. (1990) *Principles and Practice of Analytical Chemistry*, 3rd edition, Blackie Academic & Professional, Glasgow.

proton, and its signal split into two equal parts, separated by the coupling constant J_{AX}, measured in Hz. In an exactly similar way, the signal of the 'X' proton is also split by the same constant, as shown below.

For three well-separated signals, the system is called an 'AMX' system.

The number of arrangements of n neighbouring magnetic nuclei will split the signal into a multiplet of $(n + 1)$ peaks, the intensities of the multiplet peaks being given by the Pascal's triangle rule, shown in Table 8.7.

Table 8.7 Pascal triangle scheme for coupling

Neighbours	Multiplet	Intensities						
1	Doublet				1	1		
2	Triplet			1	2	1		
3	Quartet			1	3	3	1	
4	Quintet		1	4	6	4	1	
5	Sextet	1	5	10	10	5	1	
6	Septet	1	6	15	20	15	6	1

If there are two sets of non-equivalent neighbours, A and X about a central M proton, they will have their own separate coupling constants and there will be a total of $(n_A + 1) \times (n_X + 1)$ peaks.

If the chemical shifts of the proton signals are closer together $(\Delta v/J < 7)$, then the inner peak intensities increase at the expense of the outer and the system, now an 'AB' system, shows *second-order* splitting illustrated in Figure 8.18.

Equivalent protons do not couple, but non-equivalent ones do. To illustrate this, consider the important set of compounds possessing the vinyl group, e.g. vinyl chloride and styrene.

$$H_B \diagdown \diagup H_X$$
$$C = C$$
$$H_A \diagup \diagdown Y$$

If Y is an aromatic group, then the chemical shifts are

$$
\begin{array}{ll}
H_B & \delta = 5.1 \text{ ppm} \\
H_A & 5.6 \\
H_X & 6.7
\end{array}
$$

and the coupling constants are, approximately:

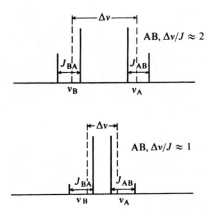

Figure 8.18 Second-order coupling between adjacent protons with similar chemical shifts. Source: Fifield, F.W. and Kealey, D. (1990) *Principles and Practice of Analytical Chemistry*, 3rd edition, Blackie Academic & Professional, Glasgow.

$$
\begin{array}{ll}
J_{AB} & 3 \text{ Hz} \\
J_{BX} & 10 \\
J_{AX} & 17
\end{array}
$$

This gives the splitting and spectrum shown in Figure 8.19. It is important that typical coupling patterns be recognised.

Coupling in ^{13}C NMR.   Since the presence of a proton on a ^{13}C carbon will split its NMR signal into a doublet, the carbon NMR spectra may become

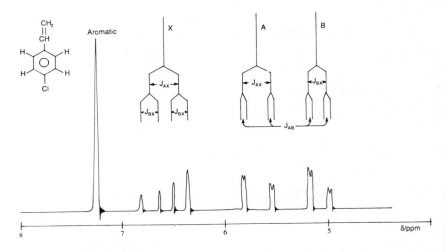

Figure 8.19 The proton NMR spectrum of *p*-chlorostyrene showing the coupling of the three protons of the vinyl group.

very complex. As before, the multiplet structure of the peaks is given by the Pascal triangle and (*n* + 1) rules. Thus a methyl carbon peak with three attached hydrogens will be a quartet with intensities (1:3:3:1) and a methine carbon with a signal hydrogen is a doublet. If there are *no* hydrogens attached, for example in a ketone carbonyl carbon R–CO–R, or the substituted carbon of a benzene ring, then the signal is a singlet.

Decoupled spectra. The complex spectra may be simplified by de-coupling the proton signals. Two methods are frequently used.

In broad-band decoupling the sample is irradiated with a spread of frequencies covering the proton NMR range. This irradiates all the proton signals and causes rapid transitions between the different orientations, so that the proton signals will appear to be an averaged orientation and their coupling with the ^{13}C nuclei is removed. This can often lead to a clear spectrum with distinct single peaks for each carbon atom.

Unfortunately, this removes all the coupling information which would help in interpretation of the spectrum. An alternative is to use off-resonance decoupling. Here, the irradiation of the proton frequencies is deliberately offset from resonance by a few hundred hertz, and consequently the coupling effects are present, through greatly reduced (Figure 8.20).

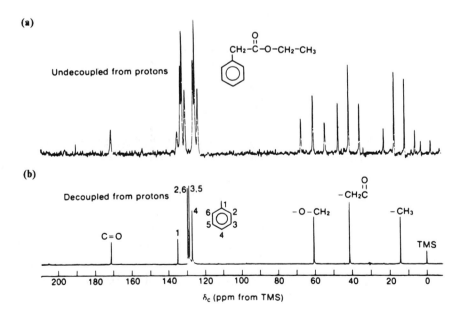

Figure 8.20 ^{13}C NMR spectra of ethyl phenylacetate: (a) undecoupled from protons and (b) decoupled from protons. Source: Moore, J.A. and Dalrymple, D.L. (1976) *Experimental Methods in Organic Chemistry*, W.B. Saunders, New York.

8.4.6 Applications of NMR

The chief application of NMR is in the identification of organic species from many chemical areas. Natural products, pharmaceuticals, paints and polymers and metabolic intermediates may all be identified by NMR.

8.5 Mass spectrometry This is a spectrometric method which does not involve the absorption or emission of electromagnetic radiation. The introduction of a sample in a molecular or atomic state into a region where it may be converted into ionic particles and fragments and then analysed by measuring the mass/charge ratio of the ions is an extremely sensitive, versatile and important analytical method.

8.5.1 Instrumentation

In order to allow molecules and ions to pass freely through the apparatus, the main flight tube system is evacuated to a very high vacuum, 10^{-4}–10^{-6} Pa so that molecular collisions are minimised (Figure 8.21).

Samples are introduced into the apparatus by controlled evaporation from a cooled solid, by allowing a slow leak through a heated capillary inlet or a pinhole, or by inserting the sample placed on the end of a probe through a vacuum lock, and then heating the probe electrically. Combinations of chromatographic techniques and thermal analysis apparatus with mass spectrometry allow separated samples and volatile evolved products to be analysed directly.

The sample must then be converted into ions. The most usual method of ionisation is by electron impact (EI). The sample is brought into contact with a high-energy electron beam whose energy is about 70 eV, or more than 6000 kJ mol^{-1}. This first causes ionisation of the molecule M (or atom) and then may break it into fragments:

$$M + e^* = M^{+\cdot} + 2e \tag{8.13}$$

$$M^{+\cdot} = Ft^+ + Nt \tag{8.14}$$

where M^+ is the molecular ion,

Ft$^+$ is a fragment ion which may also be a radical ion, and Nt is a neutral molecule or radical. For example, for propanone

$$CH_3 \cdot CO \cdot CH_3 + e^* = [CH_3 \cdot CO \cdot CH_3]^{+\cdot} \quad (m/z = 58) \tag{8.15}$$

$$[CH_3 \cdot CO \cdot CH_3]^{+\cdot} = [CH_3 \cdot CO]^+ + CH_3^\cdot \quad (m/z = 43) \tag{8.16}$$

Because EI causes severe fragmentation, so that the molecular ion may be of very low abundance, other techniques are also used.

In chemical ionisation (CI) a gas such as methane or isobutane is introduced which is ionised to primary ions (e.g. CH_4^+) which react with

(a)

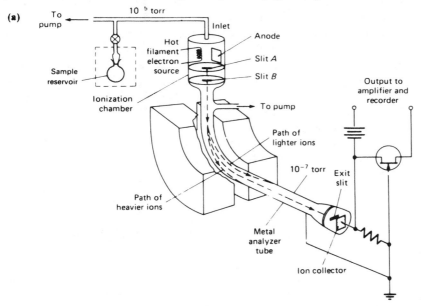

(b)

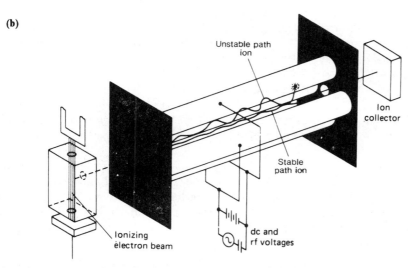

Figure 8.21 Schematic diagrams of (a) a single magnetic focusing mass spectrometer and (b) a quadrupole mass spectrometer. Source: Lichtman, D. (1964) *Res. Dev.* **15** (2), 52.

excess methane to give secondary ions such as CH_5^+. These react with the sample molecules:

$$RH + CH_5^+ = RH_2^+ + CH_4 \qquad (8.17)$$

CI produces $(M + 1)^+$ ions, and $(M - 1)^+$ ions by hydrogen abstraction and these undergo less fragmentation because of their lower energy.

In fast atom bombardment (FAB) high energy xenon or argon atoms are produced by ionisation, acceleration and subsequent transfer of energy by collision to produce a high-energy beam of neutral atoms. These are directed onto the sample in a high boiling matrix such as glycerol or carbowax. Ionisation occurs by translational energy transfer and $(M + H)^+$ ions often produced. This technique is particularly useful for high molecular weight species such as proteins.

Other ionisation methods such as field desorption, field ionisation and rf spark discharge are also used.

The molecular ions and fragment ions are next accelerated into an analyser section by an 'ion gun' having a potential difference, V, across the plates and focusing slits. The analyser may be a single magnetic field of strength B, as shown in Figure 8.21 where the ions of mass m and charge z follow a circular path of radius, R, such that, for a single-focusing mass spectrometer,

$$m/z = B^2 R^2 / (2V)$$

Double-focusing mass spectrometers have both electrostatic and magnetic sectors for greater resolution, and may allow measurement of ion masses very precisely.

Many modern systems use the quadrupole mass spectrometer, which is also described as a 'mass filter'. Four parallel poles are each charged with both a constant dc voltage and an oscillating radiofrequency voltage. The ions follow a complex path due to their interaction with the electric fields produced and only under a very particular condition of field and of m/z will an ion follow a stable oscillatory path to the detector. Other ions will be lost. By scanning the dc potential and rf field the range of ions may be detected.

The quadrupole ion trap configuration allows ionisation and analysis in the same space and permits very rapid scanning of samples from chromatography.

The separated molecular and fragment ions pass through slits into the electron multiplier and collector. The signal is amplified and processed to a recorder, cathode ray oscilloscope, or, more usually, in modern systems to a computer and data processor. Mass spectra are almost invariably presented as a plot of the relative intensity as a percentage of the most abundant peak or 'base peak'.

$$\text{Relative intensity } \% = \frac{100 \times \text{peak height}}{(\text{height of most abundant peak})} \qquad (8.18)$$

8.5.2   Isotopic composition and accurate masses

The isotopic composition of stable natural atomic species is very definitely known. Table 8.8 shows the abundance and exact masses of atoms commonly found in organic species, and isotopic compositions of other species may be found from references. These data allow us to predict both the exact mass of a species and also the probable ratios of the intensity of the molecular ion peak to those of its isotope peaks at $(M + 1)$ and $(M + 2)$.

Table 8.8   Abundance and mass of common isotopes

Atom	Isotope	Exact mass	% Natural abundance
H	1	1.0078	99.985
H	2	2.0141	0.0015
B	10	10.0129	19.6
B	11	11.0093	80.4
C	12	12.0000	98.892
C	13	13.0034	1.108
N	14	14.0031	99.635
N	15	15.0001	0.365
O	16	15.9949	99.759
O	17	16.9991	0.037
O	18	17.9992	0.204
F	19	18.9984	100
S	32	31.9721	95.0
S	33	32.9715	0.76
S	34	33.9679	4.22
Cl	35	34.9689	75.79
Cl	37	36.9659	24.20
Br	79	78.9183	50.52
Br	81	80.9163	49.48

8.5.3   Nitrogen rule

It is worth commenting that, since only nitrogen among the common elements has an *odd* valency (or oxidation number) of three, but the most abundant isotope has an *even* mass, then any molecule with an odd molecular mass must contain an odd number of nitrogens. If the molecular mass is even then the number of nitrogens is even or zero.

For an ion containing n atoms of an element with two isotopes of fractions a and b whose masses differ by d, the probable contribution of the isotopes to the successive peaks M, $M + d$, $M + 2d$, $M + 3d$ and so on is given by the binomial expansion

$$(1 + b/a)^n = 1 + n(b/a) + (n)(n - 1)(b/a)^2/2$$
$$+ (n)(n - 1)(n - 2)(b/a)^3/6 + \ldots \qquad (8.19)$$

Thus for the two chlorine isotopes Cl-35 and Cl-37, where $d = 2$ mass units, and $(b/a) = 0.32$, then a molecule with a single Cl $(n=1)$ will have two peaks

M and $(M + 2)$ in the ratio 1:0.32 or 100:32. With two chlorines ($n = 2$) there will be three peaks M, $M+2$, $M+4$ in the ratio 1:0.64:0.10 or 100:64:10 or 9:6:1.

For a molecule with 10 carbon atoms, $n = 10$, $d = 1$ and $b/a = 0.011$ the contribution of the carbons to the intensities of the molecular ion peaks will be

$$M/M + 1/M + 2 = 1:1.1:0.0055 \quad \text{or} \quad 100:11:0.55$$

Tabulated values are available both for the accurate masses and for the $M/M + 1/M + 2$ ratios for molecules containing C, H, O and N up to $m/z = 500$ and these enable molecular formulae to be suggested.

Example. The mass spectrum of a possible industrial pollutant chemical gave an accurate mass of 182.033 and the ratios of the peaks at $M/M + 1/M + 2$ were 10:8.6:1.1.

1. There are clearly an even number (or zero) nitrogen atoms because the molecular mass is even. There is no chlorine, bromine or sulphur present, because the $M + 2$ peak is small.
2. The ratios for several possible formulae are:

Formula	$(M + 1/M)\%$	$(M + 2/M)\%$
$C_7H_6N_2O_4$	8.58	1.13
$C_9H_{10}O_4$	10.04	1.25
$C_{11}H_6N_2O$	12.79	0.95
$C_{14}H_{14}$	15.35	1.10

The $M + 2$ ratios are all very similar, but the only possible match for the $M + 1$ ratio is the $C_7H_6N_2O_4$.
3. If we calculate the accurate mass to be expected for this formula we find $M = 182.0326$, close to the value found.

8.5.4 Fragmentation

There are many schemes which relate the structure of the molecules to their fragmentation mechanism (Table 8.9). Some important points may be summarised as follows:

1. The relative intensity of the molecular ion peak decreases with increasing molecular weight and with increase in chain branching for an homologous series of compounds.
2. Fragmentation to stable fragments, especially if these can be stabilised by resonance, is more likely. Benzene rings, allylic carbocations $[CH_2{=}CH{-}CN_2]^+$ and tertiary carbocations are readily formed.
3. In molecules containing heteroatoms (e.g. O, N, S) the C–C bonds next to the heteroatom are frequently cleaved, leaving the charge on the heteroatom fragment.

Table 8.9 Typical mass spectral ions and losses

Ion mass (m/z)	Possible formula (or loss)[a]	Possible compound type
M–14	(–CH$_2$)	Homologue chain
M–15	(–CH$_3$)	Methyl compound
M–18	(–H$_2$O)	Alcohol, acid
M–26	(–C$_2$H$_2$)	Aromatic hydrocarbon
M–27	(–HCN)	N-Heteroaromatic, ArNH$_2$
29	C$_2$H$_5$, CHO	Ethyl, aldehyde
30	NO	Nitro compounds
M–31	(–CH$_3$O)	Methyl ester
39	C$_3$H$_3$	Aromatic and heterocyclic
43	C$_3$H$_7$	Propyl substituent
	CH$_3$CO	Acyl group
M–44	(–CO$_2$)	Ester, anhydride
M–45	(–COOH)	Carboxylic acid
58	CH$_2$=C(OH)–CH$_3$	Ketones
M–60	(–CH$_3$COOH)	Acetate
65	C$_5$H$_5$	Aromatic hydrocarbons
76	C$_6$H$_4$	Benzene derivatives
77	C$_6$H$_5$	Monosubstituted benzene
91	C$_7$H$_7$	Benzyl compounds
105	C$_6$H$_5$ · CO	Phenyl ketone, benzoate

[a] Losses from molecular ion shown in parentheses, e.g. M–14.
Fragments may also show as losses, e.g. M–43 or 43.

4. Cleavage is often associated with the elimination of small, stable, neutral molecules such as water, CO, NH$_3$, olefines, HCl, HCN or H$_2$S.

5. The possibility of rearrangements should be considered. One which is often found is the McLafferty rearrangement which produces fragments not simply related to cleavages from the original compound.

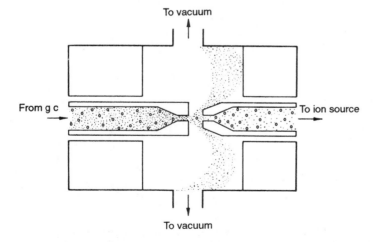

Figure 8.22 A jet separator for coupling a gas chromatograph and a mass spectrometer. Source: Fifield, F.W. and Kealey, D. (1990) *Principles and Practice of Analytical Chemistry*, 3rd edition, Blackie Academic & Professional, Glasgow.

8.5.5   *Applications of mass spectrometry*

Mass spectrometry may be used for both atomic and molecular species, and is a most sensitive technique for qualitative identification of volatile species. While it is an essential tool for the analysis of organic compounds and organic pollutants, it is also a powerful asset when combined with other analytical techniques, and the examples given will concentrate on these combined methods.

(a)

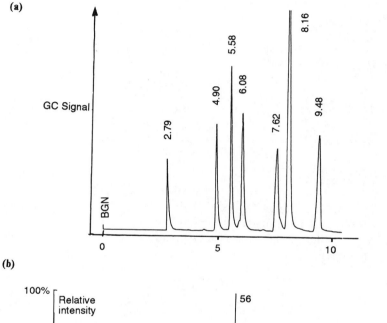

(b)

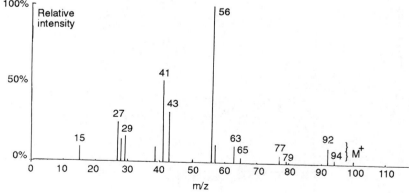

Figure 8.23 GC-MS of a solvent mixture. (a) Gas chromatogram showing the seven components. (b) Mass spectrum of the peak at 4.90 min. The two molecular ion peaks at 92 and 94 in the ratio 3:1 suggests one Cl atom. The peak at 56 is due to loss of HCl. Peaks at 27, 29, 41 and 43 are characteristic of alkyl groups. The fragmentation pattern identifies the compound as 1-chlorobutane: $ClCH_2CH_2CH_2CH_2CH_3$.

Gas chromatography coupled with mass spectrometry (GC–MS) The carrier gas stream from a gas chromatograph may be transferred through a suitable interface into a mass spectrometer. The interface is necessary because the GC operates at approximately atmospheric pressure, while the MS works at very low pressure. Sometimes the carrier gas stream is split so that some goes on to a detector (e.g. flame ionisation detector) while the rest goes to the spectrometer.

Two main interfaces are used. The jet separator shown in Figure 8.22 has two aligned jets separated by a narrow gap. If the carrier gas has a much lower molar mass (e.g. helium $M_r = 4$) than the sample molecules, more carrier gas will escape between the jets, and the pressure will be reduced and the sample concentrated.

Alternatively, a porous disc or tube interface may be used which allows only a portion of the carrier plus sample through to the mass spectrometer. Capillary columns may sometimes be connected directly.

Figure 8.23a shows a GC-MS trace where the total ion current is effectively the detector signal for the GC. Considering one particular peak, the full mass spectrum is shown in Figure 8.23b

(a)

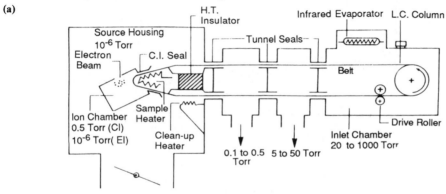

(b)

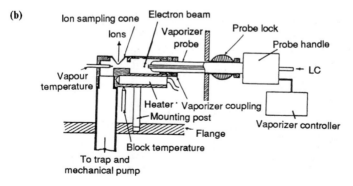

Figure 8.24 LC-MS interfaces. (a) The moving belt interface. The belt is made of high temperature polyimide. Source: Millington, D.S., *New MS Techniques for Organic and Biochemical Analysis*, VG Micromass Ltd., Cheshire. (b) A thermospray interface. Source: Vestal, M.L. (1985) *Eur. Spec. News* **63**, 22.

LC-MS.   The difficulty with combining liquid chromatography with mass spectrometry is again one of removing the bulk of the mobile phase, which in this case is liquid.

Two systems are frequently used. The moving belt interface is shown in Figure 8.24a. The sample is deposited from the LC column onto a moving belt made of high-temperature polyimide polymer and dried by infrared

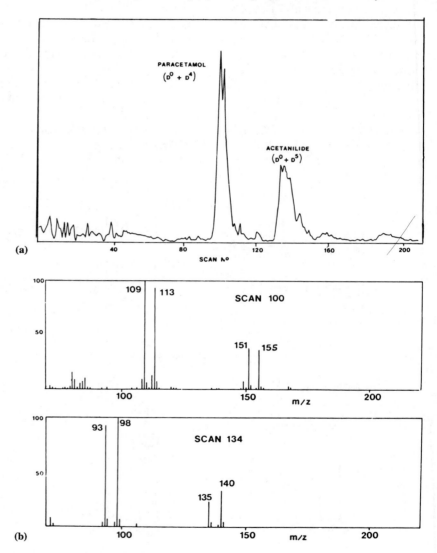

Figure 8.25 LC-MS of drugs extracted from urine. (a) Chromatogram obtained using a reverse-phase column with 50:50 methanol/water as the elutant. (b) EI mass spectra of the drugs, together with deuterium labelled analogues as internal standards. Source: Millington, D.S., *New MS Techniques for Organic and Biochemical Analysis*, VG Micromass Ltd., Cheshire.

heating. The belt passes through two sealed off compartments where the solvent is evaporated and the pressure reduced by vacuum pumps. The belt then passes through the ionisation chamber where the sample is vaporised by a heater. The belt is cleaned by a more powerful heater before its next circuit.

The thermospray system shown in Figure 8.24b converts the liquid mobile phase plus sample into a supersonic jet by rapid heating. The solvent vaporises and is rapidly pumped away, while the sample remains to be ionised in the chamber (Figure 8.25).

The references given at the end of chapter 6 should also be consulted for this chapter.   **Further reading**

Hollas, J. M. (1987) *Molecular Spectroscopy*, Wiley, Chichester.
Williams, D. H. and Fleming, I. (1989) *Spectroscopic Methods in Organic Chemistry*, 4th edition, McGraw-Hill, London.
Kemp, W. (1986) *NMR in Chemistry* Macmillan, London.

9 Measurement of ionising radiations and radionuclides

F.W. Fifield

9.1 Introduction

The effects of ionising radiations on living organisms are of clear environmental significance. In general terms the tissue damage caused is in proportion to the amount of energy deposited by the ionising radiation, although, for low levels of exposure the relationship is poorly understood. It is also clearly established that the type of radiation involved and its characteristic energy influence the hazard associated with exposure to it. A third factor which influences the hazard is a distinction between a radiation dose received from a source *external* to the body, and an *internal* one originating from radionuclides which have been incorporated into the body. In order to sustain an appropriate range of measurements, the techniques and methods must have the capability to provide the following:

1. measurement of radiation intensity and exposure from external and internal sources;
2. distinction between, and measurement of, radiations of different types;
3. determination of the characteristic energies of radionuclides;
4. distinction between natural and anthropogenic sources of radiation.

In order to understand how these requirements are achieved, it is first necessary to describe the phenomena of radioactivity and ionising radiations a little more fully. This discussion is limited to those radiations normally referred to as *alpha*, *beta*, or *gamma*. It must be remembered, however, that whereas these are of pre-eminent environmental concern in general terms, workplace monitoring in the nuclear industry could require consideration of a wider range of radiations such as neutrons and heavy ions.

9.2 Ionising radiations and radioactivity

Although ionising radiations and radionuclides can be of natural or artificial origin their characteristics, once produced, do not differ. Broadly speaking the same techniques of measurement apply to both types, and at this stage of the discussion the differences in origin may be largely overlooked. In environmental assessments the situation is different as is shown later in chapter 16.

In terms of the overall scale of energy changes observed in physical phenomena nuclear processes involve very high energies. It should be remembered therefore that terms such as *high* or *low energy* in this section are used in a comparative sense within the discussion of nuclear matters. Just as atoms and molecules exist in discrete energy states (chapter 3), so do nuclei, and a change from one state to another is accompanied by absorption or emission of a discrete amount of energy. It follows that nuclear radiations may be used as the basis of spectrometry giving qualitative and quantitative information as do atomic and molecular spectra.

9.2.1   Alpha radiation (α)

α-Radiation consists of particles which correspond to helium nuclei stripped of their electrons, $^{4}_{2}He^{2+}$. The particles interact very efficiently with their surroundings and rapidly lose their energy to the medium of passage, in which they have short ranges, e.g. < 1 mm in solids and ca. 1 cm in air. The short range has consequences for the design of measuring equipment and in the hazard that they present Any damage produced by the energy deposited is concentrated into a small volume and α-particles are more hazardous in this sense than most other radiations. On the other hand, the short range means that they only present a hazard as a surface contaminant or on injection. The particles can be generated in a variety of different nuclear reactions but are most commonly encountered as a result of the radioactive decay of a heavy nuclide ($A > 90$). This can be represented generally by equation (9.1), and exemplified by the decay of uranium-235 shown in equation (9.2).

$$^{A}_{Z}X \rightarrow {^{A-4}_{Z-2}}Y + {^{4}_{2}}He^{2+} + \gamma \text{ rays} \tag{9.1}$$

$$^{235}_{92}U \rightarrow {^{231}_{90}}Th + {^{4}_{2}}He^{2+} + \gamma \text{ rays (multiple)} \tag{9.2}$$

The γ-rays are emitted as the atomic nucleus relaxes to its ground state after the α-emission. They have a characteristic pattern which may be used together with the sharp energy spectrum of the α-particles (Figure 9.1) themselves to characterise and identify the decay reaction.

9.2.2   Beta radiation (β⁻ or β⁺)

β⁻ and β⁺ particles (negatrons and positrons) are both generated in the decay of unstable atomic nuclei. Different conditions of instability will lead to the emission of one or the other according to equations (9.3) and (9.4).

(a)

Radionuclide	Recommended half-life (years)	Important alpha particle energies (MeV)
Americium-241	433	5.442 (12.5%), 5.484 (85.2%)
Curium-244	17.8	5.763 (23.6%), 5.806 (76.4%)
Plutonium-239	24100	5.103 (11%), 5.142 (15%), 5.155 (73%)

(b)

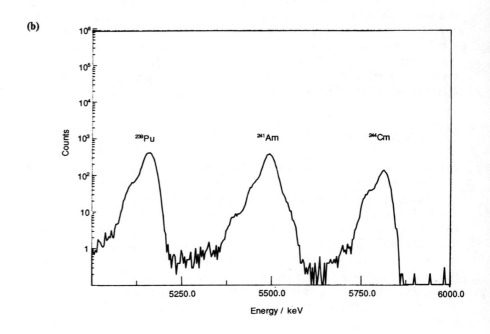

$$^A_Z X \rightarrow {}_{Z+1}^{A} Y + \beta^- \qquad (9.3)$$

$$^A_Z X \rightarrow {}_{Z-1}^{A} Y + \beta^+ \qquad (9.4)$$

Both decay sequences are usually accompanied by the emission of γ-rays as the daughter nuclei relax to the ground state although in some important cases, e.g. $^{14}_{6}C$, $^{3}_{1}H$ no γ-rays are emitted. Where γ-rays are observed they may be used spectrometrically for identification, and provide an alternative basis for measurement of the decaying nucleide.

Negatrons are in essence electrons which have an origin in a nuclear reaction. They have the same mass and charge as electrons but carry high energies consistent with their nuclear origin. With a smaller charge and mass than α-particles they interact with other materials less efficiently and have longer ranges, e.g. a few cm in solids and ca. 1 m in air. As a consequence of the nature of their process of generation (not discussed here) all particles from the same source do not have the same energy. Consequently spectra are broad (Figure 9.2) and have limited value in spectrometric

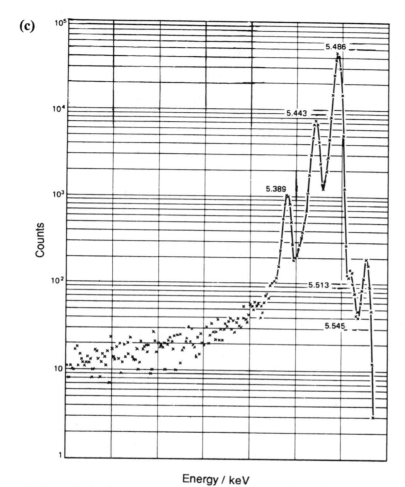

(c)

Figure 9.1 Examples of α-spectra. (a) Principal α-energies for ²⁴¹Am, ²⁴⁴Cm, ²³⁹Pu; (b) low-resolution spectrum; (c) high-resolution spectrum for ²⁴¹Am. Sources: (b) courtesy of G.A. Wells, Kingston University; (c) redrawn from Knoll, G.F. (1989) *Radiation Detection and Measurement*, 2nd edition, Wiley, Chichester.

discrimination. Liquid scintillation counting (section 9.3.5) can, however, be used for spectral resolution provided the β⁻ energies are sufficiently different.

Positrons carry a positive charge equal in magnitude to that of the negatron. They have the same mass and similar spectral characteristics. Interaction with electrons occurs readily and rapidly. Both particles are annihilated, and two photons are produced with energies of 0.51 MeV, which is the energy equivalent of the mass of the particles. Their detection and measurement is based upon the characteristic annihilation radiation

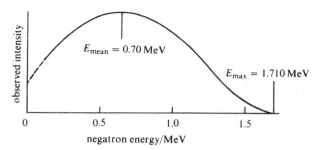

Figure 9.2 Energy spectrum of ^{32}P negatrons. Source: Fifield, F.W. and Kealey, D. (1995) *Principles and Practice of Analytical Chemistry*, 4th edition, Blackie Academic & Professional, Glasgow.

rather than on the original particle. The characteristic 0.51 MeV photons will produce a peak at that energy in a γ-ray spectrum.

9.2.3   Gamma radiation (γ)

γ-Radiation is electromagnetic radiation of short wavelength (10^{-9}–10^{-3}nm), with energies similar to the high energies carried by other nuclear radiations. With zero mass and no charge the radiation has high penetrating power and in diffuse media such as air may be detected at extreme distances from the source. The decrease in intensity represented in the inverse square law is often the limiting factor. γ-Radiation may be measured on the basis of its ionising characteristics.

On interaction with matter some γ-rays will undergo *photoelectric absorption*. In this process the total energy of a radiation quantum is transferred to an electron which is expelled from its orbital with an energy characteristic of the radiation. These electrons give rise to the sharp characteristic peaks in a γ-ray spectrum. However, other γ-rays will lose energy incrementally in a succession of collisions, and in so doing expel a series of electrons with energies below that of those released in immediate photoelectric interactions. This is known as the *Compton effect* and the spectum background produced, as *Compton scatter*. The features of a typical γ-ray spectrum can be seen in Figure 9.3.

The emission of γ-rays results from the relaxation of excited atomic nuclei to their ground states following an initial nuclear event. This leads to a pattern of γ-rays with distinctive energies which is extremely valuable in the identification and determination of environmentally distributed radionuclides. Taken together with the considerable penetrating power of the radiation this means that samples of many types can be analysed using a minimum of sample preparation. A simple example of γ-ray emission accompanying radioactive decay is given in Figure 9.4 and its spectrum in Figure 9.3.

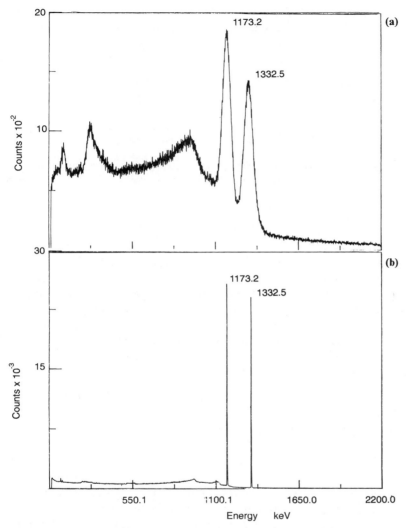

Figure 9.3 γ-Ray spectra for ^{60}Co. (a) Sodium iodide detector; (b) germanium semi-conductor detector. Courtesy of G.A. Wells, Kingston University.

9.2.4   Internal conversion

Under some circumstances some of the excess nuclear energy is transferred directly to orbital electrons which are then emitted from the atom in an *internal conversion* (IC) process. Mixed decay, with some atoms emitting γ-radiation and some atoms conversion electrons, may be met. The latter are measured by the techniques applied to β⁻-particles.

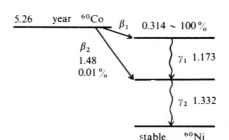

Figure 9.4 Decay scheme for ^{60}Co. Source: Fifield, F.W. and Kealey, D. (1995) *Principles and Practice of Analytical Chemistry*, 4th edition, Blackie Academic & Professional, Glasgow.

9.2.5   Radioactive decay

Radioactive decay in which an energetically unstable atomic nucleus moves to a lower energy state by the emission of ionising radiations, as outlined above, is a major source of environmental radiation. It is therefore important to note some features of decay reactions, particularly with regard to their kinetics, the concept of which has been introduced in chapter 3. Radioactive decay shows first-order kinetics, in which the rate of decay of the sources is proportional to the number of unstable atoms present. As the latter decreases so does the rate so that it takes an infinite time for complete decay. The simple decay reaction equation (9.5)

$$X(\text{unstable}) \xrightarrow{\text{decay}} Y(\text{stable}) \qquad (9.5)$$

is represented by the differential equation (9.6)

$$-\frac{dN_x}{dt} \propto N_x \qquad (9.6)$$

where N_x is the number of nuclei, and t is time. More conveniently the equation is written containing a proportionality constant, λ, which is termed the decay constant and characterises the decay reaction in terms of its rate,

$$-\frac{dN_x}{dt} = N_x \cdot \lambda_x \qquad (9.7)$$

λ is unaffected by any changes in physical or chemical conditions. The integrated form of equation (9.7) is often more easily used, i.e.

$$N_x = (N_x)_0 \cdot \exp(-\lambda x \cdot t) \qquad (9.8)$$

Where $(N_x)_0$ is the initial amount of x present, and N_x at time t. By evaluating t for $N_x = (N_x)_0/2$ the half-life, $t_{1/2}$, for the decay is obtained, and

$$t_{1/2} = \frac{0.693}{\lambda} \qquad\qquad (9.9)$$

$t_{1/2}$ characterises the decay reaction as does λ, but is more readily and directly measured. It is common to quote half-lives to represent decay rates. In practice it is usual to make radioactivity measurements under carefully standardised conditions so that a, the activity, can be substituted for N.

9.2.6 Units of radioactivity and radiation measurement

Environmental measurements of ionising radiations and radioactive materials are concerned with four key aspects.

1. the *amount* of a radioisotope present and its activity;
2. the *dose rate* of radiation and *exposure* to it;
3. *the dose* of radiation *absorbed*; and
4. the *effects* of the radiation absorbed.

Before dealing with the units used in the quantification of these factors it is necessary to consider this complex subject a little more fully.

The question of the amount of a radioactive substance and its activity is relatively straightforward. All that is required is a measurement of the radiation associated with the decay of the particular nuclide. For environmental samples where radioactive substances may be widely dispersed at low concentrations this may, however, be experimentally difficult despite the simplicity of the principles. The relationship between the amount of a radionuclide and its activity has been discussed in the immediately preceding section. Decay rate is expressed in terms of the Becquerel (Bq) which is defined as one nuclear disintegration per second (dps).

The concept of radiation dose refers to the radiation which actually falls upon a target and is a function of the intensity of the radiation (dose rate) and the length of the exposure to it. Radiation effects, and hence radiation damage are linked to the amount of energy liberated in the target, by excitation and ionisation. Radiation dose is quantified in these terms. The unit of *absorbed dose* is the Gray (Gy) which corresponds to one joule per kilogram of tissue. Typical background levels are 0.1 Gy per hour.

A further factor to be taken into account is the differing abilities of nuclear radiations to produce tissue damage whilst the total amount of energy liberated remains constant.

9.3 The detection and measurement of radiation

A high proportion of techniques used for the detection and measurement of ionising radiations have certain basic characteristics in common. In outline, a detector interacts with the radiation, generating electrical signals which are then processed electronically. The signals are in the form of a series of

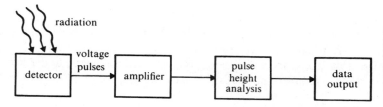

Figure 9.5 Schematic layout of radiation detection and measurement. Source: Fifield, F.W. and Kealey, D. (1995) *Principles and Practice of Analytical Chemistry*, 4th edition, Blackie Academic & Professional, Glasgow.

pulses, the number of which is related to the intensity of the radiation, and whose size may often be used as a measure of its energy. Pulse height analysis (PHA) is the basis for nuclear spectrometry. Even in the simplest instruments a degree of pulse size discrimination is used to distinguish between signal pulses and background. The outline of such instruments is represented schematically in Figure 9.5. Important types of instrument are discussed below.

Additional techniques important in radiation measurement are based upon the use of photographic film or thermoluminescent crystals. In the former the degree of 'fogging' of the film provides a measure of the radiation. The latter employs crystals which retain the energy liberated by the incident radiation until they are heated, when it is released as a pulse of UV radiation. This pulse has a size which represents the amount of energy released and hence the original radiation dose.

When making measurements of environmental radiation it is often necessary to carry out considerable sample preparation prior to the actual measurement. To this end chemical concentration and separation of radionuclides is often employed. Radiation detectors vary widely in principle and experimental configuration. In the following section general principles are discussed.

9.3.1   Gas ionisation detectors

One form of radiation measurement which is very widely employed is based upon the ionisation of a gas, typically argon. Usually a small amount, (5%), of a secondary or quench gas (CO_2, CH_4, CH_3CH_2OH or Br_2) is incorporated, whose function is to absorb electrons and restrict the duration of the resultant electrical pulse. This makes the pulses more amenable to electronic processing. The gas is used within a closed tube, exemplified in Figure 9.6. The central anode is connected to a high voltage supply and the outer cathode usually to earth. Voltages in the range 300–3000 V can be used, the selected value is dependent upon the exact nature of the tube and the filling

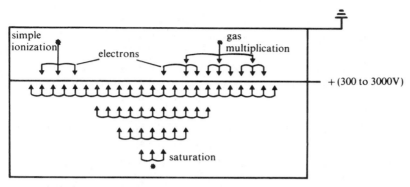

Figure 9.6 Radiation induced ionisation in an argon-filled detector. Source: Fifield, F.W. and Kealey, D. (1995) *Principles and Practice of Analytical Chemistry*, 4th edition, Blackie Academic & Professional, Glasgow.

gas. Values around 1 kV are typical. When ionising radiation penetrates the tube it produces free electrons and argon ions. The electrons are accelerated towards the anode by the potential gradient producing secondary ionisation in the process. On their discharge at the anode an electrical pulse results.

Gas ionisation detectors can be configured and used in a number of different ways. Two versions that may be encountered are the Geiger-Müller counter and the proportional counter. The former operates at relatively high voltages and produces pulses which are similar in size irrespective of the nature and energy of the ionising radiation. Simple and robust in construction they are easily made in portable forms and are very widely employed for the routine measurement of radiation levels. The latter are operated at lower voltages where a proportionality between the pulse size and the energy of the incident radiation exists. Hence they are suited for a more sophisticated employment in which an element of pulse height analysis and spectrometry is required. They are often found in X-ray spectrometers in this role.

9.3.2 *Semiconductor detectors*

Parallels may be seen between the ionisation of gases and of semiconductor crystals. Key differences are the greater densities of solids which lead to more efficient interaction with radiation and the inability of the positive species produced in the ionisation to move freely in the solid state. This latter restriction means that the positive charge centre, or *positive hole* as it is sometimes called, moves as a result of the sideways transfer of electrons under the influence of high voltages of 3–5 kV. The electrons produced in the initial ionisation are able to move freely and are discharged at the anode in a coherent group, to produce a sharp and well-defined electrical pulse. There is a strict proportionality between the energy of the incident radiation and the pulse size, which provides an excellent basis for nuclear spectrometry. Semiconductor detectors are widely used in this role for γ-ray and α-ray particle spectrometry. Detectors are fabricated from the intrinsic semiconductors silicon or germanium in hyperpure forms by the introduction or implantation of controlled amounts of impurities in order to optimise their electrical characteristics.

α-Particle spectrometry is generally carried out using silicon detectors configured in the form of a thin water. They can be operated at ambient temperatures but to ensure good resolution operation inside an evacuated chamber is needed. Sources for measurement need careful preparation and because of the short range of the particles must be mounted close to the detector. Samples with an environmental origin for e.g. uranium or plutonium can be electroplated on to stainless steel discs prior to measurement. A typical α-spectrometer is illustrated in Figure 9.7.

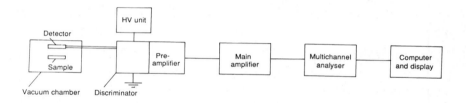

Figure 9.7 Schematic construction for an α-spectrometer.

γ-Ray and X-ray spectrometry employ detectors fabricated from germanium or silicon with the latter being most appropriate for X-rays and low energy γ-rays. In order to ensure a high level of detection for these penetrating radiations it is necessary for the detectors to be of large size. Crystals with volumes of $100 \, cm^3$ or more are now available. As for α-detectors discussed earlier, operation inside an evacuated container is needed to avoid leakage currents carried by adsorbed species. An additional problem derives from the small thermal energy that is needed to promote electrons into the conduction bands of the crystals and produce a significant background current. To avoid this, the detector must be refrigerated with liquid nitrogen to low temperature (77 K) when in use. A typical detector for a γ-ray spectrometer is shown in Figure 9.8. A γ-spectrum has already been illustrated in Figure 9.3. All high resolution γ-ray spectrometry is carried out by the use of semiconductor detectors. On the other hand for X-ray measurements the alternative of a wavelength dispersive system which provides even better resolution is available (see chapter 7).

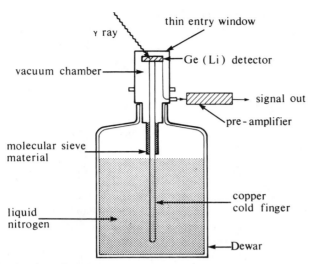

Figure 9.8 Construction of a 'dipstick' type semiconductor detector system. Source: Fifield, F.W. and Kealey, D. (1995) *Principles and Practice of Analytical Chemistry*, 4th edition, Blackie Academic & Professional, Glasgow.

9.3.3 Sodium iodide detectors

Sodium iodide absorbs energy from ionising radiations and re-emits it as UV light. By means of a photocathode and a photomultiplier equivalent electrical pulses can be produced. This provides the basis for an efficient and robust type of radiation detector which is shown schematically in Figure 9.9. The detectors are sometimes known as crystal scintillation detectors. The emission of UV radiation takes place at strain centres within the crystal.

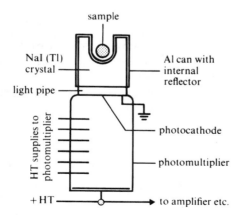

Figure 9.9 A typical scintillation detector (well crystal NaI).

Additional strain centres, accompanied by increased efficiency can be produced by incorporating a proportion (1%) of thallous iodide, TlI. Tl^+ can substitute electrically for Na^+ but, because it is a much larger ion, produces strain within the lattice.

9.3.4 Organic scintillators

Some types of organic compounds are also able to absorb energy from ionising radiations and re-emit it as UV light. The mechanism involves electronic transitions in molecular orbitals and the most commonly encountered organic scintillators are conjugated aromatic compounds. These compounds are used in two main ways. They may be incorporated into transparent polymers and used in various instrumental roles as general purpose radiation detectors. The polymer composites can be configured in many different ways enabling detectors to be produced in a variety of sizes and shapes. Organic scintillators are also extensively employed in solution as so called *liquid scintillators*.

9.3.5 Liquid scintillation counting

When suitable organic scintillators are in solution in close proximity to a radioactive source, a high efficiency for the counting process is ensured. This counting technique is particularly important for measurements, where the source is a pure β^--emitter with low particle energies. ^{14}C, ^{3}H and ^{35}S are particularly important although not unique examples. Recently, the use of liquid scintillators has been extended for the routine measurement of α-emitters as well. From the spectroscopic point of view the signals obtained are of rather poor resolution although for β^--emitters the limiting factor remains the intrinsic broad energy distribution associated with decay process, and only emitters with energies differing by factors of greater than 5 can be resolved, and then only in binary mixtures. For α-emitters the situation is more favourable and given the availability of sophisticated instrumentation greater success can be achieved (Figure 9.10).

Liquid scintillation counting is carried out using a *counting cocktail* which contains a number of different materials. A typical cocktail would contain:

1. solvent;
2. emulsifying agent;
3. primary scintillator;
4. secondary scintillator;
5. sample.

The performance of the system is highly dependent on this composition and each component needs to be considered carefully. With regard to *the solvent*,

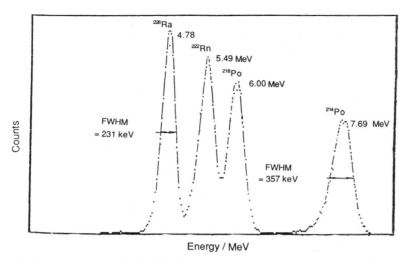

Figure 9.10 α-Spectrum obtained by liquid scintillation spectrometry. Source: redrawn from McDowell, W.J. (1992) Photon/electron-rejecting alpha liquid scintillation (PERALS) spectrometry: a review. *Radioactivity and Radiochemistry*, **3**, (2).

it must clearly maintain the scintillators in solution as well as being a good, general purpose solvent for potential samples. Toluene is the most widely used. However, samples with significant ionic character may be only water soluble. An efficient emulsifying agent will ensure good contact between the sample and scintillator. The solvent needs to be selected with care as in addition to its solvent properties it provides the initial sites for the absorption of radiation from the sample, the energy of which is transferred in radiationless processes to the primary and secondary scintillators in sequence before emission as UV light. The function of the primary scintillator is the efficient reception of the energy via its molecular orbitals and that of the second scintillator to emit it at a longer wavelength in a wavelength region where the detecting photomultipliers operate most satisfactorily. Typical scintillators are given in Figure 9.11.

Interference in liquid scintillation counting arise from processes other than interaction with ionising radiation which promote the emission of UV light quanta, those that inhibit their emission and those in which emitted light is absorbed within the cocktail. The former processes are known as *luminescence* and may arise from chemical or biochemical interactions with the sample. They can be counteracted by using special cocktails and instrumental discrimination between the single quantum produced in a luminescence process and the multiquanta signal from ionising radiations. Suppression of the emission of light is known as *chemical quenching* and generally results from the perturbation of the molecular orbitals in the

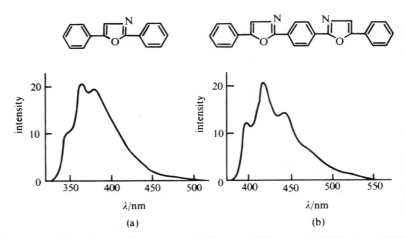

Figure 9.11 Typical primary and secondary scintillators. (a) Primary; 2,5-diphenyloxale (PPO). (b) Secondary, 1,4-bis-2-(5-phenyloxazolyl)-benzene (POPOP). Source: Fifield, F.W. and Kealey, D. (1995) *Principles and Practice of Analytical Chemistry*, 4th edition, Blackie Academic & Professional, Glasgow.

scintillant cocktail by electronegative compounds. Chlorinated solvents are typical examples and should be excluded. The absorption of light after emission is called *colour quenching* and is the result of simple absorption of light by coloured materials. Sample solutions may be bleached to reduce this effect but care must be exercised in order to assess compounds which are not overtly coloured but may still absorb UV radiation.

In summary, liquid scintillation counting is an important and widely used technique. It is, however, somewhat complex in operation and a specialist knowledge is needed for its effective and accurate employment.

9.3.6   Detection by films

The detection and measurement of ionising radiations on the basis of their interaction with films is long established and was, indeed, the key to the discovery of radioactivity by Becquerel. In the most familiar form, films with radiation sensitive emulsions are used. The location of 'fogging' revealed on development indicates the location of the radiation whilst the density of 'fogging' is a measure of the intensity. Films may be used in a conventional format with the emulsion supported on a polymer backing or less commonly as a *stripping film* where a thin layer of emulsion is coated on to the surface of a specimen to ensure close contact.

A specialised type of film which does not depend upon the photographic principles is also used. This film consists of a suitable polymer in which significant *particle track* damaged is produced. By treating with a strong alkali solution, a process known as *etching* the track can be enlarged. They can then be used to indicate the position of a radioactive source and by counting the particle tracks the amount of radioactivity may also be measured. This teachnique is especially appropriate for α-emitters or fission fragments.

9.3.7   *Concentration and separation of radionuclides*

In many environmental samples radionuclides may be widely dispersed. They may thus require pre-concentration and separation prior to instrumental measurement. Broadly speaking all of the techniques discussed in chapter 5 have been used for this purpose. Specific examples can be found in chapter 16.

Fifield, F.W. and Kealey, D. (1995) *Principles and Practice in Analytical Chemistry*, 4th edition, Blackie Academic & Professional, Glasgow.
Knoll, G.F. (1989) *Radiation Detection and Measurement*, 2nd edition, Wiley, Chichester.

Further reading

10 Electrochemical techniques

E.M. Buckley

10.1 Introduction

Analytical chemistry is primarily concerned with the characterisation and measurement of chemical systems. In environmental analysis the concern is the qualitative and quantitative evaluation of a diverse range of environmentally significant analytes. Electroanalytical techniques are one of a variety of techniques considered in this context. They complement other analytical techniques such as chromatography or spectroscopy.

10.2 Electrochemical principles

Electrochemical methods of analysis include all methods of analysis that measure current, potential and resistance, and relate them to analyte concentration. An important concept in electroanalytical chemistry is heterogeneity. The analyst must remember that the analytical measurement is made on a small portion of the sample solution, the portion in contact with the electrodes. The analyte used for electrochemical measurements is generally a solution of ionic materials in a polar solvent, for example potassium chloride in water. The ions interact with the solvent to become *solvated*, and the solvated ions may move through the solution under the influence of forces resulting from temperature gradients, concentration gradients or potential gradients. The solvated ions interact with each other so that their *activity*, *a*, is different from their concentration, *c*, by a factor called the activity coefficient, γ, which depends on the concentrations and charges of all the ions in the solution making up the *ionic strength, I*, where

$$a = c\gamma \tag{10.1}$$

and

$$I = 1/2\Sigma c_i z_i^2 \tag{10.2}$$

where c_i is the concentration of ions of type i and z_i is the charge on ions of type i. The activity is a thermodynamic property that measures the effectiveness of the ion in solution. The analytical chemist is chiefly concerned with a reliable measurement of concentration and would prefer to keep the ionic strength, I, and thus the activity coefficient, γ, constant. This can often be achieved by the addition of an electrolyte solution which will not participate in any reaction other than the adjustment of ionic strength and which is called a *total ionic strength adjustment buffer*, TISAB.

The ions in the bulk of the solution are free to move through the liquid. When a potential difference is applied, a current may flow through the solution carried by the solvated ions and dependent in magnitude on their nature and concentration. The measurement of the *conductance* (i.e. the reciprocal of the resistance) of the solution, is the technique of *conductometry*. Near the electrodes, the concentration of ions may change due to reaction. If ions are removed from the solution by plating out onto the electrode, a concentration gradient is established between the bulk of the solution and the electrode surface. Ions diffuse across this *diffusion layer* and carry the current at a rate dependent upon their concentration in the bulk solution. This is the principle underlying *voltammetric* techniques, such as *polarography*. At the interface between the electrode and the solution, an *electric double layer* is established by the ions in the solution, even if no current flows. The difference in potential between the electrode surface and the bulk of the solution is the *electrode potential*. Electrochemical cells are either *galvanic* or *electrolysis* type. Where a galvanic cell uses a spontaneous chemical reaction to produce electricity, an electrolysis cell uses electricity to drive a chemical reaction.

When a current is passed through an electrolyte solution, a reaction may occur at an electrode surface. The amount of product species produced by the reaction depends on the current, the time and the nature of the species. This is summarised in Faraday's Laws. For example, for the electrodeposition of copper:

$$Cu^{2+}(aqueous) + 2e^- = Cu(metal) \qquad (10.3)$$

$$\text{Mass of Cu deposited} = (It/F) \times \text{molar mass of copper}/2 \qquad (10.4)$$

where I is the current in amperes and t is the time in seconds.

Electrochemical nomenclature is often confusing. Important terms to remember are the *anode* and the *cathode*. The anode is the electrode at which *oxidation* takes place, and the cathode is the electrode at which *reduction* occurs. The specific theory required will be introduced as each technique is described.

10.3 Potentiometric techniques

10.3.1 Introduction and theory

Potentiometric techniques have been described as static techniques, i.e. no current flows. They are probably the most widely used of all of the electroanalytical techniques, particularly in the environmental laboratory. To make a potentiometric measurement one requires a *potentiometer*, an *indicator* electrode, a *reference* electrode and an *electrolyte* solution containing the analyte. Both electrodes are connected to a potentiometer and are immersed in the analyte solution.

The *potential difference* between the two electrodes that make up the electrochemical cell, otherwise referred to as the *emf* (electromotive force) of the cell, is recorded. By convention the reduction half-cell reaction is written or drawn at the 'right-hand electrode' and the oxidation at the 'left-hand electrode' (see chapter 3).

$$E_{(cell)} = E_{(r.h.electrode)} - E_{(l.h.electrode)} \qquad (10.5)$$

For a general reaction such as the reduction of copper ions by zinc metal,

$$Cu^{2+} + Zn = Cu + Zn^{2+} \qquad (10.6)$$

the *galvanic* cell would involve the copper metal electrode immersed in a solution of copper ions (for example, copper sulphate solution), and a zinc metal electrode immersed in a solution of zinc ions (for example, zinc sulphate solution). To stop the solutions mixing a porous disc is used, or a *salt bridge* using an electrolyte whose positive and negative ions carry equal currents (for example, potassium chloride solution) is made (Figure 10.1).

The cell reaction can be divided into two *half-cell reactions* or electrode reactions. At the right-hand electrode, reduction occurs:

$$Cu^{2+} + 2e^- = Cu \qquad (10.7)$$

At the left-hand electrode, oxidation occurs:

$$Zn - 2e^- = Zn^{2+} \qquad (10.8)$$

The free energy change, ΔG of the total reaction, as detailed in chapter 3, depends on the activities of the species involved:

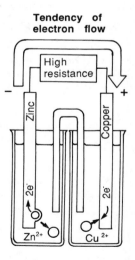

Figure 10.1 An electrochemical cell.

$$\Delta G = \Delta G^{\ominus} + RT \ln(a_{Cu} \cdot a_{Zn^{2+}}/a_{Cu^{2+}} \cdot a_{Zn}) \tag{10.9}$$

This is related to the cell emf, E,

$$\Delta G = -nFE \tag{10.10}$$

So

$$E = E^{\ominus} - (RT/2F) \ln(a_{Cu} \cdot a_{Zn^{2+}}/a_{Cu^{2+}} \cdot a_{Zn}) \tag{10.11}$$

or

$$E = [E_{Cu}] - [E_{Zn}] \tag{10.12}$$

or

$$E = [E_{Cu}^{\ominus} + (RT/2F) \ln(a_{Cu^{2+}}/a_{Cu})] - [E_{Zn}^{\ominus} + (RT/2F) \ln(a_{Zn^{2+}}/a_{Zn})] \tag{10.13}$$

This is referred to as the *Nernst Equation*. As pointed out in chapter 3, each electrode potential is referred to the *standard hydrogen electrode* (SHE) as the primary reference. This is an inert metal, generally platinum, in contact with hydrogen ions at unit activity and also with hydrogen gas at 1 atmosphere pressure. For the electrode reaction,

$$H^+ + e^- = 1/2H_2,$$

the Nernst equation is

$$E_{H_2} = E_{H_2}^{\ominus} + (RT/F) \ln(a_{H^+}/(a_{H_2})^{1/2}) \tag{10.14}$$

and we define

$$E_{H_2}^{\ominus} = 0.000 \text{ V}$$

For any electrode referred to the SHE, and having a reduction electrode reaction,

$$Ox^{m+} + ne^- = Red^{(m-n)+} \tag{10.15}$$

where Ox and Red are oxidised and reduced species respectively, the potential is

$$E_{Ox/Red} = E_{Ox/Red}^{\ominus} + (RT/nF) \ln(a_{Ox}/a_{Red}) \tag{10.16}$$

or, converting to base 10 logarithms and working at $T = 25°C$

$$E_{Ox/Red} = E_{Ox/Red}^{\ominus} + (0.0592/n) \log(a_{Ox}/a_{Red}) \tag{10.17}$$

If a straight line graph of slope $(0.0592/n)$ is obtained when electrode or cell response at 25°C is plotted against log(activity), then the electrode is said to show *Nernstian behaviour*. The $E^{\ominus}$ values as the standard reduction potentials relative to the SHE are listed in Table 10.1, and are also called the *electrochemical series*.

Table 10.1 Standard electrode potentials at 25°C, $E^{\ominus}$ (V)

Most electropositive, most reducing	
$Li^+ + e^- \rightleftharpoons Li$	-3.045
$K^+ + e^- \rightleftharpoons K$	-2.924
$Ca^{2+} + 2e^- \rightleftharpoons Ca$	-2.76
$Na^+ + e^- \rightleftharpoons Na$	-2.712
$Mg^{2+} + 2e^- \rightleftharpoons Mg$	-2.375
$Al^{3+} + 3e^- \rightleftharpoons Al$	-1.706
$Zn^{2+} + 2e^- \rightleftharpoons Zn$	-0.763
$Fe^{2+} + 2e^- \rightleftharpoons Fe$	-0.409
$Sn^{2+} + 2e^- \rightleftharpoons Sn$	-0.136
$Pb^{2+} + 2e^- \rightleftharpoons Pb$	-0.126
$2H^+ + 2e^- \rightleftharpoons H_2$	0
$AgBr + e^- \rightleftharpoons Ag + Br^-$	0.071
$AgCl + e^- \rightleftharpoons Ag + Cl^-$	0.223
$Cu^{2+} + 2e^- \rightleftharpoons Cu$	0.340
$Cu^+ + e^- \rightleftharpoons Cu$	0.522
$I_2 + 2e^- \rightleftharpoons 2I^-$	0.535
$Fe^{3+} + e^- \rightleftharpoons Fe^{2+}$	0.770
$Hg_2^{2+} + 2e^- \rightleftharpoons 2Hg$	0.799
$Ag^+ + e^- \rightleftharpoons Ag$	0.800
$2Hg^{2+} + 2e^- \rightleftharpoons Hg_2^{2+}$	0.905
$Br_2 + 2e^- \rightleftharpoons 2Br^-$	1.065
$4H^+ + O_2 + 4e^- \rightleftharpoons 2H_2O$	1.229
$Cl_2 + 2e^- \rightleftharpoons 2Cl^-$	1.358
$Au^{3+} + 3e^- \rightleftharpoons Au$	1.42
$Ce^{4+} + e^- \rightleftharpoons Ce^{3+}$	1.443
Least electropositive, least reducing	

Of course, in reality, all potentiometric electrodes do not exhibit a Nernstian response. These electrodes are still analytically useful, and the measured potential is compared to concentration by way of comparison with known standards, either using a calibration curve, the method of standard addition or other approved analytical methods as described in chapter 2.

10.3.2 Practical considerations and applications

Instrumentation. The instrumentation required is a potentiometer, a reference and an indicator electrode mounted in a sample holder with stirring facility, and a digital display or a chart recorder. Modern commercially available instruments for potentiometric measurements are essentially electronic digital voltmeters. The *signal generator* is the activity of the ion in solution, the *transducer* is the cell with the indicator electrode and reference electrode, *signal processing* normally involves amplification, and the *readout device* displays millivolts. The output can be adjusted to give a reading in a pH scale, where the instrument is called a pH meter, or as a p[ion], where the instrument is a selective ion meter. Usually the device will have a dial enabling the user to select pH output or a millivolt scale. Analytically useful instruments will also offer an expanded millivolt scale. When considering the instrumentation, precision and accuracy are important requirements. A meter that can resolve 0.1 mV is useful for most purposes including direct measurement. For titrimetry a resolution of 1 mV is usually sufficient. Accurate potential measurement requires the resistance of the meter to be very high with respect to the resistance of the cell. Slope factors and standard potentials are temperature dependent and therefore pH meters need some method of temperature adjustment. No such adjustment is offered for millivolt ranges. The analyst must be careful if it is necessary to consider temperature changes. The pH meter will also have a facility to allow for calibration with appropriate buffers. It is useful to be able to interface the instrument with a personal computer, attach a chart recorder, or even an oscilloscope is valuable if characterising and developing new electrodes. When setting up the equipment it is useful to note that, by convention, the indicator electrode is the cathode and the reference electrode is the anode. Portability and ruggedness are important considerations for the environmental analyst. Reliable, portable meters are commercially available.

Cells and reagents.   The sample for potentiometric analysis should be supplied in solution. The ionic strength is adjusted, if required, by the addition of TISAB. A TISAB may also contain a pH buffer, or in some cases complexing agents. All reagents should be prepared in clean glass or plastic ware with deionised water. The cell is usually a beaker, in some cases with thermostatic control achieved by using a water jacket. A constant, controllable stirring rate is required for all potentiometric measurements.

Electrodes.   The electrochemical cell comprises two electrodes, the indicator electrode and the reference electrode. An awareness of how the electrodes function is important so that the analyst can consider what precautions need to be taken to protect the electrode and maintain its reliability. The effect of the sample matrix on the electrode performance, pH susceptibility, temperature tolerance of the electrode, etc. should also be considered.

All commercial electrodes come with specific storage and conditioning instructions. These must be followed if the electrode is to perform well and reliably throughout its lifetime. After each measurement, and before the electrode is transferred to a new solution, it should be rinsed with deionised water and gently patted dry with tissue, to avoid dilution or contamination of the next sample.

Reference electrodes.   When making any measurement in analytical chemistry it is important to minimise the variables. Therefore, the primary requirement of a reference electrode is that it provides a constant potential and is independent of the composition of the sample solution. Absolute electrode potentials need not be measured. Instead, relative potentials are measured. As discussed above, all electrode potentials may be referred to the standard hydrogen electrode (SHE) which is defined as having a standard potential, $E^{\ominus}$, of 0.000 V. Although the results for the SHE are reproducible, it is not a very convenient electrode for practical analysis, hence a more suitable reference electrode must be used. This must satisfy the following criteria:

1. it must have a potential which is constant with respect to the SHE and independent of the solution into which it is put;
2. the filling solution in the electrode must not react with the analyte;
3. the filling solution should have ions of equal conductance to minimise any potentials at the liquid junctions (for example KCl or KNO_3).

An example of a reference electrode is illustrated in Figure 10.2. The reference electrolyte is in contact with the sample solution via a liquid junction. This is achieved using a porous diaphragm such as a glass frit, ceramic junction or other suitable porous material. A constant flow of reference electrolyte is released slowly into the sample. At this interface another potential occurs. This *liquid junction potential* is due to the different mobilities of anions and cations. To minimise the contribution to the measured potential, the following precautions should be taken. A concentrated reference electrolyte such as KCl or KNO_3, which have ions of similar mobilities, should be used, the diaphragm should be kept clean and free of precipitates and the sample solution should be stirred constantly during the measurement. The level of the reference electrode filling solution must be kept above the level of the sample solution, to avoid plugging the junction. For some applications it may be necessary to use a more specialised

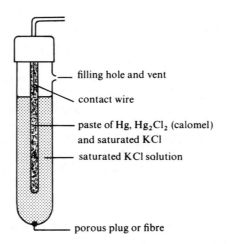

Figure 10.2 Saturated calomel reference electrode (SCE).

reference electrode, a non-aqueous reference for example. In most practical cases one of two reference electrodes are used. These are the calomel electrode and the silver/silver chloride electrode. They are both very well documented and commercially available.

The electrode reaction of the calomel electrode is:

$$Hg_2Cl_2(solid) + 2e^- = 2Hg(I) + 2Cl^-(aqueous) \qquad (10.18)$$

so that, since the activity of pure liquid mercury and of pure solid mercury(I) chloride are both 1, the Nernst equation becomes

$$E_{cal} = E_{cal}^{\ominus} - (RT/F) \ln(a_{Cl^-}) \qquad (10.19)$$

If the concentration of the chloride filling solution remains constant, the E_{cal} will also be constant, although it will depend on temperature.

The calomel electrode (CE) is made of mercury in contact with calomel (mercury(I) chloride) and filled with potassium chloride. Where the filling solution is saturated KCl, the electrode is known as the saturated calomel electrode (SCE). The potential of this electrode with respect to the SHE varies with reference electrolyte concentration, and as mentioned above, temperature. A calomel electrode with a filling solution of 3.5 M KCl will have a standard potential of 250 mV versus SHE at 25°C. The more common SCE has a standard potential of 244 mV versus SHE at 25°C. The electrode reaction of the silver/silver chloride electrode is

$$AgCl(solid) = Ag(solid) + Cl^-(aqueous) \qquad (10.20)$$

Thus,

$$E_{AgCl} = E_{AgCl}^{\ominus} - (RT/F) \ln(a_{Cl^-}) \qquad (10.21)$$

Contact with the analyte solution is made through the salt bridge of KCl, or if this might react with the analyte, through a *double junction reference electrode system* of KCl and then KNO_3. The silver/silver chloride electrode is prepared from a silver wire coated with a layer of silver chloride and immersed in a solution of potassium chloride. With a filling solution of 3.5 M KCl the Ag/AgCl electrode has a potential of 204 mV versus SHE at 25°C.

Indicator electrodes. The indicator electrode responds directly to the analyte. Ideally the potential of this electrode would be the only variable in the measurement system, and, of course, this variability would be directly proportional to the analyte concentration. The response of the indicator electrode should be fast, reversible, and governed by the Nernst equation. Reversibility is an important concept in electrochemistry. A reversible cell is one where the chemical process producing the emf is a reversible reaction. Reversible electrodes can be of three kinds.

Type I, metal/metal ion.   This simplest kind of indicator electrode consists of a clean wire of the metal whose ion is to be determined. Where the analyte is directly involved in the electrode reaction the electrode is known as an 'electrode of the first kind'. These classical metallic electrodes are important today as components of reference electrodes. Practically, the most useful are silver and mercury. A clean silver wire and calomel electrode can be used to determine silver ion concentration in solution:

$$Ag^+ + e^- = Ag(s) \qquad E^\ominus = 0.799 \text{ V} \tag{10.22}$$

$$Hg_2Cl_2(s) + 2e^- = 2Hg(I) + 2Cl^-, \qquad E^\ominus = 0.268 \text{ V} \tag{10.23}$$

These data can be substituted into the Nernst equation, and the equation rewritten at 25°C to

$$E = 0.558 + 0.059 \log[Ag^+] \tag{10.24}$$

to show that E, the measured potential, is proportional to the silver ion concentration.

Type II, metal/metal salt/metal ion.   Other metallic based electrodes include 'electrodes of the second kind'. These electrodes are constructed of metal and a sparingly soluble salt of the metal, in contact with a saturated solution of the anion of the salt. The silver/silver chloride electrode is an example of such a system. This is an anion-sensitive electrode for the chloride ion; it is also an important component of the silver/silver chloride reference electrode.

Type III, redox.   Redox electrodes are inert metal electrodes made of platinum or gold in contact with a solution containing both species of a redox couple, e.g. Fe^{2+}/Fe^{3+}.

The indicator electrodes described above are no longer routinely used in environmental analysis. Hence, the remainder of this section will concentrate on *membrane electrodes*. This group of electrodes includes a large variety of analytically useful electrodes, and is an area that currently stimulates a lot of research interest.

The ideal indicator electrode not only needs to respond rapidly to changes in analyte concentration, but it needs to respond in the presence of other ions found in the sample matrix. Therefore, an ideal indicator electrode is also selective, reproducible in response, and has a useful lifetime. The most successful indicator electrode in potentiometry is the *ion-selective electrode* (ISE). The ISE is different to the metallic electrodes described above because it has a selective element incorporated into it. This is usually the membrane itself or a component of the membrane. The membrane must have a *charge carrier*. The terms specific and selective should not be confused. The membrane potential of most ion selective electrodes is affected by more than one type of ion. The selectivity of the electrode depends on the degree of the contribution to the membrane potential from other ions present in solution. When developing a new analytical technique, or constructing a new electrode, *selectivity coefficients* for all possible interferents should be determined. This is achieved by simple experimentation and use of the Nikolsky Eisenman equation,

$$E = E^{\ominus} + (0.0592/n_A) \log_{10}(a_A + K_{A,B}{}^{pot} a_B{}^{n_A/n_B}) \qquad (10.25)$$

$K_{A,B}{}^{pot}$ is the potentiometric selectivity coefficient of the ISE for ion A in the presence of ion B. A general scheme for an ion-selective electrode is illustrated in Figure 10.3. The potential resulting from the electrode being immersed in solution of the ion of interest is called the membrane potential. The theory of membrane potentials can be complex, and depends, to some extent, on the type of membrane. Essentially it could be described as a boundary potential, formed from two potentials, each associated with one of the two membrane/solution interfaces. The boundary potential is simply the difference between these two.

ISEs can be classified according to the type of membrane used, for example a rough classification would be solid membranes, liquid membranes, and composite electrodes. This would cover most of the ISEs useful

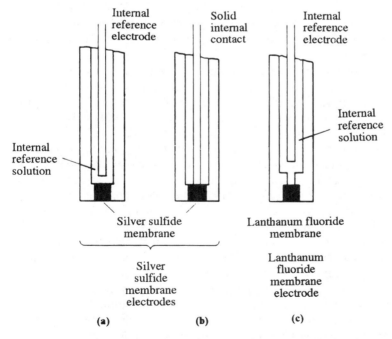

Figure 10.3 Types of crystalline ion-selective electrodes. (a) Solid-membrane electrode sensitive to Ag^+ and S^{2-}; (b) all solid-state solid membrane electrode sensitive to Ag^+ and S^{2-}; (c) membrane-configuration ion-sensing electrode for fluoride ion.

in environmental analysis. For some applications, mainly clinical, *microelectrodes* are used. Glass electrodes, solid and liquid membrane electrodes have all been miniaturised. ISEs have also been combined with semiconductor field effect transistor (FET) technology to give *ion-selective field effect transistors* (ISFETs), which operate as FETs, where the gate has been replaced by an ion-selective membrane. For the environmental analyst an advantage of the ISFET is its small size and rugged construction that lends itself easily to field work.

Solid membrane ion-selective electrodes.   The most common measurement made by an ISE is that of pH. pH electrodes are glass electrodes selective for the hydrogen ion. The glass membrane is the ion selective element and the charge carriers are the sodium ions within the interior of the membrane. Various glass compositions have been used. It is possible to buy glass electrodes that are suitable for use at high or low temperatures, and are accurate at extremes of pH. The '*alkaline error*' associated with glass electrodes is due to the response of the glass to alkali metal cations, and depends on the composition of the glass. pH electrodes are usually supplied with the external reference incorporated in an outer sleeve. The electrode is filled with a pH buffer, and generally a chloride filling solution. It is so designed that the measured potential is zero at pH 7. The pH electrode must be soaked in water for some hours (24–48 h depending on manufacturer), until a hydrated layer is formed. This layer is vital for the formation of the membrane potential.

The glass electrode is easy to use and suitable for any potentiometer or pH meter. Calibration buffers are commercially available, and should be used when fresh. Samples should be collected in borosilicate glass, and analysed as soon as possible (refer to chapter 2). Glass electrodes are also available for the analysis of sodium, potassium, lithium and silver. The potassium electrode can also be used for the determination of ammonium ions at low potassium concentrations. An example of a *solid-state membrane* electrode is the fluoride ISE (see Figure 10.3). The solid-state membrane is made of a single crystal, from a crystalline compound such as chalcogenide, or microcrystals dispersed in an inert resin such as silicone rubber. Where more than one type of crystal is compressed together to form a disc the second crystal may or may not participate in the electrochemical reaction. Sometimes silver sulphide is added to the membrane to aid conductance, because divalent ions are immobile in crystals or sometimes just to strengthen the membrane. Solid electrodes are used to determine bromide, cadmium, chloride, copper, fluoride, iodide, lead, silver and sulphide. These electrodes also have a secondary response to the counter-ion of the charge carrying species, e.g. the silver electrode is also responsive to sulphide.

The fluoride electrode is a successful example of a single crystal, solid-state ISE. It is made from a lanthanum trifluoride crystal. Within this hexagonal lattice the fluoride ions are relatively mobile. The conductivity of the crystal is enhanced by doping with europium. The fluoride ISE exhibits a Nernstian response to fluoride ions in the concentration range $1-10^{-6}$ M. The only important interferent is the hydroxide ion, due to its equivalent size. Direct measurements of free ion are reliable to 1 ppb. Standards and samples must be stored in plastic containers, and addition of total ionic strength adjustment buffer (TISAB) is recommended to maintain a constant ionic strength. This buffer may be modified to adjust the solution pH in order to minimise hydroxide interference, and also to release fluoride from any complexes it may have formed in solution.

Liquid membrane ion-selective electrodes.   The liquid membrane is made up of some water immiscible organic phase containing an ion-exchanger or a neutral ion carrier, an ionophore. The analyte ion forms an ion pair with the exchanger or a complex with the ionophore. As the response of the electrode is, to a great extent, determined by the exchanger or ionophore, they must form stronger complexes with the analyte than with any other ion likely to be present. Also, the distribution of the exchanger between the organic phase membrane and the aqueous sample solution should significantly favour the former. The ISE for nitrate is a good example of a liquid membrane incorporating an ion exchanger. In commercial nitrate electrodes the membrane is usually polyvinyl chloride (PVC). As the nitrate ion is a strongly hydrophilic anion, the exchanger should be a strongly hydrophobic cation such as a tetraalkylammonium salt. This is a particularly useful electrode in environmental analysis and the method compares favourably with colorometric methods for the determination of nitrates in soils and water. Sample solution ionic strength should be adjusted with either sulphate- or phosphate-based solutions. The choice of ionic strength adjuster depends on the application, having established that no interference from the sulphate or phosphate ions occurs in the sample concentration range. Chloride ions do interfere with this electrode, and it is therefore recommended that a double junction reference electrode is used. Liquid membrane electrodes with ion exchangers are also available for calcium, carbonate, hydrogen carbonate, chloride, perchlorate and tetrafluoroborate.

Liquid membrane electrodes with ionophores include the antibiotic, valinomycin. This particular ionophore exhibits a high selectivity for potassium in the presence of sodium, which is important in clinical analysis. The glass electrode for potassium measurements is more durable and more convenient to use but not as selective as the ionophore-containing ISE, for the analysis of potassium in natural waters.

Synthetic, neutral organic ligands are also very popular. The ionophore forms a structure, similar to valinomycin with a polar cavity and lipophilic exterior; for monovalent ions the ionophore will have 5–8 polar groups. It must be stable, but flexible enough to facilitate fast ion exchange. These electrodes have been used for the determination of barium, sodium and for lithium analysis in the primary coolant in pressurised water reactors. In some cases a neutral ionophore electrode is used for the determination of ammonia in, for example, boiler feed waters. However, usually the gas sensing electrode, see below, is more common for the determination of ammonia. Synthetic ionophores for divalent ions have also been developed with specialised ligands to enable selectivity for divalent ions in the presence of monovalent ions. A liquid membrane ISE is used for the determination of calcium in natural waters.

Gas-sensing ion-selective electrodes. A gas-sensing electrode is constructed of a gas permeable coating or membrane which separates the ISE from the analyte solution. This highly porous membrane is made from a hydrophobic polymer, such as polypropylene or polytetrafluoroethylene, and is a selective barrier which does not participate in the electrochemical reaction. The pores allow the passage of gas through the membrane, and the hydrophobic nature of the membrane itself prevents liquid entering the pores. Gas-sensing electrodes are available for analytes such as ammonia, carbon dioxide, nitrogen dioxide, hydrogen fluoride and hydrogen cyanide. For sulphur dioxide and hydrogen sulphide, the membrane is often of silicone rubber. The first electrochemical gas sensor was the Clark type oxygen electrode designed in 1956 (see section 10.3.3 on amperometric sensors). The first device using a potentiometric probe was for carbon dioxide. A glass electrode in a solution of sodium hydrogen carbonate is immersed in the sample solution. An equilibrium pressure of carbon dioxide is set up in the pores of the membrane which matches the concentration of carbon dioxide in solution and that concentration is matched in the internal solution:

$$CO_2(aq) + 2H_2O = H_3O^+ + HCO_3^{3-} \tag{10.26}$$

the pH of the internal filling solution changes accordingly. The ammonia electrode is a particularly important gas sensing electrode and works on similar principles:

$$NH_4^+ = NH_3 + H^+ \tag{10.27}$$

An important environmental application, the determination of total nitrogen, can be carried out using the normal Kjeldahl digestion, and determining ammonia in the digest with a gas-sensing electrode.

Potentiometric biosensors. Combining the selectivity and specificity of certain biological macromolecules with an ISE considerably widens the analytical scope of conventional ISEs. These sensors could also be classified as membrane electrodes, where the membrane is a biocatalytic membrane. The most common are enzyme electrodes. Tissue, bacteria and immunoglobulins have also been used. The first successful potentiometric enzyme electrode was described by Guilbault and Montalvo in 1969 for the determination of urea. The enzyme must somehow be immobilised onto the surface of the electrode; this is achieved either by chemical modification or physical entrapment. As with any biological molecule it is important to consider the biological phase when using and storing the electrode.

The biosensor is introduced to the analyte solution, the analyte, which is now called the substrate, must be transported to the electrode surface, diffuse through the membrane to the active site of the enzyme, react to form a product which is transported to the surface of the ISE and measured. It is the concentration of the product of the enzyme reaction that is determined. Hence, these electrodes are suitable for analytes not electroactive themselves, but capable of forming an electroactive product on reaction with an enzyme. The sample solution must be rapidly and constantly stirred to speed up the transport processes, and it is also important that the response of the enzyme is compatible with the response of the ISE. Although biosensors are more commonly used in clinical analysis, they are also applicable in environmental analysis, for example biosensors have been constructed to determine nitrates and nitrites and also for less common organic analytes.

10.3.3 Potentiometric titrations

These are essentially titrations where the potentiometric measurements are made during the titration, and the endpoint detected by a sharp change in potential. The indicator electrode can be selected to respond to either a reactant or product. The absolute value of the potential is not required. Reference electrode and indicator electrode are immersed in a constantly stirred solution. The potential is measured after each addition of titrant. When the endpoint is near, the increments of titrant added must be small, and enough time allowed for the indicator electrode to reach a constant potential. Auto-titrators are readily available and should be seriously considered for analytical work.

Potentiometric titrations include:

1. *neutralisation titrations* such as the determination of acidity and alkalinity, using a pH electrode (refer to chapter 4);
2. *redox titrations*, based on a redox reaction between analyte and titrant, for example using a platinum wire electrode for the determination of the ferric/ferrous ratio; also the use of redox indicators such as methylene blue, diphenylamine etc.; redox titrations also include the *iodometric titrations* for As(III), SO_2, HCN, H_2S etc.;
3. *precipitation reactions* are available for a variety of determinands such as lead, cyanide, the rare earths, etc.;
4. *complexation titrations* use among others, complexing agents such as EDTA or DTPA (refer to chapter 14) to determine analytes such as

copper or lead, etc.; complexation titrations with ISEs are not as universally applicable as those with visual indicators, but the shape of the titration curve obtained is a useful indication of possible interferences.

In general, potentiometric titrations are very useful for analytical determinations, they are precise as the data collected averages the potential over a number of readings.

10.3.4   Current developments

In this interdisciplinary age, potentiometric techniques are being combined with emerging technologies to offer exciting new possibilities. Arrays of ion-selective electrodes are being combined with sophisticated data acquisition techniques for the simultaneous determination of environmentally important analytes. The search for new and wider ranging ionophores is continuing.

10.4   Voltammetric and controlled potential techniques

10.4.1   Introduction

Voltammetric techniques have been classified as dynamic electrochemical techniques. In their operation the potential is controlled and the current is monitored. Although very important within analytical chemistry, the use of voltammetric techniques is not as widespread as it could be. The reasons for this are historical, but today, with sophisticated electronics at the disposal of the analytical chemist, they are enjoying a revival.

10.4.2   Theory

Voltammetric techniques are based on the measurement of current as a function of potential. The current is produced at an electrode surface following the oxidation or reduction of the analyte at a characteristic potential. Oxidation or reduction at the electrode surface is essentially electron-transfer (or charge-transfer). In any voltammetric technique it is the charge transfer that is being measured. The current is measured in amperes or coulomb/s, i.e. the rate of flow of charge. Voltammetric measurements are therefore measurements of the *rate of reaction*. The electrochemical reaction at the electrode surface is driven by the application of a potential to that electrode. The applied potential is the *excitation signal* and the measured current is the *resulting signal*.

As well as giving useful information about electron transfer mechanisms, redox processes and surface studies, voltammetry is an important quantitative tool. The voltammetric measurement is made in a cell filled with electrolyte in which three electrodes are immersed, the indicator electrode, the reference electrode and the auxiliary (or counter) electrode. A potential waveform is applied to the indicator electrode with respect to the reference electrode. The solution may or may not be stirred depending on the analytical technique. At some potential the redox reaction will occur, the current is measured and plotted against potential. The potential at which the reaction occurs is characteristic of the analyte, the amount of current that is measured is related to concentration. For a reversible, or Nernstian process, the reduction of an electroactive analyte may be represented by

$$Ox + ne^- = Red \tag{10.28}$$

Resulting in an equation similar to the Nernst equation:

$$E = E^\ominus + (RT/nF)\ln([Ox]/[Red]) \tag{10.29}$$

where [Ox] is the concentration of oxidised analyte at the electrode surface, and [Red] the concentration of reduced analyte. If the analyte loses electrons to the electrode it has been oxidised, if it receives electrons from the electrode it has been reduced. In order to be reduced or oxidised, the analyte must be at the electrode surface. Migration, diffusion or convection are the routes by which the analyte travels to the electrode surface. Hence, the experimental conditions must be rigorously controlled. Migration is minimised by the use of an electrolyte and the convection is controlled by stirring, and monitoring temperature effects. The theory relating current to concentration is particular to each technique. As the analyst tends to use comparison to known standards, it is not necessary to consider individual cases here. However, it is important to fully understand an electrochemical system when designing an analytical technique.

A general example of the 'absolute' relationship between current and concentration is given by the Ilkovic equation and its variations. This equation is used for polarographic analysis and considers a process based on linear diffusion to a spherical electrode:

$$i_d = 708 n D_o^{1/2} C_o^* m^{2/3} t^{1/6} \tag{10.30}$$

where i_d is the diffusion current in amperes, D_o is the diffusion coefficient in $cm^2 \, s^{-1}$. C_o^* is the bulk concentration in $mol \, cm^{-3}$, m is the mercury flow rate in $mg \, s^{-1}$ and t is the electrode drop time in seconds. Variations of the Ilkovic equation are available for other systems and electrodes. In general it is represented by

$$i_d = k C_o^* \tag{10.31}$$

10.4.3   Practical considerations and applications

Instrumentation.   The instrumentation required for voltammetric measurements is essentially a three electrode system in an electrochemical cell; a

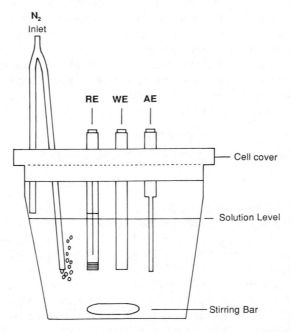

Figure 10.4 Typical configuration of a three electrode system in an electrochemical cell. RE = reference electrode; WE = working electrode; AE = auxiliary electrode.

typical configuration is illustrated in Figure 10.4. The electrodes are connected to a potentiostat, based on operational amplifiers, which applies the potential and measures the resulting current. Modern instruments are controlled by microcomputers which can assist the analyst in selecting experimental conditions and in collecting, interpreting and presenting data.

Cells and reagents.   The arrangement of the electrodes within the cell is important. The reference electrode is placed close to the working electrode and located between it and the auxiliary electrode. The cell material depends on the application, usually a glass beaker with a close fitting lid which includes ports for the electrodes and purging line. If the reaction being investigated involves a potential sweep in the negative direction then the reduction of oxygen may interfere. To avoid this, the solution is deoxygenated, usually achieved by purging the solution with an inert gas such as nitrogen, argon or helium prior to analysis, and blanketing the solution with the gas during analysis. In some cases the reference electrode may need to be isolated during the analysis, this may be achieved by using a sleeve filled with an electrolyte compatible with both the sample solution and the electrode. The cell may need to be lagged, with a water jacket for example, for temperature control. For analysis with a solid stationary electrode a conventional cell is used. For analysis using a dropping mercury electrode (*polarography*) or a hanging mercury drop electrode, special cells are commercially available, these control the drop size, the drop rate and provide a safe method of dispensing mercury. Alternatively a rotating solid electrode, or rotating cell may be used. Commercial cells are available to facilitate this. Thin layer cells are considered in section 10.5.3. Electrochemical techniques have been combined with the advantages of spectroscopy. The cells must be optically transparent and are often thin layer, known as optically transparent thin layer electrochemical cells (OTTLE).

Electrolyte solutions are a combination of solvent and supporting electrolyte. The choice of the electrolyte solution depends on the application. In general the solution must be conducting, chemically and electrochemically inert. In other words, the electrolyte solution facilitates passage of current, and over as wide a potential range as possible, it should not contribute to any chemical reactions and must not undergo any electrochemical reaction. In environmental applications, the most common electrolyte solution is water with an added salt or buffer. In some studies, usually organic electrode processes, the system may be non-aqueous. Acetonitrile or dimethyl sulphoxide are common solvents. Supporting electrolytes added include tetrabutylammonium hexafluorophosphate or tetrabutylammonium tetrafluorophosphate ($TBAPF_4$).

Electrodes.   As with electrolyte solutions, the cell geometry, the potential wave form applied, and the choice of electrode depend on the application. The most difficult choice is that of the working, or indicator electrode. The shape of the electrode is important, as is its size or any modification to the surface. The more common electrodes useful for environmental analysis are considered below.

Reference electrodes. The reference electrodes used in voltammetric techniques are the same as those used in potentiometry. The size and geometry may differ depending on the experimental conditions, but the theory is still relevant. The saturated calomel and the silver/silver chloride are the most common reference electrodes. For non-aqueous systems an electrode such as silver/silver sulphide in the appropriate solvent may be used. For quantitative analysis a reference is almost always used. In some cases where there is no suitable reference, a silver wire may be substituted as a pseudoreference and a well defined redox couple added to the system to act in the manner of an internal standard.

Auxiliary electrodes.   The function of the auxiliary electrode is to complete the circuit, allowing charge to flow through the cell. Auxiliary electrodes are usually platinum wires, foil or mesh, most commercial systems supply platinum wire electrodes. Mercury pools and graphite rods are sometimes used, but are not common for analytical measurements. Where chronoamperometric techniques are used, the auxiliary electrode may need to be isolated within the cell; this is achieved by using a glass sleeve and porous frit.

Indicator electrodes.   The potential is applied to the indicator electrode. This electrode is the site of the redox reaction, and is where the charge transfer occurs. The ideal characteristics of an indicator electrode are a wide potential range, low resistance, and a reproducible surface. The potential window of such electrodes depends on the electrode material and composition of the electrolyte.

Mercury electrodes.   The first important analytical voltammetric technique was polarography. This technique uses a dropping mercury electrode. Modern techniques such as stripping voltammetry use a hanging mercury drop, or controlled growth mercury drop. The potential window of this electrode, in an aqueous electrolyte (1 M KCl) versus SCE at 25°C, is from approximately $+0.2$ V to -1.8 V. Mercury itself is oxidised at $+0.3$ V. This wide potential range, and the easily renewable surface ensure the enduring popularity of this electrode. During analysis, scanning to negative potentials at the mercury electrode causes the reduction of oxygen that results in an interference. To prevent this from happening the solution must be purged of dissolved oxygen. Environmental applications include the analysis of many inorganic ions, particularly transition metal cations, copper, iron, cadmium, etc., in a variety of matrices. Non-aqueous systems have been described for ions readily hydrolysed such as Al^{3+}. Anions such as vanadate and iodate are also readily determined. The many organic reactions that may be investigated with this electrode include the oxidations or reductions of aldehydes, ketones, quinones, nitro containing groups, sulphides and disulphides. The pH of the electrolyte solution is particularly important for organic analysis. The information that can be obtained about adsorption phenomena at the mercury electrode is particularly useful in the study of speciation (see chapter 14). These techniques give information about the chemical 'form' of the analyte of interest as well as its total concentration. In addition to the mercury drop electrode, mercury may be deposited onto the surface of a solid electrode, electrolytically or otherwise, to produce a mercury film electrode. These electrodes may be used as detectors in flowing streams. They are also useful in trace analysis and stripping analysis.

Solid metal electrodes.   The most common solid metal electrodes are platinum, gold and silver. The platinum electrode has a potential window in the approximate range $+1$ to -0.5 V in aqueous buffer pH 7 versus SCE.

In a non-aqueous system, e.g. acetonitrile with 0.1 M TBABF$_4$, the window is extended to the range $+2$ V to -2 V versus SCE (estimated).

Carbon electrodes.   Carbon electrodes include vitreous (or glassy) carbon, and carbon paste. The glassy carbon is the most common in electroanalytical applications. The usual electrode construction is a rod of glassy carbon, sealed into an inert electrode body. A disc of the electrode material is exposed to the solution. The cleaning of these electrodes is very important, and in some cases an electrochemical or chemical treatment of the electrode surface is required. These electrodes are particularly useful for anodic studies, and also for stripping analysis.

Microelectrodes. These are electrodes constructed from the same materials as the macro-electrodes described above but with a diameter of not more than 50 μm. This means that the electrode is smaller than the diffusion layer, hence, mass transport of analyte to the electrode surface is enhanced, the signal-to-noise ratio (SNR) is increased, and measurements may be made in highly resistive media. To the environmental analyst microelectrodes should offer, in the future, the prospect of small portable systems requiring little or no sample preparation.

10.4.4   Techniques

The names of voltammetric techniques depend on the potential waveform applied. The excitation and resulting voltammograms are illustrated in Figure 10.5.

Potential waveforms. The simplest waveform is a steady increase or decrease of potential with time. This technique is known as *linear-scan* and is the waveform applied in classical poloragraphy. It is also the potential waveform applied in *cyclic voltammetry* (CV). Cyclic voltammetry, which has often been described as the electrochemical equivalent of spectroscopy, is usually the first electrochemical experiment performed on any new analyte. The applied potential sweeps backwards and forwards between two limits, the starting potential and the *switching potential*. The waveform is in the shape of a triangular linear-scan (see Figure 10.5). The solution is never stirred, hence mass transport is diffusion controlled. Experienced interpretation of a cyclic voltammogram will give information about the kinetics of the system, associated chemical reactions, the number of electrons transferred, the reversibility of a system, the diffusion characteristics, etc. Hence, the medium in which the reaction occurs is very important. At a first glance, a cyclic voltammogram will tell the environmental analyst in what potential

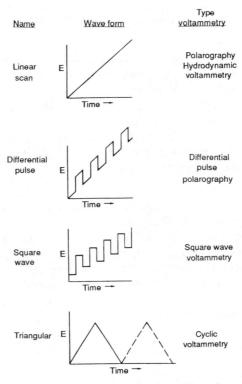

Figure 10.5 Potential excitation signals used in voltammetry.

range to look for the oxidation or reduction of the analyte. It is a very useful diagnostic tool. The medium in which the reaction occurs is particularly important in electroanalytical chemistry. Varying the pH of the solution, for example, may lead to the elucidation of associated or coupled chemical reactions. Figure 10.6 is a simple cyclic voltammogram of the ferri-ferro-cyanide couple, $Fe(III)(CN)_6^{3-}/Fe(II)(CN)_6^{4-}$. This CV illustrates a reversible single electron transfer. $E_{1/2}$, the peak position, gives qualitative information. At a constant scan rate the peak height would give quantitative information in terms of the current involved in the electron transfer. The peak to peak separation, ΔE_p, for a reversible redox reaction, gives the number of electrons transferred. Further manipulation of the conditions or the experimental parameters gives more information such as the diffusion coefficients of reactants or products.

For routine quantitative analysis of well understood analytes, pulse waveforms are more useful, and offer enhanced sensitivity and resolution. Of all of the waveforms available from commercial instrumentation, two are of particular use to the environmental analyst: *differential pulse voltammetry* and *square wave voltammetry*. The advantage of all pulse techniques is that

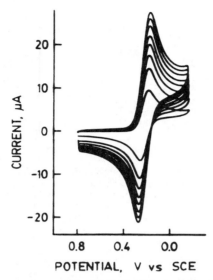

Figure 10.6 Cyclic voltammograms of 4 mM $K_3Fe(CN)_6$ in 1 M KNO_3. Pt electrode. Scan rate (v) = 20, 50, 75, 100, 125, 150, 175, and 200 mV/s.

the waveform is so designed as to discriminate against *non-Faradaic current*, the background current not due to electron transfer, hence, improving sensitivity. How this is achieved depends on the particular waveform. The enhanced resolution achieved is particularly useful when several electro-active species are being analysed simultaneously. The low detection limits make it an ideal technique for trace analysis of arsenic in milk, or trace levels of chromium(VI) in natural waters in the presence of copper(II) and iron(III). Another advantage for the environmental analyst is the specificity of the technique. Differential pulse techniques have been used to determine trace levels of arsenic(III) in the presence of arsenic(V). This discrimination of valence state is particularly important for toxicity monitoring.

The differential pulse waveform is illustrated in Figure 10.5. This wave-form is compatible with all of the electrodes discussed here, and has been applied to a huge variety of analytical problems. The development of square wave voltammetry (SWV), also illustrated, is more recently developed and is a particularly powerful technique. When square wave stripping voltam-metry at a mercury film electrode was combined with a thin layer cell for the determination of In^{3+} in natural waters, a linear calibration curve in the range 100–2000 ppb was obtained, with a limit of detection of 8 ppb. A similar technique was used to determine lead and calcium in the presence of dissolved oxygen. This significantly reduces the analysis time as no purge step is required, and could be an advantage to the analyst who wants to develop techniques useful for field work.

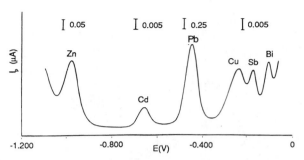

Figure 10.7 Differential pulse stripping voltammogram for the simultaneous determination of Zn^{2+}, Cd^{2+}, Pb^{2+}, Cu^{2+}, Sb^{3+} and Bi^{3+} in seawater at pH 1 containing 2 M chloride. Deposition timing, 20 min (Zn, Cd, Pb) and 40 min (Cu, Sb, Bi); pulse amplitude, 35 mV (Zn, Cd, Pb) and 10 mV (Cu, Sb, Bi); scan rate 2 mV/s (Zn, Cd, Pb) and 0.5 mV/s (Cu, Sb, Bi).

Stripping techniques. Stripping techniques are a combination of any of the techniques described in this chapter, combined with an extra preconcentration step. The analyte is deposited onto the electrode from a stirred solution, and then determined by any of the above techniques, being 'stripped' from the electrode in the process. The anodic stripping voltammetric technique (ASV) involves electrolytically depositing the analyte onto an electrode behaving as a cathode, followed by analysis using oxidation. In *cathodic stripping voltammetry* (CSV), the preconcentrated analyte is determined by reduction. Sometimes the analyte adsorbs onto the electrode at open circuit, i.e. with no potential applied (note: this is not the same as zero potential); the technique is then referred to as *adsorptive stripping analysis*. This technique combined with pulse waveforms is very useful for trace analysis. The limits of detection are sub-ppb and it is also useful for determining the chemical nature of the trace metal ion. The mercury drop electrode is most commonly associated with stripping techniques in environmental analysis. Figure 10.7 illustrates a typical application of stripping analysis combined with differential pulse voltammetry.

Amperometric sensors. Amperometry is identical in theory to voltammetry, the only difference is that, whereas in a voltammetric experiment the applied potential is scanned, in amperometric experiments the current is measured at a fixed potential. An electrochemical technique that all environmental analysts will be aware of is the measurement of dissolved oxygen. The Clark type oxygen electrode is an amperometric sensor. This is a readily portable system that can be used in the field. This 'electrode' is a complete electrochemical cell comprising of a platinum or gold disc cathode (the indicator electrode), and a silver ring shaped anode, surrounding the cathode, and coated with silver chloride (the reference electrode). The filling solution is usually 0.1 M potassium chloride. This assembly is separated from the analyte solution by an oxygen permeable membrane, in the manner of potentiometric gas electrodes. The electrode is immersed in stirred or flowing solution and the oxygen diffuses through the membrane to the platinum electrode, where it is reduced to water.

$$\tfrac{1}{2}O_2 + 2H^+ + 2e^- \rightarrow H_2O \qquad (10.32)$$

The rate of diffusion of the oxygen to the cathode is proportional to the partial pressure of oxygen in the sample. The electrode is calibrated prior to use with known oxygen saturated water solutions. Interferences include other dissolved gases such as chlorine, etc. The interest in amperometric biosensors is very encouraging and accounts for a significant amount of current research in voltammetry. The obvious example to illustrate amperometric biosensors is the enzyme electrode for the clinically important analyte, glucose. The sensor is a working electrode modified with the enzyme *glucose oxidase*. Other biological macromolecules such as an immunochemicals are also employed in this technology, as for potentiometric biosensors (section 10.3.2). Biosensors have had a huge impact in clinical analysis and are not unknown in environmental analysis. They have been used to determine phosphates in fertiliser, but are more usually restricted to specific organic pollutants for which there may be a suitable enzyme or immunochemical.

10.5 Electrochemical detection in flowing streams

10.5.1 Introduction

Voltammetry and potentiometry have both been used as detectors for flow injection analysis (FIA) and chromatography, particularly HPLC. These techniques are described in chapter 5. This section considers only the electrochemical measurement involved. The need for such measurements has also been prompted by the increase in automated analysis and continuous on-line or in-line monitoring.

The ideal detector for FIA and HPLC is one that is sensitive, has a high signal-to-noise ratio, is stable with a reproducible signal, has a linear dynamic range and a fast response time. Electrochemical detectors fulfil these criteria. The main disadvantage of the electrochemical technique as a detector for FIA or HPLC is its selectivity. They are not universal detectors usually necessary for separation techniques. To overcome this they are often used in conjunction with another detector such as ultraviolet.

10.5.2 Potentiometric measurements in flowing streams

For potentiometric measurements in flowing streams, few modifications need to be made to the measurement technique. A flow cell is required which should meet the specifications of any continuous monitoring system. The indicator and reference electrodes are the same as those used for direct potentiometric measurements with minor modifications to allow for sealing, cleaning procedures, etc. Electrode fouling is an important consideration when working with flow cells, either as detectors or for on-line monitoring. Usually analyses in flowing streams suffer less from temperature variation than direct measurement techniques. If the temperature does need to be adjusted, cells are available with temperature lagging facilities. The reagent requirements are as for direct analysis, i.e. pH or ionic strength adjustment. If these requirements are not compatible with the analysis stream or mobile phase, it may be possible to introduce additional reagents into the stream before it reaches the detector. The advantages of using potentiometric devices as detectors for FIA or chromatography are that the instrumentation and measurement technique is simple, the signal is straightforward to process, and suitable small volume cells are available. One of the main disadvantages of the ion selective electrode for chromatographic analysis is its selectivity, a more universal detector is usually required. Mobile phases with the ionic strengths required for potentiometric measurements mean it is more common for inorganic analysis.

Although potentiometric detectors for flow analysis are more widely used in the clinical laboratory than in the environmental laboratory, they are still used for determinations such as nitrate and total nitrogen in soil or water. This may well change in the near future, since this technique is the focus for current research into multi-ion analysis combining electrode arrays with FIA.

10.5.3   Voltammetric detection in flowing streams

Voltammetric techniques combined with the thin layer cell, have particular advantages as regards selectivity, sensitivity and linear dynamic range. This type of detection involves a chemical process and is not simply a physical measurement. This fact is often exploited when electrochemical cells are used as post-column reactors to convert the analyte chemically prior to detection. An important consideration when designing such a system is the composition of the mobile phase. The solvent type, its conductivity, and pH are important factors. In some cases post-column mixing is used. This is not an ideal situation; more satisfactorily the chromatographic separation is carried out using a conducting mobile phase. Amperometric detectors are the most commonly used detectors for flowing systems (Figure 10.8). The working electrode is held at a pre-selected potential with respect to the reference. The auxiliary electrode is usually incorporated into the cell as stainless steel tubing, a suitable auxiliary electrode material. A development of the single working electrode cell is the cell with multiple working electrodes. These electrodes are usually held at different potentials and

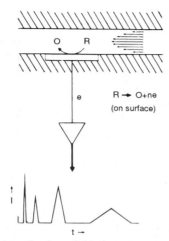

Figure 10.8 Schematic view of thin-layer amperometer detection.

monitored simultaneously. The electrodes may be in series or in parallel. If in parallel, electrochemical products from one electrode may be monitored at the other electrode. More extensive arrays of working electrodes have been developed with multi-channel potentiostats to produce the electrochemical equivalent of diode array detection.

Aluminium, iron and manganese have been simultaneously determined in natural waters by HPLC with a combination of electrochemical detection and spectrophotometric detection. The combined sensitivity and selectivity offer a powerful analytical tool. As well as the obvious inorganic environmental applications, amperometric detection is particularly important when applied to trace organics. Chlorophenols in waste water is one such example. Determination of chlorophenols with a glassy carbon electrode proves more selective than the conventional UV detector.

10.6 Other electroanalytical techniques

10.6.1 Introduction

It has been shown that the electrochemical reaction is a powerful tool that yields analytical, mechanistic and kinetic information. The various techniques can be used, either alone, or to complement one another to explore the reactions of interest in detail. The environmental scientist can exploit these techniques to give useful qualitative and quantitative information about a variety of analytes. Electroanalytical techniques such as conductometry, coulometry and electrogravimetry should not be under estimated and are briefly introduced here.

10.6.2   Conductometry

Conductometric techniques are commonly used to determine the end point of titrations. Also, the direct measurement of conductivity is a sensitive technique for measuring ion concentration. An everyday application is the measurement of the conductivity of de-ionised water to ascertain its purity and the effectiveness of the de-ioniser. Pure water has a conductivity of approximately $5 \times 10^{-8} \, \Omega^{-1} \, cm^{-1}$. Conductivity measurements are routinely used to indicate the total dissolved electrolytes in environmental water samples.

The theory of conductance measurements is based on Ohm's Law:

$$I = E/R \qquad (10.33)$$

where I represents current, E the applied potential and R the resistance. Conductance is the reciprocal of resistance and is measured in siemens (S) or reciprocal ohms. Conductivity is a characteristic property of a material, it is the conductance of a one metre cube of that material. Conductivity is the reciprocal of resistivity, and, as resistivity is usually measured in Ω cm, it is acceptable and indeed usual to quote conductivity in reciprocal ohms per cm ($\Omega^{-1} \, cm^{-1}$), instead of per metre. The conductivity cell comprises of a pair of well defined platinum electrodes of area A and distance l apart, connected to a conductivity meter. A voltage source, E, is applied across the cell between the electrodes. The field applied is El. The total current flowing can be described by:

$$I = (EkA)/l \qquad (10.34)$$

where kA/l is a proportionality constant defined as G. The conductivity meter measures G and calculates k, the conductivity. In practice it is more precise to calibrate the cell with solutions of established conductivity and determine the unknown by comparison.

10.6.3   Coulometry

Coulometric analysis is the measurement involving the quantity of electricity generated by the complete reaction of the analyte at the electrode. Two classes of coulometric techniques are common: *constant current* techniques, otherwise known as coulometric titrations, and *controlled potential* techniques. In constant current techniques a reagent is generated at one electrode which reacts stoichiometrically with the analyte and is known as the titrant. If the current is known, then it is only necessary to time the reaction and measure the number of coulombs. The following equation can then be used:

$$q = It \qquad (10.35)$$

where q is the amount of charge in coulombs, t is the time in seconds and I is the current flowing in amperes. The number of moles of electrons is defined by

$$\text{Moles e}^- = (It)/F \tag{10.36}$$

where F, the Faraday constant, is 9.648×10^4 C mol^{-1}. In constant current coulometry it is important that the electrode reaction generating the reagent is 100% efficient and that the reaction is both stoichiometric and fast. The end-points in such titrations are determined by chemical indicators, potentiometric or amperometric measurement. In controlled potential techniques the current resulting from a complete electrochemical reaction of the analyte at the electrode is monitored with respect to time. A potential step is applied to the electrode, the electrochemical reaction starts, and the current is monitored. The end of the reaction is indicated by the current decaying to zero:

$$q = \int_0^t I \, dt \tag{10.37}$$

The number of coulombs is measured by integrating the current over the time of the reaction. These techniques are limited to electroactive species and are by nature much slower than constant current techniques.

10.6.4 Electrogravimetry

Electrogravimetric analysis is not commonly used in modern environmental analysis. It involves the electrolytic deposition of the analyte onto an electrode. The mass of the deposit is used to determine analyte concentration. The technique is a combination of electrochemistry and gravimetric analysis. The cell comprises two platinum electrodes. For a reduction process, the cathode is typically a platinum gauze of large surface area. If oxidation is involved, the polarity is switched so that the platinum gauze becomes the anode. Control of the potential is important as interference may occur due to co-deposition of other electroactive substances.

Further reading

Bard, A.J. and Faulkner, L.R. (1980) *Electrochemical Methods, Fundamentals and Applications*, Wiley, New York.

Koryta, J. and Stulik, K. (1983) *Ion-Selective Electrodes*, 2nd edition, Cambridge University Press, Cambridge.

Midgley, D. and Torrance, K. (1991) *Potentiometric Water Analysis*, 2nd edition, Wiley, Chichester.

Jeffery, G.H., Bassett, J., Mendham, J. and Denney, R.C. (1989) *Vogel's Textbook of Quantitative Chemical Analysis*, 5th edition, Longman, Harlow.

Kissinger, P.T. and Heineman, W.R. (1984) *Laboratory Techniques in Electroanalytical Chemistry*, Marcel Dekker, New York.

11 Thermal methods of analysis

P.J. Haines

11.1 Introduction

Thermal methods of analysis are now used in a very large range of scientific investigations. Besides the more 'chemical' areas, such as polymers, fine organic chemicals and pharmaceuticals, they have applications to electronics, in construction, geology and engineering, in materials science and in quality control. They often give information impossible to obtain by other analytical methods. Very often, a complex material, such as a polymer composite will show definite and characteristic effects on heating which relate to its nature, composition and history. These observations are informative about its properties and working life.

11.2 Definitions

Because thermal methods have been developed by many workers, it was necessary to agree on a common terminology, and the International Confederation for Thermal Analysis and Calorimetry (ICTAC) have produced definitive publications and articles, both on the nomenclature and on the calibration methods to be used. There are still some differences in usage, but in this text we shall use the ICTAC definitions and symbols throughout.

Thermal analysis (TA) is defined as:

> a group of techniques in which a property of the sample is monitored against time or temperature while the temperature of the sample, in a specified atmosphere, is programmed.

The programme may involve heating or cooling at a fixed rate of temperature change, or holding the temperature constant, or any sequence of these. The word sample is intepreted here to mean the substance placed into the apparatus at the beginning of the experiment, and its reaction products. The adjective is thermoanalytical. The graphical results obtained are called the thermal analysis curve, or by the specific name of the method.

The property used for study may be chosen from an extensive list, shown in part in Table 11.1. Careful distinction should be made between the terms derivative and differential. Differential techniques involve the measurement of a difference in the property between the sample and a reference, for example in differential thermal analysis (DTA) where the difference in temperature between the sample and a reference is measured. Derivative techniques imply the measurement or calculation of the mathematical first derivative, usually with respect to time, for example derivative thermogravimetry (DTG) is the measurement of the rate of mass loss (dm/dt) plotted against temperature T.

Table 11.1 Thermal methods: major techniques

Technique	Abbreviation	Property	Uses
1. Thermogravimetry (thermogravimetric analysis)	TG TGA	Mass	Decompositions Dehydrations Oxidation
2. Differential thermal analysis	DTA	Temperature difference	Phase changes Reactions
3. Diferential scanning calorimetry	DSC	Power difference	Heat Capacity Phase changes Reactions Calorimetry
4. Thermomechanical analysis	TMA	Mechnical changes	Deformations
5. Dynamic mechnical analysis	DMA	Moduli	Phase Changes Polymer cure
6. Dielectric thermal analysis	DETA	Permittivity	Phase changes Polymer changes
7. Evolved gas analysis	EGA	Gases	Decompositions Catalyst and surface reactions
8. Thermoptometry	(TO)	Optical	Phase changes Surface reactions Colour changes

There are also several less frequently used techniques which are useful for particular analytical problems. These include thermosonimetry, where the sound emission from a sample is measured, thermomagnetometry, where the magnetic properties are studied and thermoluminescence which measures the light emitted as the sample is heated.

Where two or more techniques are used on the same sample at the same time the method is called simultaneous thermal analysis (STA).

11.3 General apparatus

Almost all thermal methods use a similar arrangement of apparatus, as shown in Figure 11.1. The sample is often put into a container or crucible which is then placed in contact with the sensor which is to measure the particular property. A temperature sensor must also be present at (or very near) the sample to follow its temperature as the experiment progresses.

The sensor assembly then fits into a special furnace and the atmosphere around the sample established. This is most important, because the analyst may wish to study (or to avoid!) reactions with air or other reactive gases.

The furnace is controlled by a temperature programmer and is set by the operator to raise (or lower) the temperature of the furnace at whatever rate has

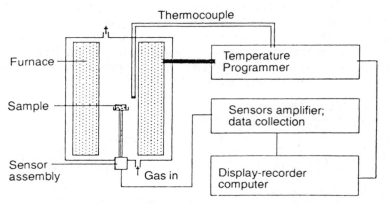

Figure 11.1 Schematic diagram of a thermal analysis system. Source: Haines, P.J. (1995) *Thermal Methods of Analysis*, Blackie Academic & Professional, Glasgow.

been chosen as most suitable. The data are collected by the sensor system, and after processing, displayed on a screen or recorder as a thermal analysis curve.

11.4 Factors affecting thermal analysis results

Many analytical methods give a result which is specific to the compound under investigation. For example, the infrared spectrum of polystyrene is characteristic of that material, and depends little on the sampling method, or instrument, or the time taken to perform the run. Many thermal methods are much less 'compound specific' and the results obtained may depend upon the conditions used during the experiment. The reasons for this will become apparent in later sections of this chapter, but may briefly be summarised as due to the dynamic nature of the processes involved. The signal generated by the sensors will depend on the extent and rate of reaction, or the extent and rate of change of the property measured. The transfer of heat by conduction, convection and by radiation around the apparatus, and the interaction of the surroundings with the sample all affect these changes.

It is therefore essential, whenever a thermal analysis experiment is reported, that the precise conditions used are included in the report. Similarly, comparison of samples should only be made when their curves are run under the same conditions, or when differences in conditions are clearly stated.

The main factors which must be considered and reported for every thermal analysis experiment are now discussed.

11.4.1   The sample

The chemical description of the sample should be given, together with its source, purity and pre-treatments. If the sample is diluted with another substance, such as an inert reference material, then the composition of the mixture should be given. Even the particle size may alter the shape of the curve, especially where a surface reaction is involved.

For example, hydrated copper sulphate, $CuSO_4 \cdot 5H_2O$, powder 50–60 μm, XYZ Co. Ltd, GPR grade, diluted 50% in alumina.

11.4.2   The crucible

The material of the sample holder or crucible should be stated, together with its shape. Sample holders are chosen because they will not interact with the sample during the course of the experiment. However, changing from aluminium to platinum, or alumina, may well alter the heat transfer because of the different thermal conductivity and changing the shape can alter the diffusion of gases to or from the sample.

11.4.3   The rate of heating (dT/dt)

Experiments may be carried out at heating rates from 0 K min^{-1}, that is under isothermal conditions, to very high rates over 100 K min^{-1}. The 'normal' rate is about 10 K min^{-1}. Similarly, experiments may be conducted with different cooling rates. Since the rates of heat transfer, of physical change and of most reactions is finite, the sample will react differently at different heating rates. There is bound to be a thermal lag between different parts of the apparatus, and the higher the rate of heating, the greater this lag is likely to be.

In order to approach equilibrium conditions most closely, we should use a very small heating rate, preferably less than $1°C \text{ min}^{-1}$. Where the average heating rate is quoted, the symbol β is often used to avoid complication.

11.4.4   The atmosphere

The atmosphere surrounding the sample and its products may greatly affect the instrumental measurements. There may be a reaction between the sample and the atmosphere and the heat transfer by the gas will also affect the results.

While a metal may be stable in argon it could form the oxide in oxygen. When a chemical equilibrium exists, for example between calcium carbonate, calcium oxide and carbon dioxide,

$$CaCO_3 \rightleftharpoons CaO + CO_2$$

then a large concentration of product will force the equilibrium back toward the reactants. Thus, calcium carbonate starts to decompose below 800°C in air containing little CO_2, but in CO_2 at 1 atmosphere pressure, the decomposition does not occur until nearly 1000°C.

We should also consider the flow rate of the gas can sweep away the products of reaction from the sample.

11.4.5   The mass of the sample

The physical properties, amount and the packing of the sample can affect the results obtained. Small samples are preferable and the average is about 10 mg. Large samples cause greater thermal lag and very small samples may give inadequate signals. Powdered samples or thin films may react more readily than large crystals or lumps.

These parameters are important for any thermal analysis experiment. As an aid to remembering the five most important, perhaps the acronym 'SCRAM' may be helpful:

1.   sample;
2.   crucible, or sample holder;
3.   rate of heating;
4.   atmosphere;
5.   mass of sample.

For particular techniques other factors must be considered. For example, in TMA and DMA the load exerted upon the sample should be reported.

11.5 Thermogravimetry   This is a technique in which the mass of the sample is monitored against time or temperature while the temperature of the sample, in a specified atmosphere, is programmed.

It should be recognised that several manufacturers and users prefer to call this technique thermogravimetric analysis (TGA). The apparatus is called a thermobalance.

In order to enhance the steps in the thermogravimetric curve, the derivative thermogravimetric (DTG) trace is frequently drawn. Remember that this is the plot of the rate of mass change, with time, dm/dt.

11.5.1   Apparatus

The major parts are:

1.   the electrobalance and its controller;
2.   the furnace and temperature sensors;
3.   the programmer or computer and
4.   the recorder, plotter or data acquisition device.

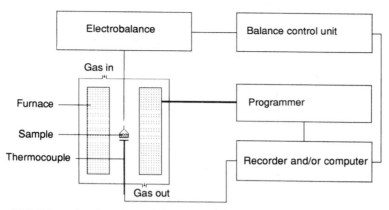

Figure 11.2 Schematic of a thermobalance system. Source: Haines, P.J. (1995) *Thermal Methods of Analysis*, Blackie Academic & Professional, Glasgow.

The balance is generally an electrobalance capable of weighing up to about 1 g with an accuracy of better than 0.1 mg. The furnace is often a wire wound ceramic tube heated electrically and controlled by the programmer or computer. The temperature, mass and derivative signals are then fed to the recorder and processed by the computer (Figure 11.2).

Samples should be small (for example, about 10 mg) and are often powdered and spread evenly in the crucible which may be platinum or alumina. The system is designed so that runs may be done in either air or other gases.

Temperature calibration. In order that the balance shall operate with as little interference as possible, the sample temperature is measured by a thermocouple placed near to, but not in contact with the sample.

Thermocouples are frequently used for measurement of sample temperature and for furnace control. For lower temperatures, chromel–alumel thermocouples can be used, while for temperatures over 1000°C, platinum–13% platinum–rhodium couples are more suitable because of their inert chemical nature.

The temperature measured by a thermocouple which is not in contact with the sample must be subject to a thermal lag which can be up to 30 K. It is very important to calibrate the thermobalance in conditions which reproduce those which will be used in actual experiments. One method for this is to use materials listed in Table 11.2 which are ferromagnetic at low temperature, but which lose their ferromagnetism at a well-defined Curie point. If the magnetic material is placed in the sample holder and a magnet placed below the sample, the magnetic force on the sample will cause an apparent increase in mass. This will disappear fairly abruptly at the Curie point as shown in Figure 11.3.

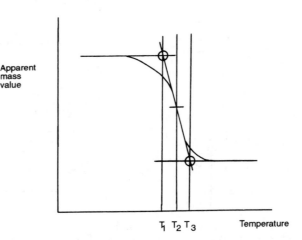

Figure 11.3 Curie point curve. For nickel, the reported values are: $T_1 = 351.4°C$; $T_2 = 352.8°C$; $T_3 = 354.4°C$. T_2 is taken as the Curie point for calibration. Source: Haines, P.J. (1995) *Thermal Methods of Analysis*, Blackie Academic & Professional, Glasgow.

Table 11.2   ICTAC calibration materials for TG

Material	Curie point (°C)
Permanorm 3	259
Nickel	353
MuMetal	381
Permanorm 5	454
Trafoperm	750

11.5.2   Applications of thermogravimetry

Calcium oxalate monohydrate ($CaC_2O_4 \cdot H_2O$).   When we heat this compound on a thermobalance, we obtain the curves shown in Figure 11.4. A single run in air at 10°C min^{-1} up to 1000°C shows three distinct changes.

1. Between room temperature and about 250°C, there is a loss of about 12% to give a stable product shown by a horizontal plateau on the TG, and a return to baseline on the DTG. Since this starts around 100°C, we might consider loss of water vapour:

$$CaC_2O_4 \cdot H_2O(s) = CaC_2O_4(s) + H_2O(v)$$

This gives a calculated loss of $100 \times 18.0/146.1 = 12.3\%$. Since this corresponds with the loss measured, this interpretation is possible.

2. Between 300 and 500°C, a loss of some 19% occurs to give a second stable product. Previous analytical methods produced calcium carbonate by heating to about 500°C so consider that:

$$CaC_2O_4(s) = CaCO_3(s) + CO(g)$$

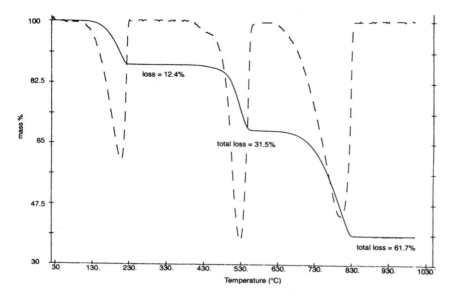

Figure 11.4 TG and DTG curves for calcium oxalate monohydrate, 12.85 mg, Pt crucible, 20 K/min, nitrogen. Source: Haines, P.J. (1995) *Thermal Methods of Analysis*, Blackie Academic & Professional, Glasgow.

The loss here should be $100 \times 28.0/146.1 = 19.2\%$ This corresponds well.

3. Above 600°C there is a final loss of about 30% to give a final product stable to the highest temperature used. The decomposition of calcium carbonate is a well established reaction with a mass loss of 30.1%:

$$CaCO_3(s) = CaO(s) + CO_2(g)$$

This may well be the final stage loss.

Words of caution! Despite the excellent agreement of the calculated and experimental mass losses, it is most unwise to deduce reaction schemes from this evidence alone! The products, both gaseous and solid should be characterised by other analytical methods, such as X-ray diffraction for solids and chemical tests for gases. In the above case, these tests have been carried out, and the reaction scheme confirmed as given.

Degradation of polymers. Polymer stability and the degradation of polymers depend on the chemical structure and physical form of the polymer or composite. There is a range from natural polymers like cellulose, which starts to char around 100°C, to polyimides which are stable up to 400°C.

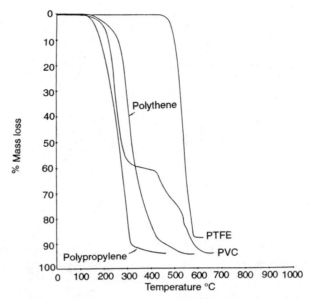

Figure 11.5 TG curves for polymer samples in air. Samples all about 1 mg, heated at 50 K/min. Source: Haines, P.J. (1995) *Thermal Methods of Analysis*, Blackie Academic & Professional, Glasgow.

The recycling, disposal and combustion of polymers may all be studied by thermal methods. The stages of degradation in air and in inert atmosphere are an important way of characterising polymeric materials (Figure 11.5).

Some polymers, like poly(methylmethacrylate), 'unzip' to the monomer, or, like poly(ethylene) to small fragments in one step. Others like poly(vinyl chloride) and cellulose lose small molecules on the way to forming a char or high boiling tarry products. This is shown by the TG curves, which also give a guide to polymer stability by the temperature at which decomposition starts.

Analysis of mixtures.   When we wish to analyse an unknown mixture of chemical substances in a real sample, we have several options open to us. We may use the range of separation techniques to isolate each component of the mixture which may then be identified and measured separately. We may use inductively coupled plasma spectrometry to determine all the elements present simultaneously. Thermal methods may be used directly on original samples from coals and soils to complex inorganic and polymeric mixtures. For example, the analysis of coals for moisture, total volatiles, fixed carbon and ash content, referred to as a 'proximate analysis', may be carried out by thermogravimetry. By heating in an inert atmosphere to vaporise moisture and volatiles successively and then, by switching over to oxygen, the 'fixed carbon' may be burnt off leaving the ash as residue. Each loss is measured on the thermobalance and the values characterise the coal.

These two techniques are considered together, since they are frequently used to study the same phenomena. They work on different principles, but respond to energy changes in samples, even if no mass change is involved. Thus an endothermic change (absorption of heat) such as melting, where mass stays the same can be detected by DSC and DTA, but not by TG.

11.6 Differential thermal analysis (DTA) and differential scanning calorimetry (DSC)

11.6.1   Differential thermal analysis (DTA)

This is a technique in which the difference in temperature between the sample and a reference material is monitored against time or temperature while the temperature of the sample, in a specified atmosphere, is programmed.

The DTA curve is generally a plot of the difference in temperature, ($\Delta T = T_S - T_R$) as ordinate against the temperature T as abscissa. An endothermic event gives a downward 'peak'.

11.6.2   Differential scanning calorimetry (DSC)

This is a technique in which the difference in heat flow (power) to a sample and to a reference is monitored against time or temperature while the temperature of the sample, in a specified atmosphere, is programmed. The heat is supplied to the sample contained in the pan, and similarly to the reference in its pan.

When the sample and reference are heated by separate heaters and the temperature difference is kept very small by supplying different amounts of power to sample and to reference, the instrument works as a power compensated DSC.

If the sample and reference are heated from the same source and the temperature difference ΔT is measured and this signal is then converted to a power difference, ΔP using the calorimetric sensitivity, this a heat flux DSC.

The DSC curve has ΔP as the ordinate and temperature as the abscissa. Since an endothermic peak involves the absorption of more power by the sample, one convention plots endothermic peaks upwards. This does cause some confusion, and in this text, we shall adopt the 'DTA' convention throughout, while indicating the sign of heat flow on the y-axis.

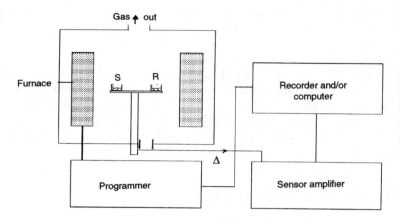

Figure 11.6 Schematic of DSC or DTA apparatus. Δ indicates the differential signal. Source: Haines, P.J. (1995) *Thermal Methods of Analysis*, Blackie Academic & Professional, Glasgow.

11.6.3   Apparatus

The major parts of the system are (Figure 11.6):

1.   the DTA or DSC sensors plus amplifier;
2.   the furnace and its temperature sensor;
3.   the programmer or computer and
4.   the recorder, plotter or data acquisition device.

The sensors are most oftem sensitive thermocouples of chromel–alumenl, gold–nickel or platinum–13% platinum/rhodium alloy. Some instruments use resistance thermometers as sensors. The furnace is a wire wound tube heated electrically. For best heat transfer, silver furnace linings are used. For temperatures below ambient, the furnace, samples and sensors may be cooled by a refrigerant placed around them. The furnace is controlled by another temperature sensor linked to the programmer and temperature and the amplified difference signal from the sensors is fed to the recorder and computer for processing.

While earlier DTA instruments used many grams of sample, modern ones need only a few milligrams (e.g. 10 mg). The reference should be as similar to the sample as possible, and is often powdered, dry alumina. Both sample and reference are placed in similar small pans of aluminium, platinum or alumina

Calibration.   It is essential to calibrate both the temperature scale and, for DSC instruments, also the power scale. ICTAC have approved a set of standard substances, which are listed in Table 11.3.

Table 11.3 Calibration materials for DSC and DTA

Material	Transition	T (°C)	ΔH (J g^{-1})
Cyclohexane	Transition	− 83	
Phenyl ether	Melting	30	
Potassium nitrate	Transition	128	
Indium	Melting	157	28.71
Tin	Melting	232	56.06
Zinc	Melting	419	111.18
Quartz	Transition	573	
Barium carbonate	Transition	810	

For a pure metal, such as 99.999% pure indium, the melting of which is shown in Figure 11.7a, the extrapolated onset temperature T_e should correspond to the correct melting point of the metal, 156.6°C. The integrated area of the peak A_S may then be used to calculate the calorimetric sensitivity constant K:

$$K = \Delta H_S \cdot m_S / A_S$$

where ΔH_S is the enthalpy of fusion of indium in J g^{-1} (28.70 J g^{-1}); m_S is the mass of sample of indium in g. If A_S is the peak area in cm^2, then K, the calorimetric sensitivity, is in J cm^{-2}. Note: If the area is calculated in K.s, or W.s then obviously the units of K will change! Where K varies with temperature, it is wise to calibrate with several materials.

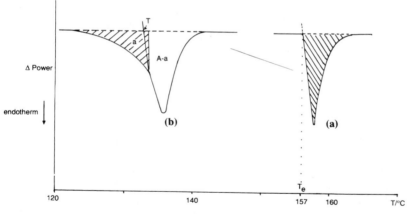

Figure 11.7 DSC curves for (a) pure indium, used for calibration and (b) impure phenacetin. Note: $a/A = F$, the fraction melted at.

Reminder! The curves which are obtained for DTA and DSC will depend on the samples and instrument conditions used:

Sample	chemical nature, purity, history
Crucible	material, shape
Rate of heating	high, low, isothermal

Atmosphere   gas
Mass of sample   volume, packing, distribution, dilution

11.6.4   Applications

These may be classified into two groups:

1. *Physical changes and measurements:* melting, crystalline and liquid crystalline phase changes, polymers transitions, heat capacity and glass transitions and others;
2. *Chemical reactions:* dehydrations, decompositions, polymer curing, oxidative attack and other reactions.

Purity determination When a material is near to 100% pure, it melts sharply, as shown by the melting of 99.999% pure indium in Figure 11.7a. If an impurity is present the melting usually occurs over a wider range and the final melting temperature is lowered as shown in Figure 11.7b. The purity of the material may be estimated by measuring the fraction F melted at each temperature T and plotting T versus $1/F$. The theory may be found in the texts given at the end of the chapter.

Poly(ethylene terephthalate). This widely used polymer shows all the characteristic transitions of polymeric materials. The DSC trace shown in

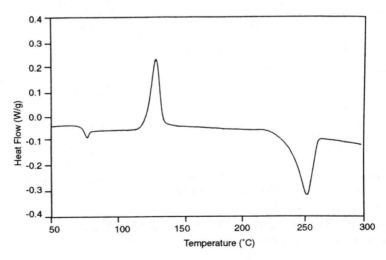

Figure 11.8 DSC curve of poly(ethylene terephthalate) (PET) quenched from melt; 10 mg, 10°C/min, flowing nitrogen. Source: Haines, P.J. (1995) *Thermal Methods of Analysis*, Blackie Academic & Professional, Glasgow.

Figure 11.8 shows the *glass transition* at around 80°C, where the polymer changes from a brittle material to a more pliable, plastic form, the *cold crystallisation* at about 130°C which is *exothermic* because the material is giving out heat as it forms crystals, and the *melting* at about 240°C which is *endothermic* since the crystals become liquid.

Kaolin The decompositions of hydrates, complexes and minerals are frequently complicated, especially when the original samples are themselves mixtures. Studies of minerals by DTA have shown that they give curves characteristic of their structure and that different minerals whose *elemental analysis* may be very similar give quite distinct DTA curves. DTA has more often been used because of the higher temperatures involved.

Figure 11.9 shows the DTA curve of kaolin (china clay) with three thermal events.

1. Around 100°C there is a small endothermic peak due to loss of moisture.
2. At about 500°C, a much larger endotherm has been shown to be due to loss of hydroxyl (–OH) water from the clay structure.
3. At 1000°C and above there is a sharp, exothermic peak which is due to the commencement of the chemical reaction forming mullite, $3Al_2O_3 \cdot 2SiO_2$.

DTA is widely used to characterise minerals and other inorganic materials. One other example may be cited where the DTA curves of *cements* were used to distinguish ordinary Portland cement (OPC) from high alumina cement (HAC) and to measure the 'degree of conversion' which the HAC has undergone. This is very important because HAC may lose its strength

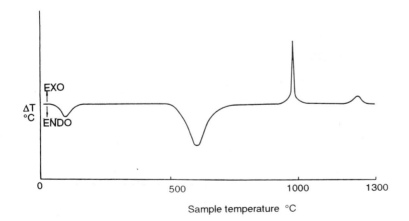

Figure 11.9 DTA curve of kaolinite. Source: Haines, P.J. (1995) *Thermal Methods of Analysis*, Blackie Academic & Professional, Glasgow.

due to an internal chemical change and structures (e.g. roof beams) may collapse catastrophically.

Oxidative stability of polyethylene. Polyethylene is a very important polymer and is used in packaging, protective sheeting, piping and moulded articles. If there is any degradation of polyethylene in the environment the desirable properties of the polymer may be affected. For example, 'blue' polyethylene water piping may become more brittle and crack. Rather than leave a sample out in the environment for many years, testing occasionally, an *accelerated ageing test* for this may be done by heating a small sample in air (or oxygen) and measuring the degradation temperature (or the degradation time at an elevated temperature such as 200°C). Figure 11.10 shows the melting of the polymer at

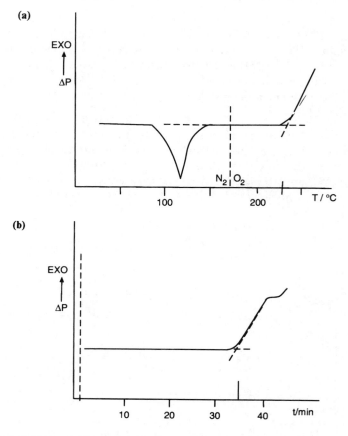

Figure 11.10 Oxidative stability of polyethylene by DSC. The dashed line represents the change from nitrogen to oxygen. (a) Scanning at 10°C/min; (b) isothermal at 200°C. Source: Haines, P.J. (1995) *Thermal Methods of Analysis*, Blackie Academic & Professional, Glasgow.

120°C followed by the exothermic onset of oxidative degradation at about 230°C.

11.7
Thermomechanical analysis (TMA) and dynamic mechanical analysis (DMA)

These are techniques is which the *mechanical properties* of the sample are monitored against time or temperature while the sample, in a specified atmosphere is programmed.

   Thermomechanical analysis (TMA) measures the deformation of the sample under a non-oscillating load. If the load is very low, the *dimension(s)* of the sample are measured and this can be called *thermodilatometry*. In *dynamic mechanical analysis* (DMA) an oscillating load is used and the storage and loss moduli are monitored.

11.7.1   Apparatus

A typical system used for TMA is shown in Figure 11.11. The *sample* is placed on an inert surface and a *probe* allowed to rest upon it, exerting a load controlled by the *force controller*. The movement of the probe due to the sample changes is measured by a *linear variable differential transformer* (LVDT) which produces an electrical signal related to the position of the

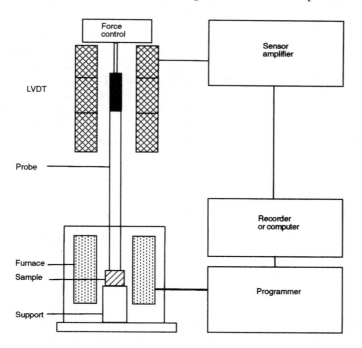

Figure 11.11 Schematic diagram of a thermomechanical analyser (TMA). Source: Haines, P.J. (1995) *Thermal Methods of Analysis*, Blackie Academic & Professional, Glasgow.

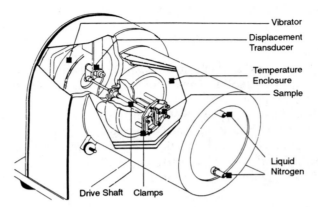

Figure 11.12 Diagram of the Rheometric Scientific DMTA. A bar sample is clamped rigidly at both ends and its centre vibrated sinusoidally. Source: Haines, P.J. (1995) *Thermal Methods of Analysis*, Blackie Academic & Professional, Glasgow.

probe. The *furnace* is similar to that used for DTA and the system may be cooled by refrigerant. The *programmer and controller* controls the rate of heating and temperatures are measured by thermocouples. Data from the LVDT and the thermocouples are stored and processed by the *computer*.

A DMA apparatus is shown in Figure 11.12. The sample is placed in one of a variety of fixing systems, depending on the mode of oscillation to be studied. For example, a bar of sample may be clamped at each end into a frame and fitted onto the system. The sample may them be oscillated by vibrating the centre point using a ceramic shaft. The *stress* depends on the power supplied to the drive shaft and the *strain* on the sample displacement. Heating, temperature measurement and control and data recording are done in a similar way to TMA. It is usual to process the data so that the *storage modulus E′*, the loss modulus *E″* and the loss tangent, tan $\delta = E''/E'$ are obtained.

Calibration. This is often done by placing a piece of calibration material (as given in Table 11.3) below the probe and measuring the temperature of transition on the TMA.

11.7.2 Applications

Expansion and transitions. If a very small load is used in TMA, the probe will move as the sample expands and the *coefficient of expansion*, α, may be measured. At transitions, such as the *glass transition*, T_g, of polymeric materials, the expansion changes and we get two intersecting lines as shown in Figure 11.13a.

When the DMA technique is used, the much higher sensitivity allows detection of more transitions at lower temperature, due to the motion of various parts of the polymer molecule. The T_g is shown by a large decrease in E' and by a peak in the tan δ curve. This is seen in Figure 11.13b.

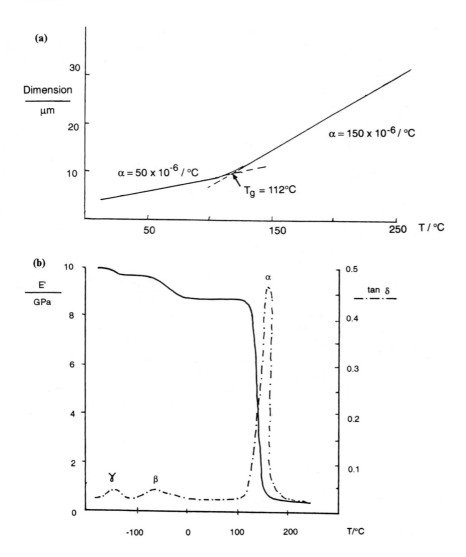

Figure 11.13 Mechanical analysis of polymers. (a) Thermomechanical analysis (TMA) of printed circuit board, 10°C/min. (b) Dynamic mechanical analysis (DMA) of printed circuit board at 7 Hz frequency, 5°C/min. The DMA shows minor transitions at low temperature. Source: Haines, P.J. (1995) *Thermal Methods of Analysis*, Blackie Academic & Professional, Glasgow.

The effects of environmental conditions, particularly moisture, on the values of the moduli and the glass transition are important when considering the end-use of polymer materials. Of course, this can be used 'in reverse' to study the effects of agressive environments on materials.

11.8 Simultaneous techniques and product analysis

If two or more techniques are applied to a sample at the same time the method is called *simultaneous*. For example, if the TG and DTA (or DSC) curves are measured together then this is *simultaneous thermal analysis* (STA). If the gases evolved from a sample reaction are measured and identified immediately by an attached mass spectrometer then this is *simultaneous* TG-MS.

The advantages of simultaneous methods are:

1. a saving in *time* from running two separate experiments;
2. the possibility of comparing runs under *exactly the same conditions*;
3. the complementary measurements allow better *identification* of events; for example the loss of moisture should be seen as a mass loss on TG and as an endotherm on DSC. Melting will only show on the DSC trace.

11.8.1   Apparatus

By making an apparatus so that *sample and reference* pans and their measuring *thermocouples* may be placed in a *thermobalance* and their masses measured as they are heated in the *furnace* under *programmed* conditions, we obtain the TG and DSC/DTA traces at the same time. Such a trace is shown in Figure 11.14.

The two traces give complementary information. Losses around 100°C are paralleled by endotherms on the DTA and correspond to loss of hydrate water. Sharp endotherms on the DTA near 300°C have no corresponding change on the TG and are phase changes of the anhydrous sample. The final decomposition on the TG is shown to be *exothermic* on the DTA.

11.8.2   Evolved gas analysis (EGA)

This is a technique in which the nature and sometimes the amount of gases or vapours evolved are monitored as the sample is heated in a specified atmosphere.

Apparatus.   Many different analytical techniques may be used for this. The simplest are chemical methods, where the evolved gases are passed into a container of a suitable absorbent and the pH, colour change, electrical conductivity or reaction are measured. For example, the evolution of CO_2

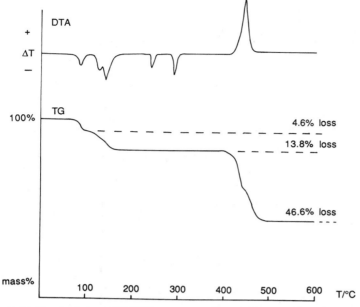

Figure 11.14 Simultaneous TG-DTA curves for barium perchlorate trihydrate $(Ba(ClO_4)_2 \cdot 3H_2O)$.

from cements may be followed by reacting it with barium hydroxide and following the conductivity. The production of HCl when poly vinyl chloride is heated may be studied by observing the pH change in an absorbent solution.

Simultaneous thermogravimetry-mass spectrometry (TG-MS). One of the most useful methods in evolved gas analysis is to pass the gases evolved from a TG (or STA) through a heated tube and an interface to reduce the pressure (as described in chapter 8) into a mass spectrometer. The use of *simultaneous* TG-MS to elucidate the stages and mechanisms of complex reactions is most important. One example which shows the potential of the method is shown in Figure 11.15

We have seen the TG of this material before in section 11.2, but this diagram shows the DTA as well for the sample heated in argon at $15°C \ min^{-1}$. The simultaneous measurement of individual ions at various mass/charge ratios shows that the first peak at 180°C corresponds largely to evolution of water, H_2O, $m/z = 18$, the second around 500°C to the evolution of carbon monoxide, CO, $m/z = 28$, with some carbon dioxide, CO_2, $m/z = 44$, possibly produced by disproportionation, and the last near 800°C to the production of CO_2.

The application of TG-MS to polymer decompositions, to the study of catalysts and of inorganic complexes complex and to minerals has proved

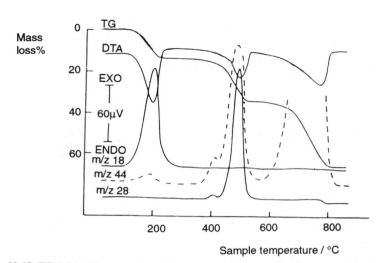

Figure 11.15 TG-DTA-MS curves for calcium oxalate monohydrate, 15 mg, 15°C/min. argon. Source: Haines, P.J. (1995) *Thermal Methods of Analysis*, Blackie Academic & Professional, Glasgow.

of great importance. The complete mass spectrum of an evolved product can be obtained and thus the product identified. When many products are evolved over the same temperature range, gas chromatography may have to be used for separation before the mass spectrometer.

Simultaneous thermogravimetry–infrared spectrometry. Since an infrared spectrum of a gaseous mixture may be obtained without reducing the pressure, and since the different gases each give their characteristic spectrum, infrared spectrometry can be a useful method for EGA. The gases from the thermal analysis apparatus (for example, a thermobalance system) are led through a heated capillary into a heated infrared gas cell or light pipe. Using a modern Fourier-transform infrared (FTIR) system, spectra may be obtained rapidly on samples with a wide range of absorbances, and the collection of evolved gas spectra during the course of a thermal analysis run can show the progress of a reaction.

The decomposition of a poly vinyl chloride–poly vinyl acetate copolymer gave three steps on the TG curve. The FTIR spectra at the first step, around 300°C showed absorbances characteristic of HCl, from the PVC and of ethanoic acid (acetic acid) from the PVAc.

11.8.3   Analysis of products and reactions

It should be recognised that thermal methods, unlike spectroscopic or elemental analysis methods are generally non-specific. An endothermic peak could be caused by a variety of changes such as a phase change, loss of solvent or a chemical reaction. Unless additional observations are made, the analysis is rather indeterminate.

Thermomicroscopy. The direct observation of the changes taking place as the sample is heated is most informative. Using a *hot-stage* and observing the sample by transmitted or reflected light using a *microscope* enables the analyst to observe phase changes, and meltings and to distinguish them from reactions. Use of *polarised light* to observe the sample gives valuable guidance to the crystal form or the liquid crystal type. Observations of the onset of 'browning' of samples such as cellulose give a clue to the stability and pretreatment of materials. When bubbling is observed during a decomposition, it suggests gas evolution which may be confirmed by EGA and with polymers it also may also indicate the nature of the final carbonised product.

The light intensity transmitted or reflected may be recorded during the heating, and we may run thermomicroscopy (or *hot-stage microscopy*) and DSC simultaneously.

X-Ray analysis. In the above sections the gaseous products from heating experiments have been analysed, but the solid materials also need to be identified.

If sufficient sample is available, it may be heated to successively higher temperatures and an X-ray powder diffraction pattern measured for each product. If this is done with a simple reaction like the oxidation of iron, the gradual decrease in the iron pattern of X-ray peaks and the accompanying increase in the X-ray pattern for the oxides of iron is easily seen.

It is possible to measure the X-ray powder diffraction pattern directly as the sample is heated and even to combine this with a DSC record.

The solid products of reaction will usually give an identification of the material, provided they are sufficiently crystalline. If a *glass* is formed, the X-ray pattern is indistinct. If the solids from each stage of the calcium oxalate hydrate experiment shown in Figure 11.15 are examined, the original material, the anhydrous product at 300°C, the calcium carbonate at 600°C and the final calcium oxide all have distinct, unique X-ray diffraction patterns.

Scanning electron microscopy (SEM). The changes which happen to the sample as it is heated often depend on the structure, crystal form and the surface nature of the sample. These are readily seen under the very high magnification of the electron microscope. The changes taking place as a sample of kaolin is heated (Figure 11.9) can be understood more readily if the solid is examined at each stage under the SEM.

11.9 Environmental applications of thermal methods

11.9.1   Geological materials

The use of TG and DTA to characterise minerals is extremely important. The DTA curves for clay minerals are easily measured and allow groups of clays to be distinguished. Borate minerals having similar chemical formulae may give different thermal analysis curves because of their different structures. Coals and oil shales can be analysed by proximate analysis methods and an estimate of the calorific value obtained by use of DSC.

Mixtures such as soils can be studied. The effects of environmental factors on the quality of soils can be investigated. Provided there is not too much clay in the sample, the TG trace will separate the moisture content as the loss around 100°C, the organic content as the loss in air up to around 500°C and the minerals remaining by any high temperature losses and the final residue, we can class soils quite accurately. Sediments deposited on the bottom of rivers, or suspended in the water can be analysed to show the mineral and organic carbon contents.

11.9.2   Recycling

The re-use of processed materials, particularly polymers, is an important method of reducing pollution both by the over-use of resources and by the inefficient disposal of waste. If the polymer waste is collected and separated into its major components, some polymers may then be recovered and recycled. Thermal methods play a role in analysing the materials and in assessing the quality of the recycled product.

If a block of polymer residue contains discrete 'chunks' of each polymer, it is possible to obtain a 'thermal spectrum' of the melting points and amounts of the most important thermoplastic polymers present.

Two products used in making plastic bottles are poly(propylene) (m.p.~176°C) and poly(ethylene terephthalate) (m.p.~250°C). By measuring the DSC curve of the recycled products the proportions of each may be found. The mechanical quality of the product may also be found by DMA.

In the recycling of polyethylene sheet, the material is re-melted and cast into small chippings for moulding. In the examination of a series of samples, coloured by different contaminants, most samples melted around 160°C and had a ΔH value of about 100 J g^{-1}. A 'blue' sample was shown to be wrong because it showed samples taken from it had *two* transitions, either at 80 and 104°C but sometimes at 125 and 162°C. This sample was rejected for re-use!

An alternative to recycling is to re-design the polymer so that it is bio-degradable. This may be done with polyethylene by including *starch* as a filler in the polymer compounding. The starch acts as a 'food' for microorganisms and even for larger creatures such as wood-lice. The effects of various environmental conditions, marine, fresh-water lake, clay and loam soils were compared with laboratory controls using thermal and other methods. The strength of the polymer matrix was measured by a TMA method and the effects on the molecular weight and structure studied by DSC. This was a long-term experiment extending over several years.

A study of the effects of chemical modifications to polymers and inorganic materials used in medical and environmental applications was done using DSC, TG, DMA and spectroscopic measurements. The changes brought about by the hydrolytic attack on the polymer were easily detected.

11.9.3   Residues

Cellulose residues from agriculture form a very abundant resource in many countries. These residues constitute a biomass fuel. The degradation and combustion of residues of straw, wood and seed residues were studied by TG and DSC and shown to be a useful resource.

11.9.4   Vaporisation studies

The environmental effects of vapours produced by industry and by synthetic chemicals are a cause for concern. While their actual effects on the biosphere and on animals must be studied by biological methods, and their concentration measured by other analytical methods, the process of vaporisation at different temperatures can be studied by DSC or TG. The rate of vaporisation depends on the vapour pressure of the material and may be measured as the rate of mass loss on a TG or by the endothermic deflection on a DSC. As the temperature is increased the vapour pressure, and hence the rate of vaporisation will increase, too. This will give a characteristic thermal analysis curve. The boiling points of pollutants may be measured using DSC and a sample pan with a pinhole aperture.

Analysis for polycyclic aromatic hydrocarbons (PAHs) by sorption onto a material or direct examination of soots has been carried out by TG-MS.

11.9.5   Flue gas treatment

The production of sulphur-containing gases during the burning of fossil fuels is a serious environmental problem. Since thermal methods can be used for the study of the combustion characteristics of coals and oils we could study the evolution and removal of sulphur gases by EGA techniques. The sulphur dioxide evolved can be measured by spectrometric, or coulometric methods. The absorption of the gases by various materials has been investigated by a modified thermobalance to show the sulphation, reduction and oxidation of the sorbent system. Reaction between sulphur dioxide and carbonate materials such as dolomite, $CaCO_3 \cdot MgCO_3$ can be followed using a thermobalance.

11.9.6   Purity

The contamination of pure materials by pollutants may be very slight, but since the DSC method can detect much less than 1% impurity and, indeed, works best at very small impurity levels, it is possible to use it to study any chemical reaction, structural change or impurity in chemical which should be pure.

11.10   Summary   Thermal methods cross the boundaries between the macroscopic industrial examination of bulk materials, the measurement of physical properties and conventional chemical analysis. They offer the analyst methods to study materials without separation, without extensive sample preparation or solution and they apply to samples from plants to polymers and from coals to complexes. However, *caution must be advised!* The results from thermal methods *do depend on the conditions used* and must often be supplemented by other analytical techniques.

Further reading   Brown, M. E. (1988) *Introduction to Thermal Analysis*, Chapman & Hall, London.
Charsley, E. L. and Warrington S. (Eds.) (1992) *Thermal Analysis—Techniques and Applications*, RSC, London.
Dodd, J. W. and Tonge, K. H. (1987) *Thermal Methods*, ACOL/Wiley, London.
Haines, P. J. (1995) *Thermal Methods of Analysis—Principles, Applications and Problems*, Blackie Academic & Professional, Glasgow.
Turi, E. A. (1981) *Thermal Characterisation of Polymeric Materials*, Academic Press, New York.
Wendlandt, W. W. (1986) *Thermal Analysis*, 3rd edition, Wiley, New York.

Chapters in general analytical treatises and textbooks

Fifield, F. W. and Kealey, D. (1990) *Analytical Chemistry*, 3rd edition, Blackie Academic & Professional, London.
Skoog, D. A. and West, D. (1980) M. *Principles of Instrumental Analysis*, Holt-Saunders, Tokyo.

Biological indicators 12
R. Manly

It is axiomatic that all living organisms, including humans, are greatly influenced by the physical and chemical attributes of their environment. Indeed, an experienced ecologist, provided with a species list of organisms from an unseen environment, can accurately predict the nature of such factors as the soil, water hardness, substratum, climate, etc. and often even geographical location. Also, for many years, geologists have used specific indicator plant species and communities to map geologic formations and locate minerals, a process known as geobotanical prospecting. Through a knowledge of the effects of physical and chemical factors on living organisms it is possible to ascertain some idea, not only of the presence, but also the levels of many substances in the environment. For example, plants often exhibit well-defined and distinguishable symptoms when particular mineral nutrients are deficient in the soils in which they are growing.

Similarly, organisms are affected by pollutants or natural substances present in excess in their environment, in ways that are often indicative of the nature and degree of such contamination. The type of effects that may be observed are some or all of the following:

1. changes in the species composition and/or dominant groups in biological communities;
2. changes in species diversity in communities;
3. increased mortality rates in populations, especially in sensitive early development stages such as eggs and larvae;
4. physiological and behavioural changes in individuals;
5. morphological and histological aberrations in individuals;
6. the build-up of pollutants or their metabolites in the tissues of individuals.

Studies of these effects have led to the development of so-called biological surveillance methods for the monitoring of environmental quality, which can sometimes replace, or at least supplement, chemical analytical methods. It is perhaps pertinent here to distinguish more precisely between the terms surveillance and monitoring. A biological survey is a static inventory of organisms, variables and processes in a chosen environment. Biological surveillance comprises a series of similar surveys carried out in the same environment over time and is therefore dynamic. If this surveillance has a specific aim, such as determining compliance with statutory limits on pollutants in that environment, then it is known as biological monitoring or biomonitoring. Notwithstanding this, for simplicity, the term biomonitoring will be used hereafter to include biological surveillance.

Biomonitoring methods can offer some significant advantages over chemical analyses. For example, most pollution monitoring programmes are based upon the collection of samples at regular intervals, as continuous sampling is usually both impractical and very expensive. Thus in a regular chemical sampling programme, particularly in a changeable environment such as the atmosphere or a river, any peaks or troughs in pollutant levels between sampling times will go undetected. In contrast, biological organisms will very often respond in a measurable way to such intermittent pollution, most notably in an adverse manner to high levels of pollutants. They also provide a time-integrated response to fluctuating levels of pollutants in the environment. Depending on the life span of the chosen organism, this response can range over a period of days (microorganisms) to one of months or years (macrophytes, macroinvertebrates and vertebrates).

In addition, a large number of pollutants may be present in any environment and there is a practical limit to the number that can be analysed in any single chemical surveillance programme. Because the biological community will respond to any harmful substance present in the environment, any changes detected can act as a warning and as a precursor to a detailed chemical analysis for increased levels of any existing pollutants or for the presence of a new or unsuspected contaminant. Furthermore, the synergistic or antagonistic effects of many pollutants cannot be detected other than by their effects on biological organisms.

There are a number of different ways in which living organisms may be used for the purposes of the biomonitoring of pollution. These can be broadly classified as shown in Table 12.1. Regardless of which of these is chosen, the collection of representative and meaningful samples is vital to the success of any survey. Many methods are available for the sampling of biological material, a detailed discussion of which is beyond the scope of this book. Some methods are prone to greater sampling error or bias than others and these must be taken into account when interpreting or extrapolating from the data obtained. Compromises may have to be reached between instigating an 'ideal' sampling programme and one that is expedient in terms of time, costs, manpower, etc. This may be particularly true when carrying out a regular monitoring programme.

The choice of sampling method may also be influenced by the desire or need to compare data collected at different times and places, often by different operatives. The requirement for a consistent sampling strategy during a monitoring programme is obvious and to this end many of the methods listed in Table 12.1, such as bioindicator and microbiological surveys, often have prescribed methods of sampling associated with them.

Table 12.1 Principal methods employed for biomonitoring

Type of surveillance	Major organisms used	Principal pollutants assessed	Advantages	Disadvantages
Community structure studies	Invertebrates, macrophytes	Organic and toxic wastes,[a] nutrient enrichment	Easy to use, low cost, no specialist equipment or knowledge required	Some specialist knowledge may be required, localised use, non-specific
Bioindicators	Invertebrates, macrophytes, algae, lichens	Organic wastes, nutrient enrichment, acidification, toxic gases	Easy to use, low cost, no special equipment required	Some specialist knowledge may be required, localised use, non-specific
Microbiological methods	Bacteria	Organic and faecal matter	Relatively low cost, directly relevant to human health	Some specialist equipment and knowledge required
Bioaccumulators	Macrophytes, macro-invertebrates, vertebrates	Toxic wastes,[a] radionuclides	Indicative of bioavailability of pollutants, relevant to human health	Time-consuming, expensive equipment and trained personnel required
Bioassays	Microorganisms, macrophytes, algae, invertebrates, lower vertebrates	Organic matter, toxic gases, toxic wastes[a]	Rapid results, relatively low cost, some continuous monitoring possible	Difficulties in extrapolating laboratory results into field situation

[a] For example, heavy metals, pesticides, fossil hydrocarbons, PCBs.

A number of biological monitoring methods have as their underlying bases the detection of changes brought about by pollution in the structure of the communities of organisms inhabiting the areas under investigation. In reality it is impractical to attempt to examine the entire community, particularly on a routine basis. Therefore, it is usual to select only certain sections of the community for study, such as the plankton or benthic organisms in aquatic ecosystems and macroinvertebrates or macrophytes in terrestrial systems.

The methods themselves are based upon the presence-absence and/or numerical abundance of individual species in a biological community. Traditionally, they have taken the form of either diversity or similarity indices and, to a much lesser extent, species abundance patterns. These all suffer from the fact that any changes observed through their use can only be subjectively, rather than objectively, linked to any specific pollutant and then only when extensive environmental data are available. The use of more sophisticated multivariate analysis techniques in recent years has enabled more objective links with influencing factors, including specific pollutants, to be made in monitoring programmes. However, the relative ease of use of many diversity and

12.2 Monitoring community structure

similarity indices, even by non-specialists, and their cost effectiveness should ensure that they continue to have a place in pollution monitoring.

12.2.1 Diversity indices

Diversity indices provide a measure of species abundance in a chosen environment in the form of a single value. They can be used to evaluate three aspects of community structure, viz. (i) the number of species or *richness*; (ii) the total number of organisms of each species present or *abundance*; (iii) the uniformity of distribution of individuals amongst the various species or *evenness*. Their value relies on the assumption that as pollution of an ecosystem increases, the loss of sensitive species reduces one or more of these factors leading to an overall reduction of diversity in the community. While this is usually the case, there are circumstances where the reverse is true. For example, there is usually an increase in the numbers and variety of plants and animals in naturally nutrient-poor (oligotrophic) lakes, such as those found in mountains or on moorland, if they become subjected to the enriching effects of mild organic pollution from animal wastes or sewage.

There are many diversity indices available, a full discussion of which is beyond the scope of this book. Some of the most commonly used indices and their methods of calculation are shown in Table 12.2. There has been much debate about the value of diversity indices in pollution monitoring, the principal advantages and disadvantages of which are summarised in Table 12.3. Despite these reservations, a number of them have been used quite extensively to monitor pollution in a range of environments.

Table 12.2 Some commonly used diversity and similarity indices

	Index	Method of calculation
Diversity indices	Shannon index (H')	$H' = \sum\limits_{i=1}^{s} P_i \log P_i$
	Simpson index (D)	$D = \dfrac{1 - N_i(N_i - 1)}{N(N - 1)}$
	Margalef index (D_{Mg})	$D_{Mg} = \dfrac{S - 1}{\log_e N}$
Similarity indices	Sorensen's index (C)	$C = \dfrac{2w}{A + B} \times 100$
	Jaccard coefficient (J)	$J = \dfrac{w}{A + B - w} \times 100$
	Pinkham and Pearson index of community similarity (P)	$P = \dfrac{1}{k} \sum \left[\dfrac{\min(X_{ia} \cdot X_{ib})}{\max(X_{ia} \cdot X_{ib})} \right]$

N_i, number of individuals of species i in a sample; N, number of individuals of all species in a sample; S, number of species recorded in a sample; P_i, proportion of individuals in ith species; w, number of species common to both samples; A, number of species in sample 1; B, number of species in sample 2; k, total number of comparisons or different taxa in the two samples being compared; X_{ia} and X_{ib}, abundance of species i in samples a and b, respectively.

Table 12.3   Relative advantages and disadvantages of diversity and similarity indices

Methods	Advantages	Disadvantages
Diversity indices	Usually simple to calculate	Diversity values vary according to the index used, sampling technique, location of sampling site and sometimes sample size
	Diversity expressed as single, quantitative value understandable to non-specialists	Interpretation of index values in relation to pollution levels not universally applicable
	No assumptions needed regarding pollution tolerance of species	Cannot distinguish between communities of pollution-tolerant or -intolerant communities
	Can be used equally with count or biomass data	Provide no information on the nature of the pollutants involved
	Species identification not necessarily required	The response of a community to pollution is not always linear and some types may increase diversity
		Relatively insensitive to anything other than extremes of pollution
Similarity indices	Usually simple to calculate	Values obtained vary according to index used
	Similarity expressed as single, quantitative value understandable to non-specialists	Affected by sample size and sometimes by species richness
	No assumptions needed regarding pollution tolerance of species	Require unpolluted site as control
		Cannot distinguish between communities of pollution-tolerant or -intolerant communities
	Species identification not necessarily required	Provide no information on the nature of the pollutants involved

12.2.2   Similarity indices

Similarity indices involve the comparison of species abundance at two different sample sites, one of which acts as a control. A number of different types have been used in pollution monitoring programmes, mainly in terrestrial environments and some of the most commonly adopted and their methods of calculation are shown in Table 12.2. As with diversity indices, their use has its advantages and disadvantages (see Table 12.3) and consequently their value has been much debated.

12.2.3   Species abundance patterns

It is not appropriate here to discuss the numerous species abundance models that have been formulated. The main advantage of their use over that of diversity and similarity indices resides in the fact that they focus on the overall pattern of diversity within a set of recorded abundance values, rather than attempting to condense all information into a single numerical value.

Despite this they have always played a minor role in pollution surveillance exercises.

12.2.4 Multivariate analyses

The lack of objectivity of the above methods with respect to influencing factors and the loss of information associated with the use of diversity and similarity indices are serious disadvantages to any pollution monitoring programme. These disadvantages can be overcome to some extent by the use of multivariate techniques for the analysis of data, although they are more complex to calculate and interpret. By grouping data in various different ways these techniques are able to detect discontinuities in communities from different sites. Thus, numerical classification of species data using any one of a number of cluster analysis methods can group together sites with similar taxonomic compositions, distinguishing them from other, less similar, groupings. Given the appropriate environmental data it may then be possible to link these differences to pollutant levels at the different sites. Even in the absence of any objective environmental data it should be at least possible to speculate and generate hypotheses to be tested in further work.

One such method of classification that has been used widely for biomonitoring purposes in recent years is Two-way indicator species analysis (TWINSPAN). In some instances it has been possible to generate simple indicator keys for use in the field using TWINSPAN (see Figure 12.1).

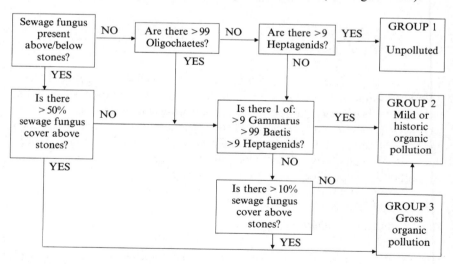

Figure 12.1 TWINSPAN key for assessment of organic pollution in streams. Source: Rutt, G.P., Pickering, T.D. and Reynolds, N.R.M. (1993) The impact of livestock-farming on Welsh streams: the development and testing of a rapid biological method for use in the assessment and control of organic pollution from farms. *Environmental Pollution* **81**, 217–228.

Ordination methods, used in conjunction with environmental data, are also very powerful ways of isolating influencing environmental factors on communities, especially when used in conjunction with cluster analyses. In the past, Principal components analysis (PCA) and reciprocal averaging/correspondence analysis (RA/CA) have been used for such purposes. However, these have tended to be replaced recently by Detrended correspondence analysis (DCA), which is an improved version of RA/CA, and requires the Fortran computer program DECORANA for its operation. DECORANA is often used in conjunction with TWINSPAN to determine the spatial affinities of groupings of species generated by the latter.

These make use of the presence or absence of certain individual indicator species or taxonomic groups known as *bioindicators* which, as their name suggests, are known to be sensitive to and respond in some detectable way to the presence of pollutants in their environment. Presence-absence (qualitative) and/or relative-absolute abundance data (quantitative), as well as physiological and morphological changes are used to ascertain the effects and distribution of pollutants in the sampling area.

12.3 Bioindicator methods

12.3.1   Biotic indices

Here data are used to assign a numerical value to individual bioindicators. The sum of the numerical values of all the bioindicator species at a given sampling site then provides an *index* value, of which there are numerous types, for the pollution at that site.

The earliest biotic index, the Saprobic system was developed in the early 1900s for use in aquatic ecosystems. Specifically, it was devised to monitor and classify the effects of organic matter in rivers and has been widely used in continental Europe. In this system four zones are recognised downstream from a major point source of discharge of organic matter:

1. the *polysaprobic zone*, extremely severe pollution;
2. the *α-mesosaprobic zone*, severe pollution;
3. the *β-mesosaprobic zone*, moderate pollution;
4. the *oligosaprobic zone*, very slight or no pollution.

The original Saprobic system was modified several times to provide a *Saprobic* or *Saprobien index* (S). By taking account of the frequency of occurrence of bioindicators and their habitat preferences a saprobic value can be obtained for each. The sum of the saprobic values for all bioindicators at a site divided by the sum of their frequency values enables a Saprobic index to be obtained for that site (Table 12.4).

Table 12.4   The Saprobic index

$$S = \frac{\sum (s \cdot h)}{\sum h}$$

where S is the Saprobic index for site; s is the Saprobic value for each indicator species; h is the frequency of occurrence of each species

s value	h value
1 Oligosaprobic	1 Occurring incidentally
2 β-mesosaprobic	2 Occurring frequently
3 α-mesosaprobic	3 Occurring abundantly
4 Polysaprobic	

Saprobic index ranges

1.0–1.5	Oligosaprobic	No pollution
1.5–2.5	β-mesosaprobic	Weak organic pollution
2.5–3.5	α-mesosaprobic	Strong organic pollution
3.5–4.0	Polysaprobic	Very strong organic pollution

The Saprobic system has been updated and refined in recent years to take account of the experience gained over the years since its introduction. The latest revision is now designated as a German Standard Method (DIN 38410 T.2) for use in water quality assessment in that country.

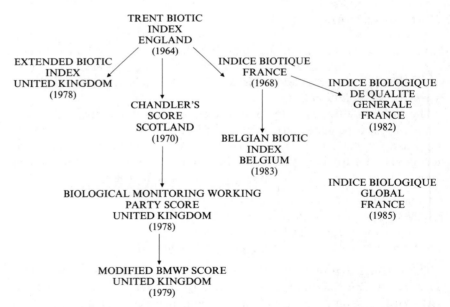

Figure 12.2 Development of important biotic index and score systems in Europe, in chronological order. Source: Metcalfe, J.L. (1989) Biological water quality assessment of running waters based on macroinvertebrate communities: history and present status in Europe. *Environmental Pollution* **60**, 101–139.

A whole range of other biotic indices has subsequently been developed since the Saprobic system was proposed, largely for use in monitoring the effects of organic matter in running waters and often for use in specific countries. Two of the most widely used of these in the UK were the *Trent Biotic Index* (TBI) and its derivative, the *Chandler Biotic Score* (CBS). The former which, as its name suggests, was originally devised for use in the River Trent in England, is based upon presence/absence data of groups of

Table 12.5 Allocation of BMWP score

Families	Score
Siphlonuridae, Heptageniidae, Leptophlebidae, Ephemerellidae, Potamanthidae, Ephemeridae	10
Taeniopterygidae, Leucridae, Capniidae, Periodidae, Periidae	
Chloroperlidae	
Aphelocheiridae	
Phryganeidae, Molannidae, Beraeidae, Odontoceridae, Leptoceridae, Gooeridae, Lepidostomatidae, Brachycentridae, Sericostomatidae	
Astacidae	8
Lestidae, Agriidse, Gomphidae, Cordulegasteridae, Aeshnidae	
Corduliidae, Libeliuidae	
Psychomyiidae, Philopotamidae	
Caenidae	7
Nemouridae	
Rhyacophilidae, Polycentropodidae, Limnephilidae	
Neritidae, Viviparidae, Ancylidae	6
Hydroptilidae	
Unionidae	
Corophiidae, Gammaridae	
Platycnemididae, Coenagriidae	
Mesovelidae, Hydrometridae, Gerridae, Nepidae, Naucoridae, Notonectidae, Pleidae, Corixidae	5
Haliplidae, Hygrobiidae, Dytiscidae, Gyrinidae, Hydrophilidae, Clambidae, Helodidae, Dryopidea, Elminthidae, Chrysomelidae, Curculionidae	
Hydropsychidae	
Tipulidae, Simuliidae	
Planariidae, Dendrocoelidae	
Baetidae	4
Sialidae	
Piscicolidae	
Valvatidae, Hydrobiidae, Lymnaeidae, Physidae, Planorbidae, Sphaeriidae	3
Glossiphoniidae, Hirudidae, Erpobdellidae	
Asellidae	
Chironomidae	2
Oligochaeta (whole class)	1

Source: National Water Council (1981) River Quality: The 1980 Survey and Future Outlook, National Water Council, London.

benthic invertebrates in relation to six key organisms found in the fauna at the sampling site. The latter was originally designed for use in Scottish upland rivers and, unlike the TBI, includes an abundance factor. Both systems have now been superseded by a range of derivatives of both (see Figure 12.2) in different countries.

In the UK, the most commonly adopted system is the *Biological Monitoring Working Party* (BMWP) *Score*. It involves a standard method of sampling at each site, which consists of disturbing the river bed by kicking for a 3-min period, while sweeping the area with a standard pond net to collect all macroinvertebrates. All species collected are identified to Family level only to accommodate wider use and eliminate problems associated

Table 12.6a   Caffrey macrophyte (higher plant) organic pollution index for Ireland

Sensitivity groupings	Macrophytes
Group A Sensitive forms	*Ranunculus* (not as below) *Callitriche hamalata*
Group B Less sensitive forms	*Ranunculus aquatilis* *Ranunculus peltatus* *Callitriche stagnalis* *Callitriche obtusangula* *Callitriche platycarpa* *Chara* spp. *Fontinalis antipyretica* *Potamogeton lucens* *Potamogeton obtusifolius* *Elodea canadensis* *Hippuris vulgaris* *Apium nodiflorum* *Rorippa nasturtium-aquaticum*
Group C Tolerant forms	*Zannichellia palustris* *Sparganium* spp. *Callitriche hermaphroditica* *Potamogeton crispus* *Potamogeton natans* *Potamogeton perfoliatus* *Nuphar lutea* *Lemna minor* *Lemna trisulca* *Enteromorpha* sp. *Scirpus lacustris* *Myriophyllum spicatum*
Group D Most tolerant forms (Pollution-favoured species)	*Potamogeton pectinatus* *Cladophora glomerata*

Source: Caffrey, J.M. (1986) Macrophytes as biological indicators of organic pollution in Irish rivers, in *Biological Indicators of Pollution*, ed. D.M.S. Richardson, Royal Irish Academy, Dublin, pp. 77–87.

Table 12.6b Index for water quality

Water quality class	Sensitivity grouping	Relative abundance
Q1: bad quality	Group A	Absent
	Group B	Absent
	Group C	Emergents sparse
	Group D	Dominant
Q2: poor quality	Group A	Absent
	Group B	Absent or sparse
	Group C	Abundant
	Group D	Dominant
Q3: doubtful quality	Group A	Absent
	Group B	Common
	Group C	Dominant
	Group D	Abundant
Q4: fair quality	Group A	Common
	Group B	Common or abundant
	Group C	Common
	Group D	Some algae
Q5: good quality	Group A	Dominant
	Group B	Abundant
	Group C	Sparse
	Group D	Absent

with misidentification and no account is taken of abundance. Each Family is then allocated a score, as listed in Table 12.5, with those regarded as least tolerant to pollution having the highest values. The individual scores are then summated to give the BMWP score. Another factor, the *Average Score Per Taxon* (ASPT) is now often calculated by dividing the BMWP score by the number of scoring taxa. Unlike the BMWP score, this has the advantage of being relatively independent of the sample size, season and even sampling method. A disadvantage of all these indices is that their use is confined to running waters and sometimes even to a particular habitat type within the river, such as riffles, i.e. areas of shallow, fast-flowing water running over a coarse pebble and gravel bed.

There are strong arguments for the use of macrophytes (higher plants) for the construction of water quality indices, not the least being the fact that they could be used in several different types of aquatic habitats. An example of such an index for use with organic pollution in Ireland is shown in Table 12.6.

In the USA, similar biotic indices have been used at various times but more widespread has been the use of the *Water Quality Index* (WQI) developed by the National Sanitation Foundation (NSF). This is not really a biotic index as it is computed using chemical and physical parameters, such as temperature and ammonia, as well as bacteriological data.

An example of a biotic index that has been developed for use in terrestrial ecosystems using lichens is the *Index of Air Purity* (IAP):

$$IAP = \Sigma_n^1 (Qf)/10 \qquad\qquad (12.1)$$

where n is the no. of species present at site; f is the frequency or cover of species; and Q is the mean no. of other species growing with these species in the area. The construction of an IAP requires information on the specific identification, numbers and frequency of epiphytic lichens in the sampling area.

All biotic index or score systems have their limitations and should be used in conjunction with other available data, such as chemical and physical parameters. However, they have the advantage of being simple to calculate and usually require minimal taxonomic expertise on the part of the operative. They also provide a single, easy to understand measure of pollution that can be used by non-specialists involved in decision-making on environmental quality issues.

12.3.2   Pollutant mapping

Another use of bioindicator organisms, for purposes other than the construction of biotic indices, is their use for the construction of species distribution maps to provide indications of pollutant distributions. Particularly well-documented examples of this include the use of a variety of plants and fungi to monitor air pollutants.

Lichens, which are composite organisms consisting of an alga(e) and a fungus living in a symbiotic relationship, are amongst the most sensitive of organisms to a variety of air pollutants, but especially to sulphur dioxide and fluorides. It has been known for over a century that lichen communities in urban and industrial areas are always impoverished and often absent. By plotting spatial and temporal variations in the distribution of species of lichen such as *Parmelia*, *Lecanora*, *Pleurococcus*, *Xanthoria* and others it has been possible to map the concentrations of some air pollutants both at local and even national level (see Figure 12.3). This is possible because information has been obtained on the tolerance ranges of specific lichen species to levels of some air pollutants, most notably sulphur dioxide (see Table 12.7). Lichen mapping is a very robust technique, in that while an expert can extract fine detail regarding certain aspects of air pollution from lichen distributions, much useful information can still be obtained by relative non-experts using cruder sampling techniques.

Mosses and microphytic fungi such as leafyeasts are also known to be sensitive to the effects of pollutants such as sulphur dioxide, ozone, fluorides and heavy metals and mapping of their distribution has been used to monitor some of these pollutants in both air and water. Comparatively little information is available regarding the dose-response relationship of most species therefore such studies are rarely quantitative in approach.

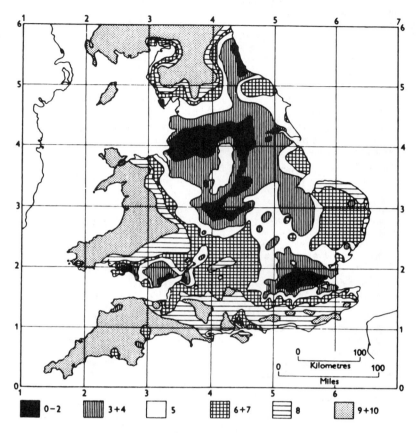

Figure 12.3 Approximate limits of lichen zones (see Table 12.7) in England and Wales. Source: Hawkes, D.L. and Rose, F. (1976) *Lichens as Pollution Monitors. Studies in Biology*, no. 66, Edward Arnold, London.

Vascular plants have also been employed to monitor pollutants through mapping studies, however, animals are rarely used for such purposes. An exception is the use of the Oribatid mite, *Humerobates rostrolamellatus*, which is known to be sensitive to sulphur dioxide. Groups of mites can be placed in containers at various sampling points over a selected area, left for a few weeks and then retrieved for a mortality count which is directly related to sulphur dioxide levels in the atmosphere.

12.3.3 Morphological and histological indicators

Presence-absence and abundance data for species are not the only measurement parameters adopted when using bioindicators. For plants these include

Table 12.7 Zone scale for the estimation of mean winter sulphur dioxide levels in England and Wales using corticolous lichens

Zone	Moderately acid bark	Basic or nutrient-enriched bark	Mean winter SO$_2$ (μg/m^3)
0	Epiphytes absent	Epiphytes absent	?
1	*Pleurococcus viridis* s.l. present but confined to the base	*Pleurococcus viridis* s.l. extends up the trunk	>170
2	*Pleurococcus viridis* s.l. extends up the trunk; *Lecanora conizaeoides* present but confined to the bases	*Lecanora conizaeoides* abundant; *L. expallens* occurs occasionally on the bases	About 150
3	*Lecanora conizaeoides* extends up the trunk; *Lepraria incana* becomes frequent on the bases	*Lecanora expallens* and *Buellia punctata* abundant; *B. canescens* appears	About 125
4	*Hypogymnia physodes* and/or *Parmelia saxatilis*, or *P. sulcata* appear on the bases but do not extend up the trunks. *Lecidea scalaris*, *Lecanora expallens* and *Chaenotheca ferruginea* often present	*Buellia canescens* common; *Physica adscendens* and *Xanthoria parietina* appear on the bases; *Physicia tribacia* appear in S	About 70
5	*Hypogymnia physodes* or *P. saxatalis* extends up the trunk to 2.5 m or more; *P. glabratula*, *P. subrudecta*, *Parmeliopsis ambigua* and *Lecanora chlarotera* appear; *Calicium viride*, *Lepraria candelaris* and *Pertusaria amara* may occur; *Ramalina farinacea* and *Evernia prunastri* if present largely confined to the bases; *Platismatia glauca* may be present on horizontal branches	*Physconia grisea*, *P. farrea*, *Buellia alboatra*, *Physcia orbicularis*, *P. tenella*, *Ramalina farinacea*, *Haematomma ochroleucum* var. *porphyrium*, *Schismatomma decolorans*, *Xanthoria candelaria*, *Opegrapha varia* and *O. vulgata* appear; *Buellia canescens* and *X. parietina* common; *Parmelia acetabulum* appear in E	About 60
6	*P. caperata* present at least on the base; rich in species of *Pertusaria* (e.g. *P. albescens*, *P. hymenea*) and *Parmelia* (e.g. *P. revoluta* (except in NE), *P. tiliacea*, *P. exasperatula* (in N)); *Graphis elegans* appearing; *Pseudevernia furfuracea* and *Alectoria fuscescens* present in upland areas	*Pertusaria albescens*, *Physconia pulverulenta*, *Physciopsis adglutinata*, *Arthopyrenia gemmata*, *Caloplaca luteoalba*, *Xanthoria polycarpa* and *Lecania cyrtella* appear; *Physconia grisea*, *Physcia orbicularis*, *Opegrapha varia* and *O. vulgata* became abundant	About 50
7	*Parmelia caperata*, *P. revoluta* (except in NE), *P. tiliacea*, *P. exasperatula* (in N) extend up the trunk; *Usnea subfloridana*, *Pertusaria hemisphaerica*, *Rinodina roboris* (in S) and *Arthonia impolita* (in E) appear	*Physcia aipolia*, *Anaptychia ciliaris*, *Bacidia rubella*, *Ramalina fastigiata*, *Candelaria concolor* and *Arthopyrenia biformis* appear	About 40
8	*Usnea ceratina*, *Parmelia perlata* or *P. reticulata* (S and W) appear; *Rinodina roboris* extends up the trunk (in S); *Normandina pulchella* and *U. rubiginea* (in S) usually present	*Physcia aipolia* abundant; *Anaptychia ciliaris* occurs in fruit; *Parmelia perlata*, *P. reticulata* (in S and W), *Gyalecta flotowii*, *Ramalina obtusata*, *R. pollinaria* and *Desmazieria evernioides* appear	About 35

Table 12.7 *Cont'd.*

Zone	Moderately acid bark	Basic or nutrient-enriched bark	Mean winter SO$_2$ (μg/m^3)
9	*Lobaria pulmonaria, L. amplissima, Pachyphiale cornea, Dimerella lutea,* or *Usnea florida* present; if these absent crustose flora well developed with often more than 25 species on larger well lit trees	*Ramalina calicaris, R. fraxinea, R. subfarinacea, Physcia leptalea, Caloplaca aurantiaca* and *C. cerina* appear	Under 30
10	*L. amplissima, L. scrobiculata, Sticta limbata, Pannaria* spp., *Usnea articulata, U. filipendulla* or *Teloschistes flavicans* present to locally abundant	As 9	'Pure'

Source: Hawkes, D. L. and Rose, F. (1976) *Lichens as Pollution Monitors. Studies in Biology*, no. 66, Edward Arnold, London.

absolute and relative growth rates, fresh/dry weight, root/shoot and leaf/ weight ratios, leaf area indices and numerous other morphological features. In animals, age/size or weight ratios, reproductive rates, functional asymmetry (uneven growth on one side of the body compared to the other) and other non-pathological morphological changes can provide tell-tale and measurable signs of pollutant-induced stress. A good example of this is the phenomenon of so-called *imposex* in the response of the dogwhelk (*Nucella lapillus*) to the presence in seawater of the anti-fouling compound tributyltin leached from marine paints. The presence of tributyl tin causes the development of a penis in the female whelk, the size and frequency of occurrence of which can be related to the degree of pollution by this compound.

In other instances these stresses manifest themselves as even more obvious histological or morphological changes. Thus, visible damage to plants, such as chlorosis or necrosis, has long been used to assess pollution. Such damage was first noticed in association with crop plants because of the economic implications, but in recent years it has come to prominence in the monitoring of the effects of acid rain, particularly in Central Europe (see Table 12.8), where the specific disease/stress symptoms can be related to the impact of acid precipitation in different areas. Included here is the infamous *Waldsterban* (literally 'death of trees') observed in large parts of German forests since 1980. Sometimes, certain susceptible varieties of plant are used to detect the presence of some air pollutants as, for example, through the evaluation of foliar injury to the ozone sensitive variety, Bel-W3, of the tobacco plant *Nicotiana tabacum*.

In animals the appearance of pathological features such as ulcers, tumours, inflammation, necrosis, parasitic infection, etc. can sometimes be linked to the appearance of pollutants. However, because the diagnosis of such changes usually requires very specialist knowledge and it is often difficult to distinguish between the normal occurrence of a disease and that

Table 12.8   Visible effects of acid rain on trees

Type of tree	Effect
Coniferous species	Yellowing needles (leaves)
	Loss of needles
	Longevity of needles shortened
	Distorted branches
	Crown (top of canopy) dieback
	Bark damage
	Root damage
Deciduous species	Leaf discoloration
	Leaf deformation
	Premature leaf loss
	Crown dieback
	Bark damage

induced by pollutants, these methods are not yet routinely used for biomonitoring purposes.

12.3.4   Detector and sentinel organisms

These are other variations in the use of pollution sensitive organisms. Detector organism refers to the monitoring of certain individual, indigenous species which exhibit a measurable response to pollution and therefore act as 'early warning' devices of the presence of pollutants in the environment. Great care has to be taken when choosing a particular species, taking into consideration its sensitivity and likelihood of exposure to the pollutant(s) concerned, its position in the community, its ecological and geographical distribution and its abundance. However, provided these factors are correctly taken into account, it is often possible to select an organism that can be used to monitor a specific pollutant, or group of pollutants, rather than pollution generally as do many other methods.

Sentinel organisms are sensitive species deliberately introduced into an environment, where they may not normally be found, to act as biological early warning systems or to delimit the extent of pollution. A good example of this is the development and use of fish alarms, in which disturbances of the normal physiology of pollution-sensitive fish species are taken to indicate a decline in water quality. The physiological parameters measured include changes in breathing and heart rates, swimming movements and in the case of the Elephant-nosed Mormyrid, the pattern of navigational pulses emitted by its electric organ.

12.3.5   Comparative methods

More recent developments in bioindicator methods, for freshwater aquatic systems at least, have retained the advantage of a single value as a pollution measure. In evaluating the extent of pollution of a water body in the United States it is becoming normal practice to first assign it to a particular ecotype and then to use established procedures, involving a set of 'metrics' (selected measurements), to compare it biologically with an undisturbed (pristine) water body of the same type. The result is expressed as the percentage similarity of the sites with respect to the sum of the metric values at each site. High percentage scores, i.e. 90% or more, indicate a relatively unpolluted site, while successively lower percentages are indicative of increasingly lower water qualities.

A similar approach is adopted in the recently proposed River Invertebrate Prediction and Classification System (RIVPACS) programme, developed by the Institute of Freshwater Ecology, for monitoring water quality in the UK. Using this programme, it is possible to compare the fauna at any running-water site under investigation with a 438-site database and therefore to obtain some measure of the environmental stress at that site, including that caused by pollution.

12.4 Microbiological monitoring

Microorganisms are vital components of all ecosystems whose diversity and abundance is greatly influenced by a wide range of environmental factors and changes in these can be indicative of certain types of pollution. For example, the presence of the characteristic 'sewage fungus' community of microorganisms (see Figure 12.4) in water bodies is indicative of pollution by high levels of organic matter. What is important to realise here is that the individual species in this community are indigenous inhabitants of many freshwater bodies, but it is their aggregation in an attached macroscopic growth that is important in this context. Conversely, the presence of so-called *intrusive* species in an ecosystem may also act as an indication of pollution. Thus, the presence of microorganisms in aquatic systems,

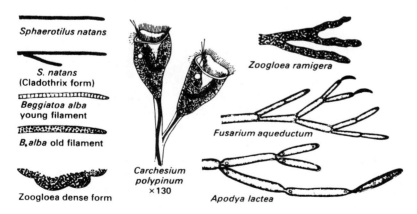

Figure 12.4 Selected characteristic species in the sewage fungus community. Source: Hynes, H.B.N. (1960) *The Biology of Polluted Waters*, Liverpool University Press, Liverpool.

particularly bacteria that normally inhabit the intestines of humans and animals, is a clear indicator of sewage pollution of water from these sources.

12.4.1   Intrusive microorganisms and faecal contamination

The presence of a number of microorganisms and invertebrate parasites in fresh and coastal waters pose a public health hazard. It is essential therefore, that where these waters are used by the public, especially for drinking purposes, they be free of these disease organisms (pathogens). Direct detection of pathogens can be an extremely expensive and time consuming business as they are often present intermittently or in low numbers. To overcome these problems cheaper and more rapid techniques have been devised to detect the presence and numbers of more common, non-pathogenic members of intestinal flora. The rationale underlying these tests is that the presence of these microorganisms is evidence of the contamination of water by faecal matter and that faecal pathogens could be present. Conversely, their absence is an acceptable, although not absolutely certain, indication that the water is safe for use by humans. These indicator organisms must satisfy the following criteria if they are to be used in such a way:

1. they must always be present when pathogens are present and must occur in greater numbers;
2. their rate of proliferation in an aquatic environment should not be greater than that of pathogens;
3. they must be more resistant to environmental stress, including disinfection, than pathogens;
4. they should be easily and unambiguously identifiable using simple and rapid techniques;
5. they should grow independently of other organisms present on artificial media.

Coliform bacteria, faecal streptococci and *Clostridium perfringens* are the most commonly used indicator organisms for faecal contamination of water. While a total count of coliforms is of some value in this context, some species are non-faecal in origin and can proliferate in water and therefore do not satisfy the above criteria. Far more satisfactory has been the use of the faecal coliform *Escherichia coli* which occurs in vast numbers as a natural inhabitant of the human intestine and whose presence is therefore a clear indication of contamination by human faeces.

Faecal streptococci can also be used as general indicators of faecal contamination of waters, but their real value lies perhaps in their use to determine possible source and temporal aspects of contamination. Faecal streptococci, although more resistant to disinfection, tend to die off more rapidly in water than do coliforms so that their presence is indicative of more recent contamination. It is also possible to distinguish between streptococci from humans and other animals such as pigs and cattle and thereby distinguish between these as the sources of contamination. Further, it has been suggested that the ratio of faecal coliforms to faecal streptococci can give some indication of the source of contamination. For example, a ratio of less than 0.6 may be indicative of storm water discharges as sources, while a ratio of greater than 4 is indicative of municipal wastes.

The prolonged longevity of the anaerobic, sulphite-reducing bacterium *Clostridium perfringens* in most water bodies can be of equivocal value as an indicator of faecal contamination. In some circumstances their persistence long after a contamination incident and possible long distance transport away from their point of origin can render their use as indicator organisms of little value. However, where transport to the laboratory is likely to be delayed beyond 12 h or toxic substances are present in the water, then their use as an alternative or supplement to faecal coliforms is recommended.

Pathogenic organisms such as *Vibrio*, *Shigella* or *Salmonella* are normally only sought when there is reason to suspect that a water body is contaminated by them. Otherwise, the elaborate and relatively expensive identification tests do not lend themselves to routine use.

Many different types of enteric viruses can be present in human faecal matter and it is possible that they may be transmitted at much lower levels than bacterial pathogens. A means of monitoring them would therefore be desirable. However, their very low level of occurrence necessitates the collection of very large water samples and, as with bacterial pathogens, methods for their detection and identification are complex and expensive and therefore not suited to routine use.

Special techniques have to be adopted when working with intrusive microorganisms, not only for the production of accurate and repeatable results, but also for safety. Samples for microbiological (bacteriological) analysis must be taken using sterile glass or plastic (non-toxic) containers with caps. Care must be taken by the sampler to avoid contamination of the sample, both during and after sampling, most notably by adopting sterile techniques wherever practicable. Analysis of samples should ideally be carried out within an hour of collection since the normal processes of reproduction, growth, death and predation will continue in the sample even after collection. Samples should never be frozen but can be stored at 4°C for up to 24–30 h if necessary. Where a disinfectant is known or suspected to be present, it should be neutralised.

Two main methods can be used for counting bacterial populations in collected water samples. They may be counted from a fixed and stained sample or else in a counting chamber, such as a haemocytometer. These methods take no account of the viability of bacterial cells. Alternatively, a total count of viable bacterial cells may be made using a variety of methods, including the growth of bacterial colonies on selected media and the development of turbidity in liquid media. However, the most commonly adopted techniques, especially for *E. coli* and other coliforms, are (i) the membrane filtration method and (ii) the multiple tube technique. In the former, a measured water sample is passed through a 0.45 µm pore diameter membrane filter which is then removed from the filter apparatus and placed face upwards on selective media formulated to encourage the growth of the desired enteric bacteria only. Individual bacteria trapped on the filter then grow into visible colonies that can be counted. The typical incubation temperature of 37°C also encourages the growth of these bacteria as opposed to an optimal range of 15–25°C characteristic for indigenous water bacteria. In the multiple tube analysis a dilution series of the water sample is established and incubated before positive and negative growth reactions are observed. On the basis of the results obtained an estimate of the numbers of bacteria in the original sample can be made.

12.5 Bioaccumulators

Many living organisms accumulate pollutants within their tissues through the process of *bioaccumulation*, whereby contaminants enter their tissues at rates greater than they can be excreted. Pollutants may be absorbed across the general body surface, through specialised structures such as lungs, gills, leaves, roots, etc. or, in the case of animals, ingested along with their normal food. Indeed, the transfer of pollutants through food chains can lead to their substantial build-up at the higher levels, such that concentration factors in the tissues can be 10^3–10^6 times that in the environment. This phenomenon is known as *biomagnification*. Such accumulation can occur throughout the lifetime of the organism without apparent adverse effects. Alternatively, the latter may manifest themselves only when certain critical levels are reached in some or all tissues or when the organism is under stress.

This ability to concentrate contaminants can be used to good effect in pollution surveillance programmes. Many pollutants may be present in water, air, soil, etc. at levels below or close to the detection limits of many chemical analytical methods. In contrast, tissues that have accumulated pollutants in the manner described above exhibit levels of pollutants often well within typical analytical detection limits. Also, the analysis of discrete samples of air, water, etc. provide only a record of the levels of pollutant present at the time they were taken, whereas those observed in a bioaccumulator organism will reflect the ambient levels present over a prolonged period of time. Furthermore, many chemical analytical methods provide information on the total amount of a given pollutant in the environment, not all chemical species of which may be available to and therefore affect living organisms. The pollutant residues present in the tissues of living organisms are clearly those available.

Not all organisms can be employed as biomonitors in this fashion. To do so they must satisfy the following criteria:

1. they must be easily identifiable;
2. they must be relatively abundant and representative of the environment in which they occur;
3. they should accumulate the pollutant(s) of interest to levels which will permit direct analysis without lethal effect;
4. there should be a clearly defined relationship between the concentration of pollutant in the tissues of the organism and those in its environment, at all sites where it occurs;
5. they should be large enough to provide adequate tissue for reliable analyses;
6. they should be easy to maintain under laboratory conditions to allow uptake experiments under controlled conditions to be undertaken;
7. they should be sedentary, so that the levels in the tissues reflect those of the environment in which they were collected;
8. they should be ubiquitous in their distribution to allow comparisons to be made between geographically separated, but similar, environments.

In reality, no single species can satisfy all these criteria and a compromise inevitably has to be reached in the choice of a species to be used as a bioaccumulator in pollution monitoring programmes.

There are two different strategies through which bioaccumulators can be exploited. *Passive biomonitoring* refers to the situation where indigenous organisms are collected from particular habitats for chemical analysis of their tissues. *Active biomonitoring* involves the placement and deliberate exposure of organisms, usually obtained from an unpolluted area, in a suspected polluted environment with a view to subsequent analysis of the tissues after what is considered to be an appropriate period. An advantage of using active biomonitoring methods is the ability to use organisms from the same genetic stock at more than one site. This reduces the potential significance to any study of differences in the rate(s) of pollutant uptake at different sites due to any genetic variability between two or more populations of the same species.

A wide range of species has been used as biomonitors, some of which are normal food items for humans and therefore have this added relevance for pollution studies. In aquatic ecosystems particular emphasis has been placed on the use of macrophytes and molluscs as bioaccumulators as they satisfy many of the criteria listed above. The use of mussels of the marine genus *Mytilus* is especially noteworthy in this context as they are the subject of a worldwide scheme known as the 'Mussel Watch'. *Mytilus* species have a very wide distribution throughout the world and Mussel Watch provides information on the levels of a range of pollutants in their tissues from widely separated locations. Fish have also been extensively used as bioaccumulators in fresh and saline waters, partly because of their use as a major food item by humans. However, their general mobility does mean that they are of limited use in pinpointing specific sources of contamination.

In terrestrial environments lichens, mosses and vascular plants have been utilised as bioaccumulators. The use of mosses perhaps deserves special mention here. It is known that living mosses accumulate metals and they have been used in both active and passive biomonitoring programmes. However, the greatest use of mosses in pollution studies has been through the deployment of so-called 'moss bags'. These are typically nylon bags filled with dead, dried moss (normally sphagnum), which are then exposed to the atmosphere for a period of time before being collected for analysis. They are particularly utilised for the monitoring of heavy metals, such as zinc, cadmium, lead, etc. as they act as very effective ion exchangers for these elements. While this technique has proved to be very effective in many studies on air pollution, it is clearly not a case of true bioaccumulation as no biological processes are involved. Indeed, artificial materials, such as plastics and resins can be employed to perform the same task.

While the great majority of bioaccumulation studies using animals have involved the use of lower vertebrates and macroinvertebrates, some have been carried out on birds and mammals, the latter including humans. However, difficulties, some of which may be of a legal or ethical nature, associated with the sampling of such relatively large organisms do impose limitations on their usefulness for such a purpose.

Despite the advantages of using bioaccumulators in pollution monitoring programmes discussed above, there are a number of important factors that must be taken into account when planning their use and when interpreting any results obtained. Biological organisms inevitably exhibit genetic variability which manifests itself as differences in morphology, physiology and biochemistry amongst individuals. This variability can be quite substantial, even amongst members of the same interbreeding population, and it may be reflected in differences in the way in which individual organisms accumulate pollutants. Age or size may also play a significant role in the rate(s) at which an organism accumulates a particular pollutant and must also be taken into account in any monitoring scheme. It is often recommended that individuals of similar size and/or age be used for analysis. Seasonal differences in accumulation may also occur and these are often related to the sex of the individual. In many organisms, especially invertebrates, the fully developed gonads represent a substantial proportion of the total body weight which may vary between the sexes and with the time of the year. For females in particular, the material and energy investment in gamete production can be enormous and can leave them vulnerable to other environmental stresses, such as those caused by the presence of pollutants. Thus, during the breeding season they may be less able to regulate their uptake of these pollutants which therefore leads to an increase in their tissue levels. Conversely, if pollutants tend to accumulate preferentially in the gonads, the release of gametes may substantially and suddenly reduce the total body burden of individuals.

It is because of these possible complications that the ability to work in the laboratory with a potential biomonitor is important. Here the dynamics of the uptake and loss of selected pollutants by the chosen species can be examined under controlled conditions. It may lead to their rejection as a biomonitor, the selection of single tissue rather than whole body analysis or some other conclusion.

Superimposed on the inherent sources of variability discussed above are those attributable to the microenvironment in which an individual lives. While environments can be characterised and classified on a broad scale, considerable variation can be present on a small scale. Thus, significant differences can occur on two adjacent beaches that have a slightly different aspect, slope, etc. Even on the same beach, a multitude of microenvironments will exist that may influence the uptake of pollutants by organisms living within them in many different ways.

The same considerations for obtaining representative and meaningful samples apply to the collection of tissues for analysis as for the calculation of the various biological indices discussed above. Particular care must be taken with samples for chemical analysis that no contamination occurs as a result of contact with the sampling apparatus, extraneous materials such as debris, storage devices, analytical equipment or any chemicals used.

Dependent on the nature of the analytical technique to be used, fresh or dried tissues may be used for analysis. The advantage of the latter is that they can be stored for prolonged periods prior to analysis and provide more accurate quantitative data, as the water content of tissues can vary enormously. Whatever the method of analysis employed, tissue samples invariably have to be subjected to a solubilisation or extraction process as they cannot be analysed directly.

A variety of methods exists for the solubilisation of tissues, usually employing oxidising mineral acids and/or other oxidising agents. Acids typically employed include nitric, sulphuric and perchloric, separately or in a great variety of combinations. Hydrogen peroxide is sometimes used in conjunction with acids.

Fresh or dried tissues may be solubilised directly in such digestion mixtures, a process known as wet-ashing. However, in some instances they may be dry-ashed at 500–550°C in a muffle furnace before treatment of the inorganic residue by acid digestion or fusion techniques in preparation for analysis. Where losses of volatile components may occur, tissues may be dry-ashed at less than 200°C in a special apparatus.

Not only is there considerable variety in the digestion medium and temperatures that can be employed for tissue or ash solubilisation but also in the apparatus in which the digestion is carried out. At its simplest, digestion may take place in heated, open tubes or else a reflux apparatus may be employed. The effectiveness of digestion is increased at elevated temperatures and pressures. To this end, samples may be sealed in digestion bombs and heated up to 275°C at pressures as high as 5000 psi. Alternatively, they may be heated up to 250°C at 1200 psi in special digestion vessels in a microwave oven. Although very accurate, these methods are expensive, time-consuming and, on a routine basis, suitable only for relatively small organic samples. The recent development of flow injection techniques in conjunction with microwave digestion will offset some of these disadvantages. Digestion procedures are used extensively for the analysis of inorganic contaminants, especially heavy metals, in tissues.

The analysis of organic pollutants, such as petroleum hydrocarbons, polychlorinated biphenols (PCBs) and pesticides requires a different approach other than solubilisation. Such molecules would not withstand attack by mineral acids or oxidising agents generally. Instead, tissues are subjected to extraction procedures employing organic solvents such as dichloromethane, methanol, pentane and hexane. Samples are best homogenised prior to extraction to break up tissues and cells and to release the contents of the latter. The homogenate may then be freeze-dried or mixed with anhydrous sodium sulphate to bind the water and then extracted using a Soxhlet or Kuderna-Danish apparatus to leach out the pollutants of interest from the tissue. Subsequently it may be necessary to employ clean-up techniques to remove potential interfering substances, such as triglycerides, from the leachate. This can be achieved by passing the crude extract through a simple column apparatus containing silica and/or activated alumina or through the use of gel-permeation chromatography.

Whatever the method of digestion or extraction employed it is essential that its efficacy be verified if it is to be of any value. This is accomplished by subjecting certified standard reference materials, of known composition and ideally of a similar nature to the sample material, to the same digestion or extraction procedure and determining the yield obtained. A range of certified standard reference biological materials is now available for this purpose.

12.6 Bioassays   This refers to the use of biological organisms under controlled conditions to determine both the short term impacts of large doses of pollutants (*acute effects*) and the long term impacts of low levels (*chronic effects*). As well as being used to study the modes of action and routes of transport of pollutants through ecosystems, the results of such studies can also be used to:

1. determine the potential impact of single pollutants or mixtures of pollutants on individuals, populations and communities;
2. determine a variety of toxic thresholds relating to their lethal and sub-lethal effects;
3. where appropriate, to determine whether a pollutant(s) is within regulatory standards;
4. contribute to the development of ameliorative measures to combat pollution;

5. ascertain the sensitivity of particular organisms to specific pollutants;
6. provide early warning of a potentially damaging pollution incident.

Numerous bioassays has been developed for use in the laboratory and in the field. A very wide range of organisms has been used, ranging from specific strains of bacteria to monitor the carcinogenicity of air or water pollutants to the use of canaries to monitor carbon monoxide in the atmosphere of coal mines. Ideally, the choice of an organism for use in a bioassay should be made with reference to the following criteria:

1. it must be sensitive to and consistent in its response to the pollutant or effect under investigation;
2. it should be widely distributed and abundant throughout the year;
3. it should have widespread importance ecologically, economically or recreationally;
4. it should be in good health and not prone to disease or parasites;
5. it should be easily maintained in the laboratory, have low genetic variability and there should be ample background data available on its biology.

In practice many of the organisms employed for bioassays meet only some of these criteria. For reasons of cost and speed, small organisms with short life cycles are often favoured. Many of the tests conducted are single-species tests despite the fact that it has generally been recognised that the comparative advantages of multi-species testing outweigh any disadvantages (Table 12.9).

Table 12.9   A comparison of the relative advantages of single-species versus multi-species bioassay testing

Bioassay method	Comparative advantages
Single-species testing	Better defined response end points Greater sensitivity to pollutant stress Less inherent variability in test system More sensitive to response parameter measured
Multi-species testing	Provide more information per test More discriminating in isolating key influencing factors Greater relevance to situation in natural ecosystems

The types of parameters measured in bioassay organisms to ascertain any effects induced by the presence of a pollutant(s) are very variable and include changes in behaviour, morphology, histology, physiology and biochemistry as summarised in Table 12.10.

Some of the tests are standard measures adopted worldwide, for example, the use of biochemical oxygen demand (BOD) and standard bottle tests to assess the polluting potential of organic effluents in water. The BOD test is a measure of the ability of a known volume of water to consume oxygen, as a result of microbial decomposition, at a temperature of 18°C over a 5-day period

Table 12.10 Summary of principal methods used for bioassays

Type of organism	Method of assay
Bacteria, fungi, protozoa	Mutagenicity BOD Nitrification studies Decomposition studies
Algae and other plants	Biostimulation and growth rates Reproductive rates Photosynthetic rate Respiratory rate Chlorophyll content Mutagenicity Morphological and histological effects
Invertebrates and vertebrates	Lethal effects Reproductive rates Development abnormalities Growth rates Feeding rates Respiratory rates Biochemical changes Morphological and histological effects Behavioural changes

in the dark (to prevent oxygen evolution through photosynthesis). The greater the amount of decomposable organic matter there is in the water sample, the greater the amount of oxygen there will be consumed (see chapter 18). Standard bottle tests involve the measurement of the growth of microscopic algae, such as *Selenastrum capricornutum*, in a sample held under standard conditions. Microorganisms are also employed during the widely adopted Ames Test to determine the mutagenicity of potential pollutants. Here, a mutant strain of the bacterium *Salmonella typhimurium* is exposed to the substance under investigation. Reversion of bacterial cultures to the unmutated form is taken as being indicative of the mutagenic potential of any suspect substance. Another test dependent on the principle of a reverse mutation makes use of the exposure of a mutant dark (non-luminescent) strain of the bacterium *Photobacterium phosphoreum* to the test sample. The restoration of luminescence in the bacterium is taken as a positive result.

Routinely adopted bioassays are not confined to using microorganisms as the test species. The water flea, *Daphnia* is used for a variety of water quality tests in which the lethal effects of possible pollutants are determined or else the sublethal effects on the growth or reproductive rates of laboratory populations are measured.

All the aforementioned tests and others can be carried out during routine surveillance programmes. For reasons of cost, complexity or time, other tests do not lend themselves so readily to routine measurement and are best reserved for basic surveys of areas and occasions where very specific pollutants and their effects are being investigated. Such tests include the monitoring of changes in behavioural activity, respiratory or other physio-logical stress, assays of enzyme activities in specific tissues of selected organisms, testicular dysfunctions in animals to name but a few.

It is recommended that the results of bioassays normally be interpreted in conjunction with other available data, especially that gathered from the field and including any relating to the prevailing chemical and physical conditions.

Part II
Specific applications

Speciation 13
G.L. Christie

Legislation governing the maximum permissible levels of a polluting element in an environmental sample such as river water, refers to total concentrations rather than the chemical form of that element. However, this total concentration provides no information concerning the fate of the element in terms of its interaction with sediments, its ability to cross biological membranes (*bioavailability*), or its resultant toxicity. Changes in *speciation* may dramatically affect the toxicity of a metal. For example, inorganic mercury species are generally unable to cross biological membranes and thus have low toxicity, but alkyl mercury species are lipid soluble and hence extremely toxic to aquatic organisms. Thus in order to assess the environmental impact of an element we must have some information concerning its chemical form (speciation).

13.1 The importance of speciation

Speciation may be defined as the different physico-chemical forms of an element which together comprise its total concentration in a given sample.

13.2 Definition of speciation

There are two main approaches to the evaluation of trace metal speciation: *experimental measurement* and *computer modelling*.

13.3 The determination of trace metal speciation

13.3.1 Computer modelling

The experimental determination of speciation is hampered by a number of problems. Sensitivity presents a major challenge since environmental samples are usually complex in nature and there may be very many species present, many of them at concentrations below the detection level of the technique employed. Although the concentrations of such species may be very low, they may constitute a significant contribution to the bioavailable, and therefore toxic, fraction of the metal. Hence it is important that they should not be overlooked. Moreover, there is the problem of disruption of the equilibrium of the system. In order to be energetically favourable, systems in nature should approximate to a *thermodynamic equilibrium*. Thus, the introduction of any analytical probe into the system may disrupt this labile equilibrium and thereby alter the very distribution one is trying to monitor. In order to circumvent these problems encountered in the experimental determination of speciation, computer modelling was introduced.

The determination of speciation using computer modelling is based on the assumption that the system under investigation is in a state of thermodynamic equilibrium. The system is then classified in terms of *components* (i.e. the chemical components present such as H^+, CO_3^{2-}, Ca^{2+} etc.) and *species* (i.e. the chemical species which may form from these components such as $CaCO_3$).

The model requires as input a complete breakdown of the chemical composition of the sample, i.e. the analytical concentrations of all of the components present together with the pH and redox potential of the system under investigation. Each program has associated with it a database of *thermodynamic stability constants*. These stability constants provide a measure of the extent to which a given complex may form. For the interaction of a metal M and a ligand L to form an ML complex, e.g.

$$Ca^{2+} + CO_3^{2-} \rightleftharpoons CaCO_3 \qquad (13.1)$$

the stability of the complex is quantified as the equilibrium concentration of the complex relative to the uncomplexed metal and ligand; i.e. for

$$M + L \rightleftharpoons ML \qquad (13.2)$$

the stability constant K is given by

$$K = \frac{\{ML\}}{\{M\}\{L\}} \qquad (13.3)$$

where { } relates to activities. In practice stability constants are measured in dilute solutions of constant ionic strength. Hence activities may be approximated to concentrations such that

$$K = \frac{[ML]}{[M][L]} \qquad (13.4)$$

The computer programs generally use stability constant data relating to the overall rather than the stepwise formation of species (see chapter 4). Thus for the combination of the metal with two ligands to form an ML_2 species, i.e. for the reaction

$$M + 2L \rightleftharpoons ML_2 \qquad (13.5)$$

the stability constant is expressed as

$$\beta = \frac{[ML_2]}{[M][L]^2} \qquad (13.6)$$

The first step in the calculation is the selection of all the relevant stability constants (i.e. for all the possible species based on the components present) from this database. By a series of mass balance equations, the computer then calculates the concentrations of all of the species present in the specified solution.

It is helpful to consider a very simple example, the speciation of a $CaCO_3$ solution, 0.001 mol dm^{-3} with no access to atmospheric gases. This particular case has served as a classic case for a wide variety of equilibrium models. Considering the simplest of all situations, a simple aqueous solution, that is, no solid species form, no redox reaction occurs and there are no adsorbent phases present. Such a situation will involve the consideration of three components, H^+, Ca^{2+}, and CO_3^{2-}. These components may combine to form the additional species $CaOH^+$, $CaCO_3^0$, $CaHCO_3^+$, H_2CO_3 and HCO_3^-. The equilibrium reactions involved and the associated stability constants (at 25°C and zero ionic strength) are shown in Table 13.1.

Table 13.1 Speciation in a solution of $CaCO_3$

Species	K	log K
$H_2O - H^+ \rightleftharpoons OH^-$	$\dfrac{[OH^-][H^+]}{[H_2O]}$	-14.0
$CO_3^{2-} + H^+ \rightleftharpoons HCO_3^-$	$\dfrac{[HCO_3^-]}{[CO_3^{2-}][H^+]}$	10.2
$CO_3^{2-} + 2H^+ \rightleftharpoons H_2CO_3$	$\dfrac{[H_2CO_3]}{[CO_3^{2-}][H^+]^2}$	16.5
$Ca^{2+} + H_2O - H^+ \rightleftharpoons CaOH^+$	$\dfrac{[CaOH^+][H^+]}{[Ca^{2+}][H_2O]}$	-12.2
$Ca^{2+} + CO_3^{2-} + H^+ \rightleftharpoons CaHCO_3^+$	$\dfrac{[CaHCO_3^+]}{[Ca^{2+}][CO_3^{2-}][H^+]}$	11.6
$Ca^{2+} + CO_3^{2-} \rightleftharpoons CaCO_3^0$	$\dfrac{[CaCO_3^0]}{[Ca^{2+}][CO_3^{2-}]}$	3.0

The equilibrium speciation profile may then be computed by solving the appropriate mass balance equations. The resulting distributions for calcium and carbonate (calculated using the computer program MINTEQA2) at pH 6, 9 and 12 are shown in Tables 13.2 and 13.3.

Table 13.2 Percentage distribution of calcium at pH 6, 9 and 12

Species	% Total metal		
	pH 6	pH 9	pH 12
$Ca^{2+}_{(aq)}$	100	95	61
$CaCO_{3(aq)}$	0	5	28
CaOH	0	0	11

Table 13.3 Percentage distribution of carbonate at pH 6, 9 and 12

Species	% of total carbonate		
	pH 6	pH 9	pH 12
CO_3^{2-}	0	5	71
H_2CO_3	32	0	0
HCO_3^-	68	90	1
$CaCO_{3(aq)}$	0	5	28

These results illustrate the effect of pH on the speciation of this simple carbonate solution. At pH 6, the calcium is present entirely as the free (aquated) metal ion and carbonate as its protonated forms (H_2CO_3 and HCO_3^-). As the pH is increased the ligand is deprotonated and calcium is able to compete with protons for the binding sites on the carbonate ligand. By pH 12, some 28% of the metal is present as $CaCO_3$. In addition, at high pH hydrolysis occurs with the formation of CaOH.

This example illustrates the type of information obtained from speciation calculations. The output data usually takes the form of these percentage distribution tables for each of the components present.

A wide range of sophisticated codes is available for such calculations. These codes differ in their mathematical method of calculation and also in their levels of sophistication; the simplest codes merely calculating the speciation of an aqueous solution. More elaborate codes have the ability to handle solid species; some have additional facilities such as the modelling of adsorption phenomena and the simulation of the mixing of two solutions.

Modelling of solids.   Solid species are incorporated into the model in the form of *solubility products*. Data relating to all possible solid species which may be present are contained within the thermodynamic database. If during the calculation the solubility product of a given solid is exceeded, the solid in question is specified as precipitating. Many computer codes have further levels of sophistication in that species may be specified as being present as solids which are not allowed to dissolve, as solids which are not allowed to precipitate even though the K_{sp} may be exceeded (i.e. a supersaturated solution), or may be excluded from the calculation altogether. This provides some crude allowance for kinetics in that a saturated species which is kinetically unlikely to precipitate may be excluded from the calculations.

Modelling of adsorption phenomena.   Many models also allow for *adsorption* phenomena to be considered. The inclusion of adsorption reactions is important since such surface reactions (e.g. the adsorption of trace metals on an iron oxide surface) may result in the removal of a toxic species from solution and thus affect the speciation of the system. Generally the approach is to treat the adsorption reaction as the formation of a 'species'; the adsorption site is treated as one component and the adsorbing species as the other component. The interaction of the adsorbing component and the adsorption site is then quantified by a pseudo-stability constant. Programs such as MINTEQA2 have further levels of sophistication in that they incorporate a variety of models for surface reactions including Freundlich, Langmuir and ion exchange models. The major drawback to this approach is the lack of reliable stability constant data characterising these adsorption reactions.

Modelling of complexation by humic and fulvic acids.   The complexation of metal ions by simple organic ligands such as citrate, oxalate, EDTA, etc. can easily be modelled since such interactions are usually well-characterised and reliable stability constant data are readily available. However, the interaction of metal ions with the macromolecular organic ligands such as *humic* and *fulvic* acids, which are of prime importance in modelling environmental samples, is more problematic. These potential ligands have high molecular weights (500–1500 for fulvic acids, 1500–300 000 for humic acids) and have several different types of functional groups. Furthermore, the functional groups cannot be considered independently of each other since these molecules develop a surface charge which produces electrostatic interactions, affecting the acidity of the functional groups. Thus, it is impossible to characterise such molecules in the same way that simple organic ligands may be treated; one cannot measure protonation and metal–ligand stability constants for humic and fulvic acids. However, modelling approaches have been developed which allow the effect of the presence of these organic molecules on metal–ligand speciation to be considered.

One such approach is the 'discrete functional group model'. In this case, humic and fulvic acids are considered to behave like simple organic ligands and are treated as an assemblage of non-interacting simple organic ligands each with a well-defined protonation and metal–ligand stability constant. In a more sophisticated version of this model, the computer generates a random assemblage of these ligands; analysis of C, H, N and O, the molecular weight and known functional groups are used to generate a random molecular structure that complies with the analytical data. Models have also been developed which allow for some consideration of the interaction between metal binding sites on these macromolecular organic ligands. The development of an electrostatic charge on humic and fulvic acids leads to a decrease in the tendency of protons to dissociate from functional groups. The 'electrostatic discrete functional group models' attempt to incorporate the effects of such phenomena into the model by the introduction of an electrostatic correction factor into the protonation and metal-ligand formation constants.

Drawbacks to computer modelling.   The main shortcoming to the computational approach is the reliability of the data used. The term GIGO (garbage in, garbage out) is often used in reference to computer modelling. The results of the model are only as good as the parameters on which they depend and any shortcomings in either the analytical concentrations of the system under investigation or in the thermodynamic data will result in a very misleading speciation profile. However, the major advantage of this approach is that changes to the model can easily be made and thus the effect of say a 10% error in one of the component concentrations can very readily be investigated. Similarly, stability constant data may be varied over a range in order to see if this has any effect on the speciation profile. This type of operation also serves to identify those species which are of major importance (i.e. those which occur at significant concentrations). It is recommended that the source data relating to the stability constants for these species be carefully checked and, if necessary, re-measured.

13.3.2   Experimental determination of speciation

The overriding problems associated with the experimental determination of speciation are sensitivity and the disruption of the labile equilibria. *Sensitivity* is a particular problem with environmental samples where the concentrations of the individual species are very low indeed and in many cases below the detection limit of the technique employed. Thus in many cases investigating the speciation of the system experimentally provides a '*fractionation*' profile (i.e. splitting the total metal concentration into free and complexed fractions) rather than a complete breakdown of the individual species present. All experimental procedures are invasive and thus likely to disrupt the thermodynamic equilibrium of the system. The nature of this disruption may vary with the technique used and hence the speciation profile obtained is to some extent method dependent. In view of this it is important to consider any speciation profile in conjunction with the method of determination.

Anodic stripping voltammetry (ASV).   This is undoubtedly the most widely used experimental technique in the determination of trace metal speciation. For details of the theory of this technique the reader is referred to chapter 10. The application of this technique to the determination of trace metal speciation is based on the fact that it allows you to differentiate between labile (i.e. weakly complexed) and total metal. The electrochemical response represents the *ASV-labile* metal fraction and reflects the hydrated metal ion content together with contributions from any metal complexes present which rapidly dissociate in the diffusion layer:

$$ML_n \underset{k_1}{\overset{k_{-1}}{\rightleftharpoons}} M + nL \qquad (13.7)$$

The amount of metal deposited at the electrode during deposition depends on the rate of dissociation (k_{-1}) of the metal complex ML_n.

The contribution of the trace metal to the ASV peak depends only on the rate of dissociation and the thickness of the diffusion layer (the thicker the layer, the greater the contribution to the ASV peak height). The ASV labile fraction is determined at the natural sample pH or after pH adjustment with buffer. The total metal concentration may be determined by ASV after first treating the sample with either nitric-perchloric acid, digestion with persulphate, or UV irradiation. The ASV labile metal fraction is generally less than the total metal and is considered to represent the bioavailable fraction of the metal.

The electrode systems usually used are a hanging mercury drop (HMDE) or a thin mercury film (MFE) on a glassy carbon support. Such systems give a sensitivity of 10^{-10} mol dm^{-3} for Cu, Cd, Pb and Zn and allow simultaneous determination of all four metals.

There are a number of additional problems associated with use of ASV in trace metal determination. The presence of surface active compounds such as humic and fulvic acids or organic pollutants may lead to interference by their adsorption onto the mercury electrode surface. These may shift the peak potential or alter the ASV peak height. In addition, adsorption-desorption processes at the mercury electrode can lead to the production of tensammetric waves which may be mistaken for metal-ion waves and thus give rise to anomalously high results. Hence, it is necessary to pretreat the sample with activated charcoal or subject it to UV irradiation (which destroys metal–organic compounds without affecting inorganic metal species) to prevent such interference. A further problem is the fact that analysis must be done at relatively high ionic strength (0.02 mol dm^{-3}) which necessitates the addition of electrolyte to most environmental samples.

Ion exchange. Ion-exchange methods have been widely used for trace metal speciation. This experimental approach has the attraction that it avoids the problems of contamination associated with any of the alternative techniques. The profile obtained does, however, vary considerably depending on the exchange resin used. A commonly used ion-exchange resin is Chelex-100 which complexes metal ions via the nitrogen and oxygen donor atoms of the iminoacetate functional groups. This resin binds ionic metal strongly, but the pore size (1.5 nm) means that colloidal particles are not retained by the column. Thus, such systems provide a fractionation between ionic and colloidal metal. However, although this loosely corresponds to the toxic and non-toxic fractions of the metal, the binding of heavy metals to biological membranes is thought to involve sulphur donor groups and not nitrogen and oxygen groups. Thus Chelex-100 may not be the most suitable material for the speciation of heavy metal ions. The use of thiol resins may provide a more realistic fractionation of the metal. Conventional cation- and anion-exchange resins have also been used in the study of metal speciation. Sequential passage of the sample through anion- and cation-exchange resins facilitates the fractionation into anion-exchangeable, cation-exchangeable and neutral (i.e. not retained by either column) metal.

Ultrafiltration.   Ultrafiltration using a membrane of known pore size may be used to fractionate trace metals in environmental samples. The method involves the application of a constant pressure of inert gas to a filtration cell, which contains the sample, an agitation mechanism and a supported membrane disc. A range of membranes (usually a thin film of polymeric hydrous gel supported on a porous polyethylene or cellulose ester matrix) with pore size 1–15 nm are available. These allow the separation of colloidal and non-colloidal metal. This method only provides a very limited fractionation corresponding to the molecular weight cutoff (200; 500; 10 000; 50 000; 100 000 or 300 000) of the membrane used. A series of fractionations could be achieved by using several membranes of different pore size. If a correlation between pore size and bioavailability can be established, this type of fractionation may provide some useful information with respect to toxicity. However, this approach does suffer from a number of drawbacks. Contamination (the filters themselves are often contaminated with trace metals and potential complexing agents) and loss of sample (adsorption onto the membrane surface may be significant and since the ratio of membrane surface to sample size is high, significant sample losses may occur) are serious problems. In addition, the change in equilibrium resulting from the ultrafiltration process may lead to the dissociation of colloidal complexes.

Dialysis.   This method is based on the same principles as ultrafiltration, employing a dialysis membrane (made of cellulose acetate, collodion or gelatine) of known pore size (typically 1–5 nm) facilitating separation of colloidal and non-colloidal material. Fractionation of the species is based on their differential rates of diffusion through the pores of the dialysis membrane. The dialysis cell, usually containing water, is surrounded by the sample under investigation, e.g. a dialysis bag containing 100 cm^3 of pure water is suspended in the river under investigation for a period of several weeks. Movement of metal species across the membrane occurs as a result of the concentration gradient. The problems associated with ultrafiltration also apply to this method with the added disadvantage that it is a much slower process; it generally takes in excess of 24 h for equilibrium to be attained. A modified procedure whereby the metal species in the dialysis chamber are continuously removed by adsorption onto a suitable material (e.g. via passage through a column containing a suitable chelating resin) may significantly speed up the process and thus provide a more effective separation.

Electrophoresis.   Both *flat bed* and *capillary electrophoresis* may be used to separate species on the basis of the magnitude and sign of their charge and, to a lesser extent, their size.

In flat bed electrophoresis, an electrical potential is applied across a strip of a porous conducting polymer. The ends of the strip are immersed in a buffer solution to control the pH and conductance of the system. The sample is applied to the centre line of the electrophoresis bed. Species migrate under the applied potential according to their sign (anions and cations in opposite directions and neutral species remaining close to the point of application), for example hydrated metal cations move strongly towards the cathode.

High performance capillary electrophoresis (HPCE) utilises a length of fused silica capillary tubing in place of the polymer bed. The potential applied between the ends of the tube affords a separation of charged species as in the flat bed system. Ultraviolet, fluorescence or mass spectrometry detection systems are used. The latter affords much more information regarding the nature of the species.

Ion-selective electrodes.   Ion-selective electrodes may be used to determine the concentration of the free (aquated) metal ion since when immersed in a test solution these probes develop a potential proportional to the log of the activity of a specific hydrated ion. Commercial electrodes are available for the detection of anions (e.g. for halides, NO_3^-, CN^-, SCN^-, S^{2-}); cations (e.g. for H^+, Na^+, K^+, Ca^{2+}, Cd^{2+}, Cu^{2+}, Pb^{2+}), and gases (e.g. for NH_3, O_2, CO_2, NO_2). However, their detection limits (10^{-6}–10^{-7} mol dm^{-3}) mean their usage is usually restricted to the measurement of the relatively higher concentrations of species occurring in polluted waters. In addition, interference due to the presence of ions other than the one being monitored may present a major problem.

Ion-selective electrodes may be used to distinguish between 'free' and 'complexed' metal, making measurement before and after the addition of an agent to promote the dissociation of complexes containing the element under investigation.

Coupled techniques.   This is a hybrid approach using at least two analytical techniques: the first to separate the species and the second to identify and quantify it. The two techniques must be coupled by an interface and hence the detection system must be compatible with the separation process. The separation is usually achieved by some form of chromatography, either gas chromatography (GC), high performance liquid chromatography (HPLC) or supercritical fluid chromatography. GC provides superior separation, especially if capillary columns are used, but the sample must be volatile and thermally stable. Mercury, tin and lead alkyl compounds can be directly determined by GC. In addition, species which contain Sn, Pb, Hg, As, Sb,

Bi, Se or Te may be separated from the matrix by conversion into chemically stable and volatile hydrides (e.g. by treatment with $NaBH_4$) which can then be separated by GC. HPLC is usually used for non-volatile compounds and supercritical fluid chromatography is a relatively new technique which is particularly useful for volatile compounds that thermally degrade.

Although the technology exists to use HPLC to separate metal species, conventional detectors (refractive index, ultraviolet, fluorescence, electrochemical, flame ionisation and infrared) generally lack the sensitivity, selectivity and applicability required. An alternative approach is to use atomic spectroscopy as a detection system. In this way the separatory power of chromatographic techniques such as HPLC and GC is coupled with the sensitivity and selectivity of atomic spectroscopy for detection. Various coupled systems have been developed including *gas chromatography-atomic absorption spectroscopy* (GC-AAS); *gas chromatography-microwave induced plasma atomic emission spectroscopy* (GC-MIP-AES) and *high performance liquid chromatography-atomic absorption spectrometry* (HPLC-AES). However, the coupling of HPLC or GC with inductively coupled plasma-mass spectrometry (GC-ICP-MS or HPLC-ICP-MS) probably represents the best way forward in this field, generating the most information concerning the nature of the various species present and its high sensitivity means sample preparation steps are simplified and pre-concentration is not usually required. Such systems are usually based on quadruple mass spectrometry. The ICP operating at atmospheric pressure and optimised for ion detection is placed on its side facing a sampling cone. The mass spectrometer operates at reduced pressure, hence a two- or three-stage differentially pumped interface is needed to transfer ions from the plasma to the mass spectrometer. Such systems have been used to detect organolead compounds at a level of 0.7 pg s^{-1}.

The use of atomic spectroscopy as a detection system has the added advantage that it is element specific. Hence, if two species co-elute, only that which contains the element of interest will be detected. Thus a much less effective chromatographic separation may be tolerated.

The use of coupled techniques is not without problems, primarily concerning the interface in the form of sample nebulisation and transport. HPLC can be coupled to ICP relatively easily via a short length of tubing (the length must be kept to a minimum in order to maintain resolution). However, since the ICP nebuliser is very ineffective, only 1–5% of the analyte is transferred to the plasma in the absence of further modifications (giving detection levels of only 100 μg dm^{-3}). Ultrasound nebulisers, thermospray vaporisers, glass frit nebulisers and direct injection nebulisation have all been investigated in an effort to improve detection levels. The latter method appears to provide the best results but is difficult to use because of its fragility. A further problem is the low tolerance of atomic spectroscopic techniques to commonly used HPLC solvents. This may be overcome to some extent by decreasing the solvent loading using microbore HPLC or cooling the nebulisation spray chamber, or the use of a special plasma torch which can operate with low argon flow rates. The use of capillary columns in GC facilitates the use of low mobile phase flow rates (typically 0.5 ml min^{-1}) which is advantageous with respect to interfacing with the detection system (a low flow rate is desirable to increase the residence time in the detection system and hence the sensitivity). The main disadvantage of this type of column is that it can only cope with a small sample size and therefore sensitivity is limited. Various adaptations to the detection systems have been made to further increase the residence time, e.g. the use of a flame heated alumina tube in AAS detection.

Other methods.   A number of other techniques are available which have some application to the experimental investigation of speciation.

Liquid-liquid extraction may be useful in the separation of neutral and charged species. This approach involves shaking an aqueous sample with an organic solvent, e.g. octanol or hexane. Neutral species will be extracted into the organic layer but charged complexes remain in the aqueous phase.

NMR spectroscopy may be used to investigate metal speciation either by looking at the NMR spectra of the metal nuclei or magnetic nuclei in the ligands. In the latter case, proton and ^{13}C NMR are commonly used but ^{31}P, ^{14}N, ^{15}N and ^{19}F are also applicable. However, sensitivity generally precludes the use of NMR spectroscopy in the study of speciation in environmental samples.

Gel chromatography may be useful in the study of metal speciation in polluted waters or in sewage, affording a separation on the basis of molecular size. However, the relatively small sample size and large eluent volume preclude its use in systems where metal concentrations are low.

Radiotracer studies are inapplicable to speciation work because of the long equilibration times (days, months or even years) between the ionic tracer and non-ionic species.

The choice of technique for the investigation of trace metal speciation depends on a number of factors. A prime requirement of any speciation method is that it should involve little sample manipulation and offer the minimum opportunity for sample contamination. Other considerations include the concentration of the metals in the sample (many of the techniques discussed will be inapplicable if the levels are very low); the complexity of the system (in particular, the presence of substances which will interfere with the method of detection used) and the level of information required (the more straightforward methods provide only a crude fractionation profile). The more sophisticated techniques such as HPLC-ICP-MS may provide greater sensitivity and provide more information regarding the nature of the species present but they are very expensive to set up and require a greater degree of analytical expertise.

13.4 Concluding remarks

The most important factor to bear in mind when considering the results from any experimental study of speciation is that they are method-dependent. Moreover, any perturbation of the labile equilibrium during the investigation will mean that the results do not necessarily reflect the speciation profile in the original sample. Hence all results should be considered in conjunction with the methods used to obtain them.

Computer modelling does not suffer from the problems associated with the experimental investigations but here again the results must be interpreted with care. The results of computer modelling studies are wholly dependent on the data used in the calculations. Thus, it is of vital importance to have accurate data regarding the chemical composition of the system (this will inevitably involve chemical analysis and many of the problems associated with experimental study of speciation may apply). Moreover, the thermodynamic stability constant data must be carefully reviewed. Pivotal formation constants should be critically assessed for their accuracy, and if necessary re-determined. The big advantage of the computational approach is that, unlike the tedious and time-consuming experimental investigations of speciation, it is very quick and can easily be repeated. Hence it is very straightforward to examine the consequences of inaccuracies in the input data by repeating the calculations, systematically varying the input data over an appropriate range.

The ideal approach to the determination of speciation is probably to use a combination of approaches. Computer modelling is most effective when used in conjunction with experimental studies. Initial investigations into the speciation of very simple, well characterised systems is required in order to validate the computer model (i.e. there is close agreement between the experimentally determined and computed speciation profiles). The model may then be used to investigate the speciation of more complex systems. A comparison of the information from this approach and that obtained using one of the experimental techniques may provide the most information about the speciation of the system.

Further reading   Ure, A.J. and Davidson, C.M. (Eds.) (1995) *Chemical Speciation in the Environment*, Blackie Academic & Professional, Glasgow.

Morrison, G.M.P., Batley, G.E. and Florence, T.M. (1989) Metal speciation and toxicity, *Chemistry in Britain* **August**, 791.

Jenne, E.A. (Ed.) (1979) *Chemical Modelling in Aqueous Systems*, ACS Symposium Series 93, Am. Chem. Soc., Washington, DC.

Florence, T.M. (1982) Speciation of trace elements in water, *Analyst* **29**, 345.

The analysis of atmospheric samples 14

C.K. Laird

The atmosphere is a mixture of three principal gases, nitrogen, oxygen and argon, which together make up more than 99.9% of dry air. The atmosphere also contains variable amounts of water vapour and a large number of minor and trace gaseous components and many aerosol and particulate species. Environmental effects are associated with changes in concentration of one or more of the minor components or with the presence of specific pollutants. Some atmospheric components of environmental interest and typical concentration ranges in tropospheric air are given in Table 14.1.

Table 14.1 Some compounds of environmental importance and typical concentration ranges in tropospheric air

Carbon dioxide	344 ppm
Ozone	<1 to >100 ppb
Carbon monoxide	<100 ppb to >20 ppm
Methane	1.68 ppm
Non-methane hydrocarbons	1 ppt to >1 ppb
Nitric oxide	5 ppt to 1 ppb
Nitrogen dioxide	<1 to >150 ppb
Nitrous oxide	310 ppb
Sulphur dioxide	<1 to >100 ppb
$CFCl_3$ (Freon 11)	200 ppt
CF_2Cl_2 (Freon 12)	350 ppt

The values given are volume concentrations or mixing ratios, where 1 ppm is 1 volume in 10^6, 1 ppb is 1 in 10^9, etc. Provided that none of the components of a gas mixture condenses over the range involved, volume concentrations are independent of temperature and pressure. It is therefore convenient to calibrate gas analysers and measure atmospheric concentrations as mixing ratios. However, environmental effects are often related to the mass concentration of pollutant, and concentrations may alternatively be expressed as mass per unit volume (usually mg m^{-3}) of air, under specified conditions of temperature and pressure. Concentrations of aerosols and particulate materials are expressed as mass per unit volume of air, again under specified conditions.

14.2 Atmospheric analyses

Most analyses of atmospheric samples come into one of three general categories: measurements to monitor one or more components of ambient air; measurements of emissions from industrial or other processes; and measurements of specific toxic substances, particularly in indoor or workplace atmospheres. The purpose of the measurements may influence the choice of analytical technique.

14.2.1 Measurements of atmospheric composition

The composition of the atmosphere is subject to short-term and long-term changes. For example, the global concentrations of several species such as methane, nitrous oxide and carbon dioxide, are increasing at rates up to about 0.8% per year. Ozone in urban air varies on a daily cycle, while stratospheric ozone varies seasonally and is also subject to a long-term decline. Measurements are needed to monitor these changes, and measurements of specific atmospheric components are required for verification or refinement of theoretical models of atmospheric physics and chemistry. In each case, the analytical system must be capable of producing an accurate and precise record of the concentration of the target species. It may be required to operate for long periods in remote locations or under adverse conditions for example in mobile sampling platforms such as vehicles, balloons or aircraft. As atmospheric models are refined and the role of minor components is elucidated, it has become necessary to measure increasingly labile species, at increasingly small concentrations.

14.2.2 Emission measurements

Many industrial plants and other processes are required to operate under licences which limit the type and amount of material which may be emitted to the atmosphere. Emission measurements are then necessary to control the plant and to ensure compliance with legal limits. Limits may be set for single processes or for a plant such as an oil refinery. Measurements may be made in the gas leaving stacks or exhausts, or remote monitoring techniques may be used to make integrated measurements of emissions from an entire plant. Pollutants are very rapidly diluted once they leave an exhaust or stack. The concentration range for measurement of a given pollutant in a stack is therefore likely to be several orders of magnitude higher than for measurement of the pollutant in ambient air.

14.2.3 Indoor and workplace atmospheres

For monitoring of indoor and workplace atmospheres it is important to ensure either that the concentration of specific pollutants, or that the exposure of personnel over a given time period, does not exceed specified limits. The analyser may be required to be self-contained, and small enough to be worn by an operator. Indication of excessive concentrations by audible or visual alarm may be more important than recording the actual concentration of the pollutant.

14.3.1 Gas chromatography

14.3 Techniques for gas analysis

Gas chromatography (GC) is one of the most powerful and versatile techniques for separation of the components of mixtures of gases or volatile liquids. The mechanism for GC separation of gaseous mixtures usually involves selective adsorption of the components of interest on solids such as alumina, silica gel, molecular sieves or proprietary adsorbents such as Tenax, Porapak or various 'carbon molecular sieves' (gas-solid chromatography, GSC). Porous layer open tubular (PLOT) capillary columns are available with a variety of adsorbent coatings which are suitable for GSC separation of mixtures of permanent gases and vapours, but precise injection of the small volumes of gas sample required for operation of these columns is difficult, and many atmospheric analyses are carried out using packed (3 mm or 6 mm OD) GC columns. Conventional GC detectors such as the flame ionisation detector (FID) and electron capture detector (ECD) are used, but GC separation may also be combined with detector systems developed to measure low concentrations of specific atmospheric components.

14.3.2 Spectrometric methods

Spectrometric methods are widely used for gas analysis, and the availability of high-power light sources, particularly lasers, together with requirements to measure labile and trace components in the atmosphere has led to the development of a variety of novel analytical techniques.

Infrared and ultraviolet photometric analysers. All gases with a covalent bond, except simple non-polar diatomics such as O_2, N_2 or Cl_2, have characteristic absorption spectra in the infrared region, and many gases also absorb in the ultraviolet. The absorption of radiation is governed by the Beer–Lambert law (chapter 8). Absorption depends on the number of absorbing molecules in the light path and is therefore sensitive to changes in temperature and pressure. Certain gas molecules emit radiation after stimulation to an excited electronic state. Fluorescence is the production of radiation by stimulation by higher-energy incident radiation. When the molecules are stimulated to emit radiation by a chemical reaction the process is known as chemiluminescence.

Both absorption and emission of radiation can be used for gas analysis. Analysers based on emission of radiation by fluorescence or chemiluminescence are inherently more sensitive than the absorption type. However, not all the potential sensitivity may be achieved in practice as some of the excited molecules may lose energy by non-radiative processes, and particularly by collisions with other molecules (quenching). Quenching may be minimised by operation at reduced pressure but this affects the complexity and cost of the analyser.

An infrared or ultraviolet absorption gas analyser consists of a radiation source such as a heated wire (IR), or a lamp or gas discharge tube (UV), a cell to contain the sample and fitted with transparent windows through which the light beam passes, and a detector. The analyser is made specific to a particular gas by selection of radiation in a wavelength region where the gas of interest is known to absorb. Wavelength selection may be either dispersive, where broad-band radiation is passed through the sample and the measurement wavelength is selected after dispersion of the emergent beam by a grating or prism, or non-dispersive, where the measurement wavelength is selected before the light is passed through the sample. Either a narrow band radiation source of appropriate wavelength, or a broad-band source with a suitable filter, may be used. Most gas analysers are of the non-dispersive type (NDIR or NDUV). Although some non-dispersive analysers incorporate continuously variable filters to allow measurement of several gases, the majority are constructed with filters selected for measurement of only one gas. Analysis of atmospheric samples usually involves measurement of low concentrations of specific components of complex mixtures and suitable analysers commonly incorporate refinements to improve the sensitivity and selectivity of the measurement. A typical dual-beam non-dispersive infrared (NDIR) analyser is shown in Figure 14.1.

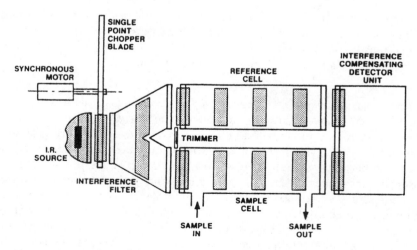

Figure 14.1 Dual-beam non-dispersive infrared gas analyser (Rotork Analysis).

Infrared radiation from the source is passed through an interference filter which selects the analytical wavelength corresponding to an absorption band of the molecule to be measured. The radiation is split into two beams of equal intensity. A rotating chopper allows the beams to pass intermittently but simultaneously through sample and reference cells. The detector consists of two sealed absorption chambers separated by a thin metal diaphragm. The diaphragm, with an adjacent perforated metal plate, forms an electrical capacitor. The chambers are filled with the gas to be detected so that energy characteristic of the gas to be measured is selectively absorbed.

The reference cell contains a non-absorbing gas. When the sample cell is also filled with non-absorbing gas, equal energy enters both sides of the detector. The presence of the component to be measured in the sample cell causes some of the radiation in the beam passing through that cell to be absorbed. This results in an imbalance in the energy entering the two sides of the detector, causing the diaphragm to deflect and so change the capacitance. The resulting AC electrical signal is processed to give the concentration of the species to be measured. NDIR or NDUV analysers with sample cells with 100–200 mm path lengths can typically measure down to ppm levels, but sensitivities may be improved by use of multiple pass sample cells to give effective path lengths of 100 m or more.

Fourier-transform spectrometry. Fourier-transform spectrometry (chapter 8) is a technique for non-dispersive measurement of absorption spectra. It is most commonly applied in the infrared region (Fourier-transform infrared, FTIR). In principle, FTIR spectrometry is widely applicable to atmospheric measurements and has several advantages over dispersive spectral measurement techniques. However, FTIR spectrometers are optically and mechanically complex, and FTIR spectrometry is used where its properties can be most fully exploited. In FTIR spectrometry the entire frequency range is monitored simultaneously. Several components of an atmosphere can therefore be measured together, and FTIR spectrometers have been mounted on space vehicles and telescopes for remote sensing of earth and planetary atmospheres. Spectral scan rates for a given resolution can be higher in FTIR than dispersive infrared spectrometry. The availability of rapid scan rates makes FTIR spectrometry suitable for reaction kinetics studies and enables FTIR spectrometers to be used as GC detectors.

Correlation spectrometry. Correlation spectrometry is a non-dispersive measurement technique in which molecules of the gas to be analysed are themselves used to specify the analytical wavelengths. In a gas filter correlation spectrometer (Figure 14.2) two similar cells, one containing the gas to be analysed, and the other a non-absorbing gas, are placed alternately

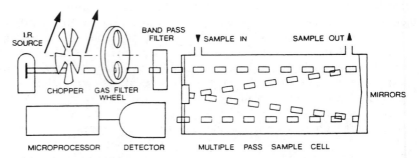

Figure 14.2 Gas filter correlation analyser.

in the light beam. The concentration of gas in the filter is set so that the light is strongly absorbed at the wavelengths of its absorption spectrum. The rotating filter wheel then acts as a chopper which modulates the radiation received by the detector only at wavelengths corresponding to the line absorption spectrum of the specifying gas. The sample gas is passed through a sample cell placed in the beam as shown. The presence of the specifying gas in the sample causes additional absorption in the radiation received by the detector. However, because radiation is strongly absorbed by the gas filter at the specifying wavelengths, the signal change due to the sample is larger when the non-absorbing cell is in the beam than when the specifying cell is in the beam. Thus the presence of the specifying gas in the sample cell results in a decrease in the amplitude of the modulated signal. The gas filter technique allows correlation with the fine line spectrum of the specifying gas. It enables measurements to be made in the presence of species such as water vapour which has a particularly complex infrared spectrum.

Tunable diode laser spectrometry. Tunable diode laser spectrometry is similar in principle to conventional spectrometry in that absorption of infrared radiation by the species of interest is measured. However, the diode laser used as the light source has a very narrow line width (of the order of 10^{-4} cm^{-1}), enabling individual vibration lines of the target molecule to be resolved. This increases the sensitivity and selectivity of the technique. The laser wavelength may be 'tuned' over a range of about 30–100 cm^{-1} by adjustment of the diode temperature or (more usually) current. The laser wavelength can also be modulated at a frequency of several kilohertz. Modulation enables electronic signal processing techniques to be used to stabilise the system and improve the signal-to-noise ratio. Absorptions as small as 1 part in 10^5 can typically be measured. With the sample in a multipass, 100-m path-length cell, this corresponds to concentrations less than 1 ppb for most molecules.

Disadvantages of the technique are the cost and complexity of the equipment, and the need for highly detailed spectral information on the determinand before the analytical wavelength can be selected. Although diode lasers of different chemical composition are available to cover the entire mid-infrared wavelength range from about 3 to 30 µm, obtaining a suitable diode for a particular application can be difficult.

LIDAR and DIAL. LIDAR (light detection and ranging) is the optical analogue of RADAR. A pulse of light from a high-power laser is emitted into the atmosphere. The light is scattered by aerosols (Mie scattering) or molecules (Rayleigh scattering) along the light path, with the Mie effect being the dominant scattering mechanism in the atmosphere. A proportion of the light is reflected back along the transmittance path and the reflected pulse is received by a telescope and measured by a photodetector. The backscattered light contains information on the position and nature of scattering and absorbing species along the light path. The position of absorbing species can be calculated by measuring the time taken for the pulse to return to the detector. To obtain information on the chemical composition of the atmosphere the LIDAR system is usually operated in a differential mode, known as DIAL (differential LIDAR, Figure 14.3). Laser light pulses at two adjacent wavelengths are emitted into the atmosphere, one wavelength being chosen to correspond to an absorption line of the target molecule, the other is not absorbed by the target species. Scattering effects vary only weakly with wavelength and are assumed to be the same for both. Signals for the two wavelengths returned from parts of the atmosphere not containing the target species are therefore identical. However, one wavelength is selectively absorbed by the target molecules, resulting in differences in the signal returned at the two wavelengths from the parts of the atmosphere along the transmittance path containing the target pollutant. The signal can be processed to give range-resolved information on the concentration of the target pollutant.

Because scattering effects are weak, DIAL is usually used at UV and visible wavelengths where spectra are simpler and absorption coefficients are higher than in the infrared region. It has been used for near-field monitoring of emissions of NO, NO_2, SO_2, O_3, CO_2 and some organic molecules. DIAL measurements in the 2.9–3.5-µm region have been used for monitoring loss rates of gaseous hydrocarbons from refineries and petrochemical plants, and a ship-borne DIAL system has been used to monitor hydrocarbon vapour losses from oil tankers loading at sea. Quantitative measurements of atmospheric gas concentrations may be made at ranges up to a few kilometres, with a range resolution of 10 m. The equipment is complex but the ability to operate remotely and to provide range-resolved measurements of pollutant concentrations is valuable.

DOAS and COSPEC. In differential absorption spectrometry measurements are made at adjacent wavelengths, respectively on and off an

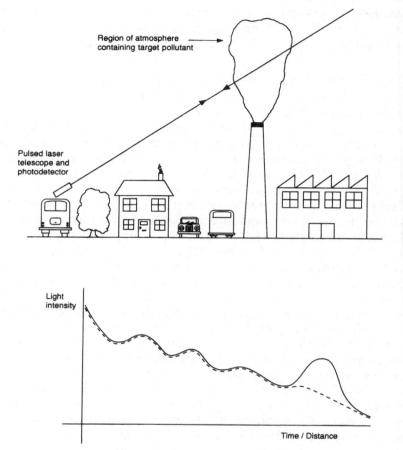

Region of atmosphere containing target pollutant

Pulsed laser telescope and photodetector

Light intensity

Time / Distance

Figure 14.3 Range-resolved remote sensing by differential LIDAR (DIAL). (——) λ_1 not absorbed by target pollutant; (– – –) λ_2 absorbed by target pollutant.

absorption band of the target molecule. The output signal is a function of the ratio of the intensities of the measurements and is thus stabilised against drift in the detector and opacity changes in the sample. For atmospheric measurements using the differential absorption technique (DOAS), high power light sources, such as tunable dye lasers or broadband sources such as xenon lamps are used. Path lengths may be up to 10 km in clean air or 500 m in heavily polluted air. The technique is extremely sensitive for molecules with high absorption coefficients, for example ppt levels of NO_2 have been measured.

In atmospheric measurements by correlation spectrometry (COSPEC) the sun is used as the light source, and sky emission is measured from the ground with a high-resolution spectrometer. Pollutants in the atmosphere absorb radiation and modify the spectrum, enabling their concentration to be determined. In practice, because of interference from water vapour in the infrared, COSPEC is limited to the visible and UV regions. It has been used for measurement of SO_2 and NO_2, and is useful for *in situ* measurements of plume concentrations within a few kilometres of an emission source. A disadvantage of the technique is the need to correct for sensitivity changes or baseline drift caused by changes in the skylight spectrum or intensity. Detection limits are in the ppm range.

Photoacoustic spectrometry (PAS).   In photoacoustic spectrometry (Figure 14.4) the gas sample is contained in a resonant, sealed cell, and is irradiated with pulsed IR or UV radiation. Some of the radiant energy is absorbed by the gas and is immediately released as heat, causing the temperature of the gas to rise and the pressure in the cell to increase. The radiation is modulated by a rotating chopper. Absorption of radiation by the sample gas results in a series of pressure pulses which are detected by microphones placed in the walls of the cell. Specificity to a particular gas is achieved by irradiation of the sample with light of wavelength at which the gas of interest is known to absorb. Either a narrow band source such as a laser of appropriate wavelength, or a broad band source together with a filter to select the analytical wavelength, may be used. The spectrometer shown may be used for measurement of up to five gases by selection of appropriate filters.

Like other spectrometric techniques, PAS involves absorption of radiant energy. However, because the energy absorption is measured directly, as a thermal effect, rather than indirectly as the intensity of attenuated or emitted radiation, there are important differences from conventional spectrometry. The requirement to make measurements in a closed cell means that PAS is a batch, rather than a continuous measurement process. The photoacoustic signal depends on the absorption coefficient of the sample and the incident light intensity. Unlike other spectrometric techniques, the signal is independent of absorption path length, and so PAS can be used for point measurements in a gas sample. The PAS technique is applicable to a wide range of gases. It is highly sensitive, and the response is linear over a wide concentration range (typically 4 or 5 orders of magnitude). Highest sensitivities are achieved with laser sources. Ppb levels of organic molecules such as toluene, benzene and vinyl chloride have been detected using a CO_2 laser. Examples of detection limits for the instrument in Figure 14.4 are 150 ppb for CO, 3 ppm for CO_2, and 60 ppb for formaldehyde.

14.3.3   Electrochemical sensors

A basic electrochemical device (chapter 10) consists of a cell with two electrodes immersed in or in contact with an electrolyte. When a gas is

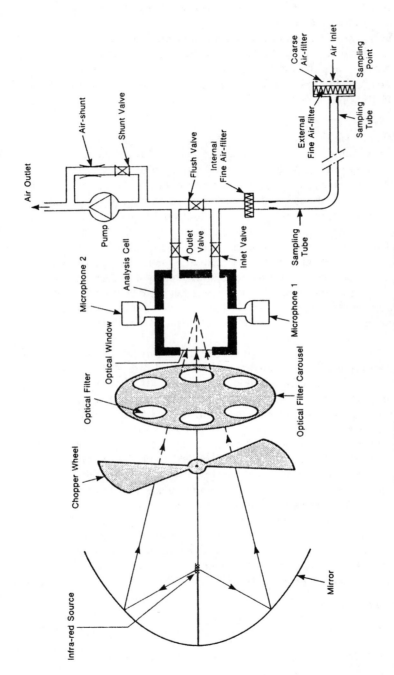

Figure 14.4 Photoacoustic gas analyser (Bruel and Kjaer).

present which takes part in a reaction at one of the electrode/electrolyte interfaces, or within the electrolyte, a current or voltage may be generated which is related to the concentration of that gas. Any reaction taking place at the sensing electrode must be balanced by a corresponding reaction at the counter electrode. To improve stability and minimise polarisation effects, a third reference electrode is often introduced. The sensor may be made selective to a particular gas by incorporation of a suitable membrane through which the gas diffuses to reach the sensing electrode.

Electrochemical sensors are available for a variety of gases, for example O_2, H_2, CO, NO, NO_2, Cl_2, HCN and SO_2. Electrochemical gas analysers are usually relatively inexpensive and have the potential for miniaturisation and for operation in harsh environments. However, the response of a sensor to the target species in a given environment may not be easy to predict, and the possibility of interference, particularly from changing moisture levels and other reactive components, must be carefully checked.

14.3.4 Chemical methods and detector tubes

Most gases of environmental interest are chemically active and can be determined by methods which typically involve selective absorption from a known volume of air sample in a suitable solution, followed by chemical or physical measurement of an absorbed species. For example sulphur dioxide may be determined by absorption in hydrogen peroxide solution followed by titration or conductimetry. Chemical methods may be used for continuous monitoring by carrying out the reaction in a flow system. Sensitivities and selectivities depend on the particular reaction and measurement technique, and therefore on the species analysed. Response times also depend on the reaction used, but are generally longer than for spectrometric analysers.

Gas detector tubes are based on chemical methods in which the presence of the test gas causes a colour change in a suitable reagent. The tube contains the reagent system necessary to provide a complete analysis. All the classical types of reactions, including oxidation–reduction, acid–base, addition, and substitution, are used. A known, predetermined volume of air is drawn through the tube. The presence of the test compound is indicated by a colour change in the tube packing, and the concentration can be deduced from the length of the stain using supplied calibration data.

Detector tubes are useful for quick determination of the concentration of many toxic substances. They are inexpensive, versatile, portable and simple to use. However, although tubes are available for detection of some 200 substances, the number of chemical reagent systems is considerably fewer, indicating that the same chemical can be used to detect a number of different compounds. For example, the same two stage reaction between benzene, formaldehyde and sulphuric acid can be used to detect either compound, but other aromatics and aldehydes will also react. Lack of specificity is a particular problem when the test atmosphere may contain unknown compounds or if it contains a mixture of compounds which react in the tube at different sensitivities.

14.4 Sampling   *14.4.1   Ambient air*

The concentration of atmospheric pollutants varies both with time for samples taken at a fixed point, and with position in an air mass. Variations are largest near to an emission source, but air masses are rarely well mixed and most pollutant concentrations show some fine structure. Sampling involves choices of where, when and how often to sample, and how many samples to take. An ideal sampling system would allow instantaneous measurement of the pollutant concentration at any point, but this can seldom be achieved in practice. The choice of sampling system is thus particularly important for atmospheric samples as it may affect not only the reliability of the analysis but also influence the information which can be obtained from the measurements. There are essentially two possibilities: extractive sampling or *in situ* monitoring.

Extractive sampling includes methods where the air sample is taken through an instrumental analyser, or where samples are collected in gas sampling vessels or bags for later laboratory analysis. Minor components of air may be sampled by selective adsorption on suitable materials. The sorbent, often a proprietary porous polymer such as Tenax or Porapak, is packed into a tube through which a known volume of air can be pumped or allowed to diffuse, or is contained in a badge which can be exposed to the air or worn by an operator. The adsorbed substances are subsequently desorbed by solvent extraction or heating, and are analysed by any suitable laboratory technique. It may be possible to determine several components in a single adsorbed sample.

In situ monitoring involves the use of sensors or analysers for local determination of pollutant concentrations. Examples include combustible gas detectors and personal monitors. Many spectrometric techniques, including DOAS, TDL and LIDAR are suitable for *in situ* monitoring. A collimated beam of light from a laser or high-intensity source is allowed to traverse a known distance in the atmosphere. The light absorption by the target species is measured either by a remotely sited detector or is returned by a retroreflector to a detector close to the light source. Provided that the emission intensity, the absorption coefficient and the absorption path length are known, the concentration of absorbing species can be calculated. However, in most cases the concentration calculated is the average along the absorption path, i.e. it is not possible to distinguish between a uniformly mixed air mass and one which contains narrow bands or plumes of pollutants. An exception to this is DIAL which provides range-resolved pollutant measurements and is therefore a true remote sensing technique (Figure 14.3).

In many cases the nature of the target species limits the choice of sampling techniques. For example, it may be difficult to ensure that labile components of a sample are not lost by reaction or adsorption on the walls of the sampling containers or analysis system, and the most reactive species can only be monitored *in situ*. In contrast, measurement of individual hydrocarbons in air requires extractive sampling followed by GC separation and analysis.

Air monitoring often involves measurements of several components of an air mass to study the interaction of different pollutants. In this case it is important to ensure that the measurements of different components can be assigned to particular points in time or position. Assuming that sampling of several components is started simultaneously at a given point, two times are involved in correlating the results, the response time and the sampling time for measurement of each component. The response time is the time taken for the sample to pass through the sampling system to the collecting or measuring device. For a continuous analyser the response time depends on the design of the instrument and sampling system and on the flow rate of the air sample. The sampling time is the time necessary to collect the actual sample for analysis, i.e. to fill a sample cell or sampling bag, or to take the measurement in a continuous analyser. The value obtained from the analysis is the concentration of the target species in the atmosphere, averaged over the sampling time. It is therefore not possible to monitor changes in analyte concentration which are short in comparison to the sampling time. For batch sampling by filling a bag or cylinder, sampling may be started instantaneously and the response time is then effectively zero, but the sampling time, i.e. the time taken to fill the container, may be several minutes or even hours. For continuous analysers, both response and sampling times may be relatively short, but may become important if pollutant concentrations in the air mass change rapidly, or if the air mass and sampling platform are moving relative to one another, for example if measurements are made from an aircraft.

14.4.2 Emissions

Emissions may be measured either *in situ* or by extractive sampling. *In situ* measurements are made by sensors placed in the gas stream or by spectrometric methods where the light beam traverses the stack or duct containing the gas. For extractive sampling a probe is inserted into the duct and gas is pumped to the analyser or sample collection device. A typical absorption train for extractive sampling of flue gas is shown in Figure 14.5. Concentrations of pollutants in the gas leaving a stack or exhaust are

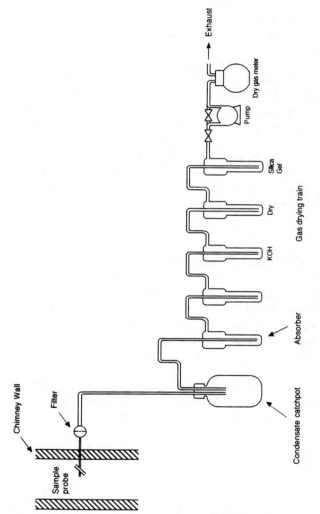

Figure 14.5 Gas absorption train for emissions sampling.

typically several orders of magnitude higher than in ambient air so less sensitive analysers can be used. For example, the concentration of sulphur dioxide in the flue gas from coal fired generating plant may be >1000 ppm, but this has typically diluted to ppb levels by the time the plume reaches ground level. Non-dispersive infrared analysers may be used for measuring emissions leaving the stack, but flame-photometric or pulsed fluorescent analysers may be required for measurements in ambient air. The gas leaving stacks or exhausts is often either toxic or corrosive, and the materials of sample lines and parts of the analyser contacted by the gas sample must be carefully chosen to avoid loss of reactive components of the gas. Trace heating of sample lines and analysers may be necessary to prevent condensation. As an alternative to trace heating, the sample may be diluted with a fixed ratio of air, sufficient to ensure that the dew point of the diluted sample does not fall below ambient temperature, but in this case more sensitive, and thus more expensive, analysers may be needed to measure the diluted sample.

Gases emitted from stacks or exhausts often contain varying amounts of excess air, either added deliberately as part of the process, or adventitiously from air inleakage, and exhausts from combustion processes contain high concentrations of water vapour. To allow meaningful comparison of emission data from different sources, the oxygen concentration in the emission gas must be measured, and the results for emission of pollutants expressed at standard conditions of oxygen and moisture concentration. As the gas is often dried before measurement, it is usually convenient to express the emission values for dry gas conditions, at a specific oxygen concentration.

Most gas analyses involve the use of a sensor or detector which is calibrated by measuring its response to a series of standard gas mixtures containing known concentrations of the determinand. Sources of stable, low-level calibration gas mixtures are thus extremely important as the accuracy of the analysis depends on the accuracy with which the concentration of the standard mixtures is known. Gas mixtures may be prepared either statically by successive addition of the components to a cylinder while the pressure or weight is measured, or dynamically by dilution of the pollutant or a concentrated standard mixture into a stream of the base gas to give a stream of mixed gas containing a known concentration of the test pollutant.

14.5 Calibration of gas analysers

Preparation of reliable static gas mixtures may present difficulties at the low concentration levels of most pollutants. It is usually necessary to prepare the final mixture in stages by dilution of more concentrated mixtures. Many pollutants are reactive, and it is difficult to ensure that components of the mixture are not lost by adsorption on the cylinder walls during preparation or storage.

Dynamic systems based on permeation or diffusion are often preferable to static mixtures for calibration. A permeation or diffusion tube is a device containing a reservoir of the test compound which is allowed to permeate through a membrane or diffuse along a capillary. The permeation or diffusion rate depends on the vapour pressure of the test compound, the temperature and the geometry of the diffusion tube or the nature of the permeation membrane. The emission rate for given device can be determined by recording the weight loss over a period of time. Provided that the device is held at constant temperature, after an initial induction period the rate of weight loss settles to a constant value that is maintained until the reservoir of test material is depleted.

Permeation or diffusion tubes are available or can be constructed to give sources of known emission rates for a wide variety of pollutants including SO_2, HCl and NO. Permeation or diffusion tube calibrators consist of an accurately temperature-controlled chamber to hold the source and through which pure base gas is passed at a known, controlled flow rate, together with plumbing and flow controllers to enable a range of concentrations to be generated in a stream of base gas which can be fed to the gas analyser. The base gas may be air or an inert gas such as nitrogen, and the calibrator may include an air pump and filters or scrubbers to ensure that the incoming gas is clean.

14.6 Analysis of the major air pollutants and oxygen

14.6.1 Carbon monoxide

Methods for measurement of carbon monoxide include gas-filter correlation spectrometry, tunable diode laser spectrometry and electrochemistry. Ambient levels of CO may also be determined by gas chromatography, with separation on a molecular sieve or porous polymer column. However, detection is indirect since the common GC detectors either do not respond to CO or are not sufficiently sensitive to measure ambient levels. The effluent from the GC column may be mixed with hydrogen and passed over a heated catalyst to convert the CO to methane (methanation) followed by measurement of the methane by FID. Alternatively, the column effluent may be passed through a heated bed of mercuric oxide. Carbon monoxide liberates mercury vapour which is determined in the outlet gas by atomic absorption spectrometry.

$$HgO + CO \rightarrow Hg_{vap} + CO_2 \tag{14.1}$$

Carbon monoxide levels down to about 10 ppb may be determined. The gas chromatograph with mercury vapour detector, known as a reducing gas analyser, may also be used to measure ambient concentrations of other gases which reduce HgO, such as hydrogen, ethene (ethylene) and ethyne (acetylene).

14.6.2 Nitrogen oxides

Nitric oxide (NO) may be measured either by non-dispersive ultraviolet (NDUV) spectrometry or by reaction with ozone to produce nitrogen dioxide (NO_2):

$$NO + O_3 = NO_2^* + O_2 = NO_2 + h\nu \qquad (14.2)$$

About 10% of the NO_2 is produced in an electronically excited state and undergoes transition to the ground state emitting light in the range 590–2600 nm. The intensity of the light emission is proportional to the mass flow rate of NO through the reaction chamber and is measured by a photomultiplier.

The reaction is specific for NO, but NO_x ($NO + NO_2$) can be measured by decomposition of the NO_2, either by UV light or by passing the sample through a stainless steel tube heated to 600–800°C. Under these conditions, most nitrogen compounds (but not N_2O) are converted to NO which is measured as above. The NO_2 concentration is then given by difference.

The flow system of a nitrogen oxides analyser is shown in Figure 14.6. Ozone is generated from ambient air by the action of UV light, and a controlled flow of ozonised air is passed to the reaction chamber for reaction with NO in the air sample. The detection limit for NO is about 1 ppb.

Luminol (3-aminophthalhydrazide) in alkaline solution reacts with gaseous NO_2 at atmospheric pressure to produce intense chemiluminescence with maximum emission intensity at 425 nm. This forms the basis for highly sensitive chemical NO_2 analysers. The detection limit is below 30 ppt. Organic nitrates such as peroxyacetyl nitrate (PAN) interfere, but are usually present at much lower concentrations than NO_2. Ozone may also interfere at low NO_2 concentrations. The luminol system may be used for NO_x measurements by catalytic oxidation of NO to NO_2.

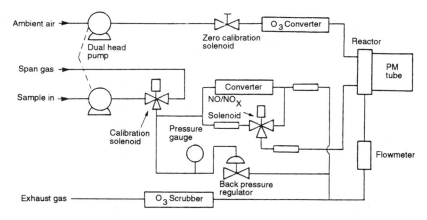

Figure 14.6 Chemiluminescent NO/NO_x analyser (Beckman).

Nitrogen dioxide may be sampled by diffusion tubes. The tubes, 71 mm long by 12 mm internal diameter, are open at one end and sealed at the other with a cap containing a disc coated with triethanolamine. During the sampling period, air diffuses along the tube and NO_2 is absorbed by the triethanolamine. Nitrogen dioxide is later determined colorimetrically by the Greiss/Saltzman reaction in which the nitrite ion diazotises sulphanilamide in orthophosphoric acid solution. The diazonium salt is coupled with N-(1-naphthyl)-ethylenediamine dihydrochloride to give a purple–red azo dye. The diffusion tube method is relatively simple and has been widely used for NO_2 surveys, for example in urban areas or near to roads, when the tubes are exposed for periods of 1–2 weeks. Diffusion tube sampling is not suitable for short-term measurements.

14.6.3   Ozone

Continuous ozone analysers are based either on UV photometry or on the flameless chemiluminescent reaction of ozone with ethylene:

$$O_3 + C_2H_4 \rightarrow \text{ozonide} \rightarrow CH_2O^* \tag{14.3}$$

$$CH_2O^* \rightarrow CH_2O + h\nu \tag{14.4}$$

A chemiluminescent ozone analyser is shown in Figure 14.7. Light emission from the reaction, centred at 430 nm, is measured by a photomultiplier and the resulting amplified signal is a measure of the concentration of ozone in the sample stream. As the light emission is a direct function of ambient

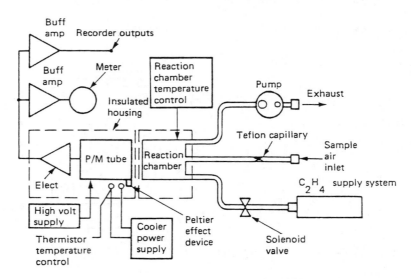

Figure 14.7 Chemiluminescent ozone analyser (Columbia Scientific Industries Corporation).

temperature, the reaction chamber is regulated to 50°C. The photomultiplier is also contained in a thermoelectrically cooled housing maintained at 25°C to minimise short- and long-term drift. The instrument can measure ozone levels in the range 0.1–1000 ppb.

14.6.4 Sulphur dioxide

Sulphur dioxide may be measured by a variety of techniques including coulometry, flame photometry, chemical methods and spectrometry. Continuous SO_2 analysers are commonly based either on flame photometry or fluorescence. In the flame photometric detector the air sample is passed at a controlled flow rate through a fuel-rich hydrogen–air flame. Sulphur compounds are reduced in the hot inner flame zone to form thermally excited S_2 dimers. As the S_2 dimer reverts to its ground state it emits light in a series of bands in the range 300–450 nm with the most intense bands at 384.0 and 394.1 nm. The 394 nm emission is monitored by a photomultiplier. The flame photometric analyser is reasonably simple and specific, but has two disadvantages. The response is non-linear, being proportional to $[S]^n$, where $[S]$ is the sulphur concentration and n is a constant, with a value between 1.5 and 2 depending on flame conditions. The detection limit is typically about 1 ppb, which although satisfactory for polluted air, is too high to allow measurement of SO_2 in clean air remote from emission sources. Commercial analysers often incorporate a linearising circuit to give an output signal which is proportional to the mass flow rate of sulphur over a concentration range of two or three orders of magnitude. It has been shown that the detection limit may be improved by about an order of magnitude by doping the hydrogen fuel with SF_6.

Fluorescence sulphur dioxide analysers (Figure 14.8) are based on the reaction

$$SO_2 + h\nu_1 \rightarrow SO_2^* \rightarrow SO_2 + h\nu_2 \ (h\nu_1 > h\nu_2) \qquad (14.5)$$

The air sample is irradiated by a pulsed UV source in the region of 214 nm and the fluorescence induced is measured in the range 240–420 nm. Unlike

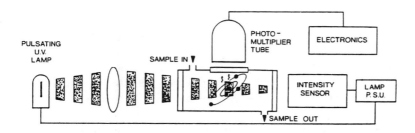

Figure 14.8 Pulsed fluorescence sulphur dioxide analyser.

flame photometric analysers which measure total sulphur, the fluorescence reaction is specific to SO_2. The detection limit is about 1 ppb.

Sub-ppb levels of SO_2 may be monitored continuously by a chemical analyser. The air sample is bubbled through mercury(I) nitrate solution. Sulphur dioxide promotes the disproportionation of Hg(I) according to the reaction:

$$2SO_2 + 2H_2O + Hg_2^{2+} \rightarrow Hg(SO_3)_2^{2-} + Hg^0 + 4H^+ \qquad (14.6)$$

The mercury vapour released is passed to a measuring cell where it is monitored with high sensitivity by atomic absorption spectrometry using a mercury vapour monitor. The detection limit is about 0.2 ppb and the calibration curve is linear up to about 100 ppb of SO_2 in air.

14.6.5   Oxygen

Oxygen is not normally a pollutant, but determination of oxygen concentrations is often necessary, for example to correct emission measurements to standard conditions, or for control of emissions from combustion processes. Continuous oxygen analysers are either based on electrochemistry or make use of the fact that oxygen, alone among common gases, is paramagnetic.

Electrochemical oxygen sensors may be either amperometric (polarographic) or galvanic (fuel cell) type. For gas sensing a micro fuel cell with a lead anode and a cathode in the form of a perforated, gold-plated convex disk is used. The electrolyte is aqueous potassium hydroxide solution. The cathode is covered with a PTFE membrane which is exposed to the test gas. Oxygen diffuses through the membrane, enabling the following reactions to take place:

$$O_2 + 4e^- + 2H_2O \rightarrow 4OH^- \qquad \text{(cathode)} \qquad (14.7)$$

$$Pb + 2OH^- \rightarrow PbO + H_2O + 2e^- \quad \text{(anode)} \qquad (14.8)$$

$$Pb + O_2 \rightarrow PbO \qquad \text{(overall)} \qquad (14.9)$$

The cell current is a measure of the oxygen concentration in the test gas.

The high temperature ceramic oxygen sensor is an important electrochemical device for *in situ* measurement of oxygen in combustion and exhaust gases. It utilises the fact that, at temperatures between 600 and about 1200°C, zirconia (ZrO_2) shows high mobility for oxygen ions. Thus if a zirconia tube at high temperature is exposed to gases of different oxygen concentration on its inner and outer surfaces, a potential is developed which is a function of the ratio of the oxygen concentrations in contact with the surfaces. The sensor (Figure 14.9) consists of a tube or thimble of zirconia stabilised with calcium. Platinum electrodes are fixed to the inner and outer surfaces. Reference gas of constant O_2 concentration (usually dry air) is

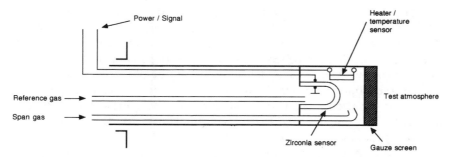

Figure 14.9 Zirconia potentiometric oxygen sensor (Hartmann and Braun).

passed over the inner surface. The voltage developed between the electrodes is then a measure of the oxygen concentration in the atmosphere surrounding the probe. The zirconia probe is widely used in combustion gas analysers and is used to measure the air-to-fuel (λ) ratio for combustion and emission control in catalyst-equipped internal combustion engines. The measuring range is 0.25–25 vol.%

The paramagnetic properties of oxygen are exploited for gas analysis in two main ways, magnetodynamic and thermalmagnetic instruments. In a magnetodynamic oxygen analyser (Figure 14.10) two spheres of glass or quartz, filled with nitrogen, are mounted on the ends of a bar to form a

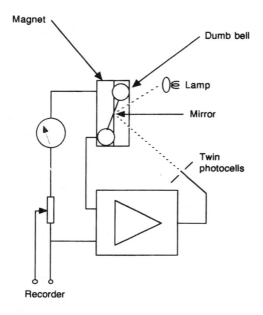

Figure 14.10 Magnetodynamic oxygen analyser (Servomex).

dumb-bell. The dumb-bell is mounted horizontally on a vertical torsion suspension and is placed in a non-uniform magnetic field between the poles of a powerful permanent magnet. The test gas surrounds the dumb-bell. Oxygen in the test gas is drawn into the magnetic field tending to displace the spheres which are repelled from the strongest parts of the field, rotating the suspension. The angular deflection of the dumb-bell is detected by a mirror and photocell assembly and control circuitry is used to produce a compensating electromagnetic torque to restore the dumb-bell to its original position. The compensation current is proportional to the oxygen concentration in the test gas.

The measuring cell of a thermalmagnetic ('magnetic wind') analyser is shown in Figure 14.11. The sample gas flows through an annular chamber, connected by a horizontal cross-tube. Two identical platinum heating and sensing coils are wound on the outside of the cross-tube, and form two arms of a Wheatsone bridge circuit, the bridge being completed by two external resistances. The coils are heated by the bridge current. One winding is placed between the poles of a powerful magnet. When a gas containing oxygen enters the cell, oxygen is drawn into the magnetic field, and tends to flow through the cross-tube. The flow velocity through the cross-tube causes heat to be transferred from one platinum coil to the other, altering the

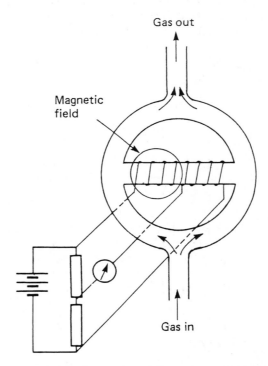

Figure 14.11 Thermalmagnetic ('magnetic wind') oxygen analyser (Taylor Analytics).

resistance and producing an out-of-balance voltage in the bridge circuit, proportional to the oxygen concentration in the test gas.

Compared to gases such as nitrogen, hydrogen and carbon dioxide, which are weakly diamagnetic, oxygen is strongly paramagnetic, and the magnetic properties of atmospheric samples are usually almost entirely dependent on the oxygen concentration. However, the magnetic susceptibility of oxygen is temperature dependent, and additionally the magnetic wind analyser is affected by changes in thermal conductivity of the test gas mixture. Nitric oxide is paramagnetic, with a susceptibility about half that of oxygen, and may interfere in the measurement. Both magnetodynamic and magnetic wind instruments can typically measure oxygen concentrations from 100% to about 100 ppm. The magnetic wind instrument is suitable for corrosive samples.

14.7.1 Methane, non-methane hydrocarbons and non-methane organics

14.7 Analysis of some secondary air pollutants

The average concentration of methane in the atmosphere is about 1.6 ppm, but local sources produce areas of higher concentration. Methane can be separated by gas chromatography and measured by flame ionisation detector. However, analysis of methane in ambient air is usually carried out either to detect long-term changes or to measure local sources, such as emissions from land-fill sites. Both require measurement of small variations on a relatively large background, and the analyser must be carefully designed and controlled to achieve the required accuracy and precision. Non-methane hydrocarbons (NMHC) are several orders of magnitude lower concentration than methane.

Determination of individual hydrocarbons in air requires selective removal of the hydrocarbons by adsorption or cryotrapping, followed by release of the concentrated sample for GC separation and measurement, usually with a flame ionisation detector. Urban air may contain up to several hundred different hydrocarbons and organic species. Identification of individual hydrocarbons and other non-methane organics is difficult. Use of coupled techniques such as GC-MS or GC-FTIR are possibilities, and some information on the type of hydrocarbon species in the sample may be obtained by use of a photoionisation detector (PID). The PID is non-destructive and may be used alone, or installed in a gas chromatograph in series with an FID. Unlike the FID which responds to all hydrocarbons, the PID responds to alkenes, alkynes and aromatic species, but not to alkanes.

Total hydrocarbons in air or emissions may be determined by infrared spectrometry or by passing the air sample directly to a flame ionisation detector, without GC separation. In this case the FID is operated as a 'carbon counter', i.e. the response of the analyser is assumed to be directly proportional to the mass of carbon passing through the detector per second, and concentrations are expressed as methane equivalents.

Tunable diode laser spectrometry has been used for remote monitoring of methane and other emissions from chemical plant or waste disposal sites. Infrared spectrometry, with the light beam passing across the carriageway at exhaust-pipe level, has been used to give an almost instantaneous indication of emission levels of CO, CO_2, and hydrocarbons from individual motor vehicles using the road.

14.7.2   Chlorofluorocarbons

Chlorofluorocarbons are man-made chemicals which have been released into the atmosphere over the last few decades and now occur at concentrations from about 300 to less than 1 ppt. They are chemically stable and may be separated and determined, together with some other trace atmospheric constituents such as trichloroethylene and other halogenated solvents, by gas chromatography with electron capture detection.

14.7.3   Organic nitrates

Organic nitrate species, particularly peroxyacetyl nitrate (PAN), are the lachrymatory constituents of photochemical smog. They are highly reactive but may be determined by gas chromatography with electron capture detection. Continuous analysers for PAN are based on catalytic oxidation to NO_2 followed by chemiluminescent determination by reaction with luminol.

14.7.4   Nitrous oxide

Nitrous oxide (N_2O) is the most abundant oxide of nitrogen in the atmosphere, at about 310 ppb. It is relatively unreactive, is not detected by chemiluminescent analysers and does not contribute to NO_x figures. It may be determined by infrared spectrometry, by gas chromatography with electron capture detection and by photoacoustic spectrometry.

14.7.5   Combustible gases

Combustible gases may be detected by sensors which measure the heat output resulting from catalytic oxidation of the flammable gas molecules to carbon dioxide and water at a solid surface. The most suitable metals for promoting oxidation of molecules containing C–GH bonds are those in Group 8 of the Periodic Table, and particularly platinum and palladium. By use of a catalyst, the temperature at which the reaction takes place is much reduced compared to gas phase oxidation.

Pellistors are solid state catalytic sensors. A bead of porous ceramic material such as alumina about 2–3 mm diameter is formed over a coil of fine (50 µm) platinum wire. The catalyst material is deposited on the outside of the bead. The platinum wire coil acts as both an electrical heater and temperature sensor. The sensor is mounted in a protective open-topped housing so that gas flow to the sensor is largely diffusion controlled. An inactive sensor, i.e. one not coated with catalyst, may also be mounted in the housing. The two sensors are connected in the measuring arms of a Wheatstone bridge circuit. The out-of-balance voltage is a measure of the combustible gas concentration. The sensitivity and selectivity of pellistors are influenced by the choice of catalyst and the operating temperature. The surface treatment of the sensor is also important and pellistors may incorporate an additional diffusion layer to give selectivity to specific gases.

The atmosphere may contain entrained or suspended solids of all sizes from molecules to particles of sand, grit or gravel, and liquid aerosols dispersed as fog, mist or rain. Particles and aerosols are directly responsible for environmental effects and may also play an important role in atmospheric chemistry. Removal of gaseous pollutants from the atmosphere often involves interaction with cloud water, rain or dust particles. The physical and chemical properties of particles and aerosols vary with particle size. The diameters of aerosols and smoke particles of most importance in the atmosphere are in the 0.01–100 µm range. However, characterisation of particles and aerosols is complex, and can only be briefly discussed here. For example, particles may occur in a variety of shapes from almost perfectly spherical droplets to irregular solids or fibres. Particle sizes cannot be measured directly, and must be inferred from measurement of a related property. The 'diameter' of irregular particles may be strongly dependent on the measurement method employed. Three parameters are important in characterising particulate and aerosol samples; the size ranges of the particles, the concentration or the number of particles in each size range, and the chemical composition or biological effect of the particles.

14.8 Sampling and analysis of particles and aerosols

Sampling of aerosols and particulates involves similar considerations to gas sampling, but the sample must normally be considered to be inhomogeneous, and cannot be treated as a homogeneous fluid. In addition to obtaining a representative sample from the gas volume, it is important to ensure that the size distribution in the sample is preserved.

14.8.1 Emissions sampling

Particulate samples for emission monitoring are usually extracted by insertion of a sampling probe into the stack through which the gas is flowing. The gas usually contains particles in a wide size range, and to maintain the size distribution in the sample it is essential to sample under isokinetic conditions, that is to ensure that the velocity of the sample stream withdrawn through the probe is the same as the velocity of the gas in the duct. The effect of sampling velocity on particle size distribution is illustrated in Figure 14.12. Isokinetic sampling is usually the most important consideration in emissions monitoring and the effect of sample velocity only becomes unimportant for particles less than approximately 5 µm diameter. The diameter and profile of the probe tip must be chosen to minimise the effect of the probe on the sample flow, and the sampling system must be carefully designed to avoid losses.

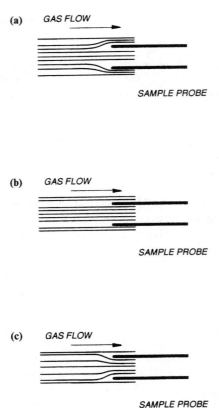

Figure 14.12 Effect of sampling velocity on particle size distribution. (a) Sub-isokinetic sampling. The sampling velocity is lower than the stack velocity. Gas tends to 'back-up' at the probe tip. Light particles are carried round the probe with the gas, while the momentum of heavier particles causes them to enter the probe. The sample is biased towards heavier particles. (b) Iso-kinetic sampling. The sampling and stack velocities are equal, and the size distribution of particles in the stack gas is preserved in the sample. (c) Super iso-kinetic sampling. The sampling velocity is greater than the stack velocity. The gas is 'sucked in' to the probe. Lighter particles are carried in with the gas, while the inertia of heavier particles causes them to miss the probe. The sample is biased towards light particles.

The size distribution of particles in a gas stream is affected by bends or tees in the ductwork and may vary across the diameter of the duct carrying the sample. To ensure representative sampling, the sampling point must be in a straight run of ducting, as far as possible from flow disturbances. Samples must be taken at defined transverse points along two diameters of the ducting at 90° to one another. The number of samples required for characterisation of the emissions varies with the location of the sampling point and the diameter of the duct, and is usually specified in the regulations governing the process.

14.8.2   *Particle and aerosol characterisation and collection*

Continuous measurements of the number and size range of atmospheric particles and aerosols may be made by methods based on light scattering. The optical density of an air sample is linearly related to particle concentration, and measurement of optical density or opacity can therefore be used to determine the mass emission rate of particles in a gas stream. An opacity monitor or transmissometer may simply consist of a light beam from a visible source such as a tungsten lamp which shines through the gas sample, or across a duct, and is measured by a photodetector. More complex instruments split the light into measuring and reference beams to give more accurate measurement and to allow compensation for noise and drift for example due to gradual degradation of the light source.

Cloud water droplet concentrations and sizes may be measured by

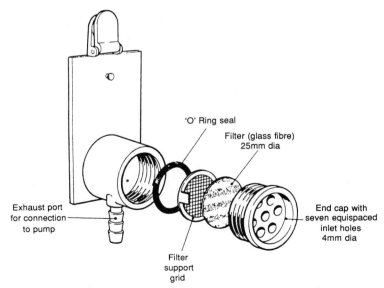

Figure 14.13 Personal filter sampling head for gravimetric dust sampling (UKAEA).

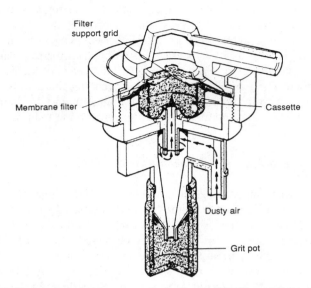

Figure 14.14 Cyclone sampler for personal monitoring (Health and Safety Executive).

detectors in which a laser light beam is focused at a point in the sample. The scattered light is measured as a function of angle relative to the incident beam. In either case, the intensity of the scattered light enables the diameters of the scattering particles to be deduced. Particulates and aerosols may be collected by deposition or impaction, filtering or centrifugation. A filter sampler for personal monitoring of total inhalable particulates is shown in Figure 14.13.

A personal cyclone sampler, which separates airborne particles into 'inhalable' and 'respirable' fractions is shown in Figure 14.14.

14.8.3   Particle and aerosol composition

Particles and aerosol samples may be examined by any appropriate technique, such as microscopy, or chemical analytical methods for solids or liquids.

Further reading   Cheremisinoff, P.N. (Ed.) (1981) *Air/Particulate Sampling and Analysis*, Ann Arbor Science, Michigan.
Guiochon, G. and Guillemin, C.L. (1988) Quantitative Gas Chromatography for Laboratory Analyses and On-Line Process Control, *Journal of Chromatography Library*, Vol. 42, Elsevier, Amsterdam
Lodge, J.P., Jr. (Ed.) (1989) *Methods of Air Sampling and Analysis*, 3rd edition, Lewis Publishers, Chelsea, MI.

Newman, L. (1992) *Measurement Challenges in Atmospheric Chemistry, ACS Advances in Chemistry Series 232*, American Chemical Society.

Sigrist, M.W. (Ed.) (1994) *Air Monitoring by Spectroscopic Techniques*, Wiley, Chichester.

Thain, W. (1980) *Monitoring Toxic Gases in the Atmosphere for Hygiene and Pollution Control*, Pergamon, Oxford.

Torvela, H. (1994) *Measurement of Atmospheric Emissions*, Springer-Verlag, London.

15 Trace elements

N.I. Ward

15.1 Trace elements in the environment

There are 90 naturally occurring elements on the earth, the magnitude and chemical form of which can widely vary both between or within environmental, geological, biological, or marine systems. The elemental composition of the earth's crust is predominantly O, Si, Al, Fe, Ca, Na, K, and Mg, whereas the human body is H, O, C, N, Ca, P, K and Cl. An example of an element that varies widely in magnitude between the various systems is aluminium. As one of the main constituents of the earth's crust, Al in rocks commonly ranges from 0.45 to 10%. However, the content of this element in plants is generally lower than 200 mg kg^{-1} (dry weight), human bone 3.6 mg kg^{-1}, human blood serum 1 to 5 µg dm^{-3}, and seawater 2 µg dm^{-3}. Similarly, the relative amounts of the major elemental constituents in seawater follow the order $Na^+ > Mg^{2+} > Ca^{2+} > K^+$, while in river water the order is $Ca^{2+} > Na^+ > Mg^{2+} > K^+$. Elemental distribution is therefore very complex and dependent upon many physical and chemical factors, including energetic radiation, weathering, pH, redox properties, solubility, etc.

In biological systems elements can be basically grouped into three categories: *major* elements consisting of C, H, N, O; *minor* elements comprising Ca, Cl, Mg, P, K, Na; and the remainder are referred to as *trace* elements. The proportion of these three elemental groups in relation to the total mass of an organism represents 96%, 3.6%, and 1%, respectively. The term 'trace elements' relates to the time when minute amounts of a number of elements were found in biological systems but they could not be precisely quantified using classical methods of analysis. Although more sensitive and specific analytical techniques have been developed this terminology has remained in popular usage. Many modern multi-element methods, such as inductively coupled plasma mass spectrometry (ICP-MS), and neutron activation analysis (NAA) are now capable of measuring elemental levels as low as <0.01 µg kg^{-1}, and so a new phrase which is commonly used is *ultratrace* elements. There is no general acceptance of what the concentration intervals should be, however, trace element levels are basically <100 mg kg^{-1} (µg g^{-1}) to 0.01 mg kg^{-1}, and ultratrace levels <0.01 µg g^{-1} or <10 µg kg^{-1} (ng g^{-1}).

In biological systems the major elements act as structural components while the minor elements maintain electrolyte balance processes. Many of the trace elements are important for the growth, development, and health of living organisms. From a biological viewpoint, trace elements are most conveniently classified into three groups: *essential, non-essential,* and *toxic.*

The trace elements essential for plants are those elements which cannot be substituted by others in their specific biochemical roles and that have a direct influence on the organism so that it can neither grow nor complete some metabolic cycle. In general B, Co, Cu, Fe, Mn, Mo, Si and Zn are known to be essential for plants, whilst Al, As, Br, F, I, Li, Ni, Rb, Se, Sr, Ti and V are known to be essential for some groups or species, but there is a need for confirmation of the evidence. In general, the above trace elements are involved in key metabolic processes, such as respiration, photosynthesis, and fixation or assimilation of nutrients; they act as metalloenzymes in electron transfer systems (Cu, Fe, Mn, Zn); or catalyse valence changes in the substrate (Cu, Co, Fe, Mo). Both deficiencies or excesses of these elements can result in metabolic disorders of plants.

In human and animal systems, trace elements are defined as being essential if depletion consistently results in a deficiency syndrome and repletion specifically reverses the abnormalities. The trace elements fulfilling these requirements are As, Co, Cr, Cu, F, Fe, I, Mn, Mo, Ni, Se, Si, Sn, V and Zn. In addition to the essential elements, there are several others which are always found in body tissues and fluids but for which no proof of essentiality has been established. These elements are often referred to as non-essential, e.g. Li, B, Ge, Rb, Sr. Some elements, Cd, Hg and Pb, are prominently classified as toxic. This is because of their detrimental effects even at low levels. It should be noted, however, that all trace elements are predominantly toxic when their levels exceed the limits of safe exposure. These limits vary widely from one element to another.

The physiological roles of the essential trace elements relate to their association with enzyme systems, either as metalloenzymes or metal–enzyme complexes or their involvement in oxidation–reduction and transport processes. Some of their other important roles include membrane permeability, nerve conduction, muscle contraction, and involvement in the synthesis and structural stabilisation of proteins and nucleic acids.

The overall role of trace elements in living organisms is directly related to the interactions within all environmental, geological, biological, or marine systems. For example, the trace element composition of soil may significantly influence the elemental composition of the vegetation, which in turn influences that of animal or human tissues or fluids via the food chain. Figure 15.1 shows a general schematic of the movement of trace elements within the environment.

The earth's crust is the primordial source of all the natural trace elements of the various environmental, geological, biological and marine systems.

15.2 Natural levels and chemical forms

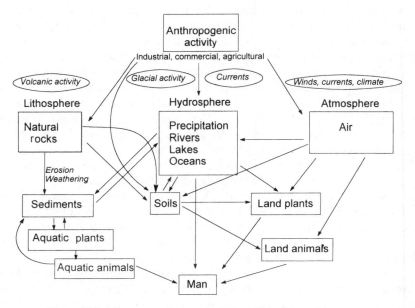

Figure 15.1 Movement of trace elements within the environment.

Throughout the ages the various types of rocks were continually being subjected to various physical and chemical processes which resulted in the formation of soil or sediment layers. Subsequently soils and sediments became the terrestrial and aquatic substratum for plants which absorbed, translocated, and in some cases bioaccumulated elements required for their sustenance. As important parts of the food chain, the edible parts of plants provides the route for trace elements to become incorporated into the body components of animals and humans. Along this natural pathway water is a significant source of trace element input. Water, through natural weathering and leaching processes, and by dissolving and reacting with materials, mobilises and distributes trace elements. In regard to atmospheric input, it would appear that volcanic activity, forest fires, soil dusts, and salt sprays contribute to both aerosols and particulates which are the major sources of global cycling of trace elements.

The natural geological and biological alterations of the earth's surface are slow. Man's input to the trace element composition of the biosphere has been very dramatic, resulting in widespread trace element changes in environmental, biological, and marine systems. These changes have in themselves provided a challenge for environmental analytical chemists who are confronted with the ever increasing demands for more sensitive methods for monitoring and distinguishing between natural components and pollutants.

In order to evaluate the impact of trace element pollution, it is necessary to establish the natural elemental levels and chemical forms of the earth's environment. Tables 15.1–15.4 show the typical natural trace element concentrations of fresh-, river- and seawater; surface soils; plant and plant foodstuffs; and human tissues and fluids.

Table 15.1 Typical natural trace element concentrations of fresh-, river- and seawater

Trace element	Concentration (μg dm^{-3})		
	Fresh	River	Sea
Li	10	12	200
B	10	10	5000
Al	10	50	2
Ti	5	10	1
V	0.5	1	2.5
Cr	1	1	0.05
Mn	10	7	0.2
Fe	500	40	2
Co	0.05	0.2	0.02
Ni	0.5	0.3	0.5
Cu	3	5	2
Zn	15	20	10
As	0.5	2	3
Se	0.2	0.2	0.1
Br	15	20	7000
Rb	1	20	120
Sr	70	60	8000
Mo	0.5	1	10
Cd	0.03	0.02	0.1
Sb	0.2	0.3	0.2
I	2	10	60
Cs	0.02	0.04	0.1
Hg	0.07	0.007	0.03
Pb	1	3	0.03

Table 15.2 Typical natural trace element concentrations of surface soils

Trace element	Concentration (μg g^{-1} dry weight)	
	Mean	Range
Li	25	<5–120
B	30	8–140
Al%	3	0.5–4.5
Ti%	0.3	0.02–1
V	85	15–360
Cr	60	5–1100
Mn	450	7–2000
Fe%	5	3–>10
Co	8	0.2–50
Ni	20	1–120
Cu	15	6–60
Zn	60	17–125
As	2.5	0.4–70
Se	0.3	0.02–2.50

Table 15.2 *Contd.*

Trace element	Concentration (μg g^{-1} dry weight)	
	Mean	Range
Br	45	5–240
Rb	75	<20–140
Sr	280	10–3500
Mo	1.5	0.2–12
Cd	0.25	0.01–2.5
Sb	0.75	0.05–3
I	1.5	0.1–16
Cs	2	0.3–25
Hg	0.05	0.004–0.70
Pb	20	1.5–80

Table 15.3 Typical natural trace element concentrations of plant and plant foodstuffs

Trace element	Concentration[a] (μg g^{-1}) (dry weight)	
	Plant	Plant foodstuffs[b]
Li	0.5 (<0.01–143)	0.3 (0.06–10)
B	5 (1–30)	5 (0.8–10)
Al	200 (6–3500)	15 (3–140)
Ti	2 (0.15–80)	1 (0.1–5)
V	0.5 (0.1–2.5)	0.01 (0.001–0.7)
Cr	0.2 (0.02–0.2)	0.05 (0.01–14)
Mn	80 (20–240)	15 (1.3–90)
Fe	120 (30–920)	60 (6–130)
Co	0.08 (0.03–0.6)	0.07 (0.008–0.2)
Ni	1 (0.1–5)	0.8 (0.06–4)
Cu	5 (1–12)	4 (0.08–9)
Zn	30 (12–60)	25 (1.2–45)
As	0.15 (0.009–1.5)	0.08 (0.003–0.3)
Se	0.06 (0.002–0.88)	0.03 (0.003–0.15)
Br	35 (5–120)	5 (0.2–40)
Rb	55 (44–130)	15 (1–55)
Sr	220 (6–1500)	25 (0.06–150)
Mo	0.3 (0.03–8)	0.5 (0.04–2.5)
Cd	0.1 (0.02–0.5)	0.08 (0.008–0.3)
Sb	0.06 (0.001–10)	0.01 (0.001–0.25)
I	0.1 (0.03–12)	0.1 (0.005–12)
Cs	0.1 (0.03–0.4)	0.007 (0.001–0.05)
Hg	0.01 (0.001–0.04)	0.003 (0.002–0.04)
Pb	1 (0.3–10)	0.7 (0.05–4)

[a] Mean (range).
[b] Lettuce, cabbage, beans, corn, cereals.

The earth's crust contains very few free trace elements. Most are found as soluble simple salts (Na, K, Rb), cationic constituents in aluminosilicates (Li, Be, Cs), insoluble carbonates or sulphates (Mg, Ca, Sr, Ba), oxides (B, Al, Si, Sc, Ti, V, Cr, Mn, lanthanides), or sulphides (Fe, Co, Ni, Cu, Zn, Ga, Ge, As, Se, Mo, Cd, Sn, Sb, Hg, Tl, Pb, Bi).

Most trace elements in waters (ground and surface) are present mainly as

Table 15.4 Typical natural trace element concentrations of human tissues[a] and fluids[b]

Trace element	Concentration (μg g^{-1})[a] (dry weight) or (μg dm^{-3})[b]				
	Kidney[a]	Liver[a]	Hair[a]	Milk[b]	Blood serum[b]
Li	0.01	0.01	0.2	2	0.8
B	0.5	1	1	5	1
Al	5	5	2	15	1
Ti	0.5	0.5	2	10	5
V	0.2	0.1	0.05	0.8	0.05
Cr	0.1	0.1	0.8	1.5	0.1
Mn	1	5	1.5	5	0.4
Fe	320	600	40	500	1090
Co	0.1	0.1	0.35	0.5	0.1
Ni	0.5	0.1	0.8	12	0.3
Cu	14	20	20	280	1000
Zn	150	250	180	1500	900
As	0.01	0.03	0.1	0.5	1
Se	0.5	2	2	18	90
Br	5	0.2	10	1520	500
Rb	5	5	0.05	100	160
Sr	0.05	0.05	0.1	10	5
Mo	0.3	1	0.1	1	0.5
Cd	15	0.8	0.2	1	0.1
Sb	0.05	0.05	0.2	1	0.3
I	0.2	0.2	0.2	80	50
Cs	0.01	0.1	0.1	0.5	0.7
Hg	0.1	0.1	0.4	2	1
Pb	3	5	1	10	3

suspended colloids or fixed by organic and mineral substances. Many trace elements do not exist in soluble forms for very long in natural waters and therefore they accumulate in bottom sediments or plankton. The chemical forms of a trace element in both fresh- and seawater depends on the nature of other species present, the pH, and the nature of the metal ions and bonded functional groups of the species formed. Table 15.5 shows some of the major chemical species found in natural waters.

Soils are considered as sinks for trace elements, and therefore they play an important role in the environmental cycling of elements. The mineral constituents of soils are normally directly related to the parent rock and the type of weathering processes. The principal components of soil are inorganic materials: sand, silt and clay. The most important are the clay particles with typical diameters of <0.002 mm, and consisting of a wide range of secondary silicate and aluminosilicate minerals. Clay minerals may contain low levels of trace elements as structural components but their surface properties (area and electrical charge) play a vital role in regulating the buffer and sink properties of soils. Other important soil minerals are the oxides and hydroxides of Fe and Mn (which determine the colour of many soils and have a high sorption capacity for trace elements); carbonates, especially calcite, which has a major influence on the pH of soils; and in some cases,

Table 15.5   Major chemical species found in natural waters

Trace element	Chemical species
Li	Li^+
B	$B(OH)_3$, $[B(OH)_4]^-$
Al	$Al(H_2O)_6^{3+} \rightarrow [Al(OH)_4]^-$
Cr	$Cr(OH)_2(H_2O)_4^+$, CrO_4^{2-}
Mn	Mn^{2+}, $Mn^{2+}SO_4^{2-}$, $MnCl^+$
Fe	$[Fe(OH)_2]^+$, $[Fe(OH)_4]^-$
Co	Co^{2+}
Ni	Ni^{2+}
Cu	Cu^{2+}, $Cu(OH)^+$, $Cu^{2+}SO_4^{2-}$, $Cu^{2+}CO_3^{2-}$
Zn	$Zn(OH)^+$, $Zn(OH)_2$, $ZnCl^+$, $ZnCl_2$, $ZnCO_3$, Zn^{2+}
Se	$Se(IV)$, $Se(VI)$
Rb	Rb^+
Mo	MoO_4^{2-}
Cd	$CdCl^+$, $CdCl_2$, $CdCl_3^-$, Cd^{2+}
Cs	Cs^+
Hg	$HgCl_2$, $HgCl_3^-$, $HgCl_4^{2-}$, $HgOHCl$, $Hg(OH)_2$
Pb	Pb^{2+}, $PbCO_3$, $PbCl^+$, $PbCl_2$, $PbCl_3^-$, $Pb(OH)^+$, $Pb(OH)_2$, $Pb(OH)_3^-$, $Pb_3(OH)_4^{2+}$, $Pb_4(OH)_4^{4+}$

phosphates, sulphides, sulphates, and chlorides. Soil organic material is around 2–5% of the total soil mass and plays an important role in regulating water-holding capacity of the soil, its ion-exchange capacity, and binding of metal ions. In particular, the most stable compounds in soils are humic substances, such as humic acid and fulvic acid which have a great affinity for interacting with metal ions. The interstitial spaces between soil particles contain the soil water, from which plants obtain their nutrients, especially trace elements and water. In fact, one of the most important properties of soil and the soil solution is pH which directly influences the fertility of the soil and nutrient availability to plants. Table 15.6 shows the major chemical species found in soils and soil solutions following weathering processes.

In the aquatic and marine environments, sediments are also an important part of the earth's crust. Sediments are river borne clay, silt, sand, organic material and a mixture of other materials. Ocean sediments are mainly fine-grained alumino silicates. The trace elements found in significant amounts in sediments are Cd, Co, Cr, Cu, Fe, Mn, Mo, Ni, and Pb. Most of these trace elements are sorbed on to clays and $Mn(IV)$ or $Fe(III)$ hydrous oxides, or occur as low soluble organic complexes. Bacteria in the top sediment–water interface can strongly influence the chemical form and toxicity of various trace elements. Biomethylation processes in such sediments can convert Hg^{2+} or AsO_4^{3-} (arsenate) into CH_3Hg^+, $(CH_2)_2Hg$ or $CH_3AsO(OH)_2$ (methylarsenic acid), $(CH_3)_2AsO(OH)$ (dimethylarsenic acid), which then become concentrated in fish and thereby enter the food chain. Other chemical forms in sediments directly depend on pH, the redox conditions, and the nature of other ligands.

Table 15.6 Major chemical forms or species found in soils and soil solutions[a]

Trace element	Chemical form or species	
	Soils	Soil solution
V	Cationic/anionic oxides hydroxy oxides	VO^{2+}, VO_4^{3-}, VO_3^-
Cr	Chromite $FeCr_2O_4$	$Cr(OH)^{2+}$, CrO_4^{2-}, CrO_3^{3-}
Mn	Oxides, hydroxides (Al, Li)$MnO_2(OH)_2$	Mn^{2+}, Mn^{3+}, Mn^{4+}, $MnOH^+$, $MnOH^{2+}$, $MnHCO_3^+$, MnO_4^{2-}
Co	Fe minerals, Mn oxides	Co^{2+}, Co^{3+}, $Co(OH)_3^-$
Ni	Pentlandite (Fe, Ni)$_9S_8$ Garnierite (Ni, Mg)$_3Si_2O_5(OH)_4$	Ni^{2+}, $NiOH^+$, $NNiO_2^-$, $Ni(OH)_3^-$
Cu	Chalcocite Cu_2S, Chalcopyrite $CuFeS_2$, Bornite Cu_5FeS_4	Cu^{2+}, $Cu(OH)^+$, $Cu(OH)_2^{2+}$, $CuCO_3$, $Cu(OH)_4^{2-}$, $Cu(CO_3)_2^{2-}$ Cu^{2+}-humic or fulvic acids
Zn	Sphalerite ZnS	Zn^{2+}, $ZnCl^+$, $ZnOH^+$, $Zn(OH)_2$, $ZnCO_3$, $[ZnCl_4]^{2-}$
As	Arsenates AsO_4^{3-}	AsO_2^-, AsO_4^{3-}, $H_2AsO_3^-$
Cd	CdS	Cd^{2+}, $CdCl^+$, $CdOH^+$, $CdHCO_3^+$, $CdCl_3^-$, $CdCl_4^{2-}$, $Cd(OH)_3^-$, $Cd(OH)_4^{2-}$
Hg	HgS	Hg_2^+, Hg^{2+}, $HgOH^+$, $HgCl^-$, $HgCl_4^{2-}$, CH_3Hg^+, $CH_3CH_2Hg^+$
Pb	PbS	$PbOH^+$, $Pb_4(OH)_4^{4+}$, Pb–organic complexes

[a] Soil solution following weathering processes.

The chemical composition of plants is generally related to the elemental content of nutrient solutions or soils. Absorption processes are very complex; the main pathway of trace elements to plants is via the roots. Foliar uptake can occur but this is only a major pathway in relation to aerial sources of pollutants. Root uptake is dependent upon the dissolved chemical forms in the soil solutions (ionic, chelated, complexed), pH, the presence of other ions, redox potentials and temperature. There is a wide variability in the bioaccumulation of trace elements among different plant species. Some elements such as B, Cd, Rb, Cs are readily taken up, whereas Fe and Se are only slightly available to plants. Trace element absorption by plant roots is also influenced by mycorrhizal fungi, which enhance uptake from the soil solution in exchange for carbohydrates from the host plant. Evolutionary changes have also resulted in metal tolerant plant species which are able to accumulate very high concentrations of specific metals (Ni, Zn, Cr, Co, Se, Cu, Hg). The trace element balance of many plant species is also dependent upon the antagonistic or synergistic interactions between chemical elements. Calcium, P, Mg and K are known to be the main antagonistic elements against the absorption and metabolism of various trace elements (B, Co, Cr, Cu, F, Fe, Mn, Ni, Sr, Zn). In some cases antagonistic interactions are beneficial to plants. For example, the use of Si complexes in plant root hairs to prevent the uptake of Al^{3+}.

The trace element composition of animal and human tissues or fluids is directly related to the trace element content and bioavailability in the soil or sediment–plant–animal–human food chain. Many common components of food in a given diet will influence both the amount of trace elements and their chemical form. For example, carbohydrates, free fatty acids, fibre, protein, phytate, organic acids, etc. can affect trace element utilisation in the body. Also complex interelement interactions will affect both absorption and metabolism of specific trace elements, e.g. excess dietary Ca, Pb and Zn will affect Zn, Fe, and Cu, respectively.

One of the major even-increasing problems is the influence of chemical pollution, especially heavy metal contaminants, upon the complex interactions of the biosphere. The relationship between animal or human health and exposure to elements through air, water and food is an important area of environmental analytical research. Before assessing the requirements for trace element analysis, it is necessary to identify the sources of elemental contamination and the effect of environmental pollution.

15.3 Trace element contamination and pollution

All trace elements occur to a varying extent within all components of the environment. The term pollution of the environment refers to an increase of a trace element relative to the natural occurrence of that element. This increase may relate to natural events, but in general is associated with human activities. In particular, industrial and agricultural practices result in a more widespread trace element contamination of the environment than does natural occurrence.

15.3.1 Air particulates

Air pollution is generally related to gaseous emissions into the atmosphere (carbon monoxide, carbon dioxide, hydrocarbons, halocarbons, sulphur dioxide, hydrogen sulphide, nitrogen oxides, NO, NO_2, N_2O, and ammonia). Trace element or heavy metal contamination can result primarily through atmospheric particles or particulates. The main sources (Table 15.7) are coal and fuel power generation plants, metal processing and smelting, transportational combustion, waste incineration, and aerosol sprays (halocarbons). The major industrial emissions are non-ferrous metal smelters, petroleum refining, cement production, chloro-alkali production, aluminium smelters, and chemical production. In general, most trace elements or heavy metals can be released through industrial emissions, although the major examples are Ca, Mg, Ba, Cd, Pb, F, Br, Ti, V, Mn, Fe, Cu, Cr, Ni, and Zn.

Metal air pollution is of major concern in that it is global and contributes to contamination of all the components of the environment. In most cases the maximum pollution levels are within a few kilometres of the emission source, but small particulate and aerosol pollutants can contaminate all areas of the earth. Several studies have shown a slow accumulation of lead in both the Arctic and Antarctic regions since the introduction of lead alkyl

Table 15.7 Trace elements released into the atmosphere from human activities

Activity	Be	F	Al	Sc	Ti	V	Cr	Mn	Fe	Co	Ni	Cu	Zn	As	Se	Br	Sr	Mo	Ag	Cd	Sn	Sb	Hg	Ti	Pb	Bi	Th	U
Coal and fuel power generation	X	X	X	X	X	X	X	X	X		X	X	X	X	X		X	X				X	X	X	X		X	X
Metal processing and smelting	X	X	X	X		X	X	X	X	X	X	X	X	X	X			X	X	X	X	X	X	X	X	X	X	X
Transport					X	X	X	X		X	X	X	X			X									X			
Waste incineration			X	X	X		X			X	X	X	X	X					X	X	X	X	X		X			
Aerosol sprays			X									X	X							X		X						

additives to petrol in the early 1920s. The physical properties of particulates in air and their transfer to vegetation and soil surfaces is complex. The major processes involved are those of Brownian motion, sedimentation, impaction, and interception. The importance of any of these processes in the actual deposition of a particulate depends upon a variety of factors including windspeed, particle size, temperature, humidity and surface roughness.

A recent problem of metal particulate air pollution is their role in the oxidation of sulphur dioxide and the formation of acidic aerosols involved in global acid rain. Many metal ions, such as Mn(II), Fe(III), Cu(II), Cr(III), Al(III) and Pb(II) act as catalysts in the reaction $2SO_2 + 2H_2O + O_2 \rightarrow 2H_2SO_4$. Trace elements also contribute to the formation of photochemical smogs (Fe, V, Ca, Pb, Br, Cl). A mixture of metal particulates and acidic aerosols can directly affect human health through the increased incidence of eye and respiratory tract irritation, bronchitis and asthma. Other problems occur with regard to an increase in the deterioration of building materials (especially limestone and marble), and a reduction in plant quality through reduced plant photorespiration.

As a result of increased industrial and transportational emissions the release of metal particulates into the environment is now under strict control with the introduction of air quality regulations. Transportational emissions are being reduced by the use of catalytic converters which can only be used on vehicles using non-leaded fuels. Industrial emissions are controlled by centrifugal separators, cyclones and filters (particles >0.01 µm), electrostatic precipitators (for small diameter particles >0.005 µm), and wet scrubbers (particles 0.2–10 µm).

15.3.2   Natural waters

A wide range of human activities contributes to the trace element pollution of the aquatic environment. The major activities include mining and ore processing, coal and fuel combustion, industrial processing (chemical, metal, alloys, chloro-alkali, petroleum), agricultural (fertilisers, pesticides and herbicides), domestic and agricultural effluents or sewage, transportational (urban and motorway run-off), and nuclear activity (Table 15.8). Elemental or heavy metal input can be from atmospheric fallout, leaching or dumping from the lithosphere, or direct into the aquatic environment (including ground, surface, river, lakes, estuarine, oceans, etc.).

The impact of water pollution depends upon the magnitude of trace element input (both as single or multiple element(s)), duration of input, physical and chemical form and associated ligands or chemicals. All of these inter-related factors will determine the elemental concentrations in the water systems, and their relative availability, transport, and toxicity. The most important factor is the chemical form in which the element exists in solution.

Table 15.8 Trace elements released into the aquatic environment from human activities

Activity	Be	F	Al	Sc	Ti	V	Cr	Mn	Fe	Co	Ni	Cu	Zn	As	Se	Br	Sr	Mo	Ag	Cd	Sn	Sb	Hg	Tl	Pb	Bi	Th	U
Mining and ore processing	X	X	X			X	X	X	X	X	X	X	X	X	X			X	X	X	X	X	X	X	X		X	X
Coal and fuel combustion	X	X		X	X	X	X	X	X		X	X	X	X	X		X	X		X	X	X	X		X		X	X
Industrial processing	X	X	X	X	X	X	X	X	X	X	X	X	X	X	X	X	X	X	X	X	X	X	X	X	X	X	X	X
Agricultural		X					X	X				X	X	X					X	X	X	X	X		X			
Domestic and agricultural effluent/sewage		X	X		X		X	X	X		X	X	X	X			X	X	X	X	X	X	X		X			
Transport						X	X	X			X	X	X			X									X			
Nuclear industry										X									X	X							X	X

This depends upon pH, solubility, temperature, the nature of other chemical species, and many basic factors of solution chemistry.

Many trace element contaminants entering the aquatic environment can have dramatic effects on the bioavailability and toxicity of biological processes. In particular, bio-amplification by plankton, or bio-transformation by bacteria in the water–sediment interface can strongly influence elemental toxicity throughout the remaining food chain. For example, Hg, As, Sn, and Pb undergo biomethylation in the water–sediment interface, resulting in the production of more toxic species which are concentrated in shellfish or fish.

During the past 50 years there has been a rapid increase in the number of major trace element related water pollution incidents. Examples include mercury in Minimata, Japan (methylmercury from a chemical plant where it is used as a catalyst in the production of vinyl chloride and acetaldehyde); cadmium in the Jinstu River, Japan (traced to effluents from zinc mining) which produced Itai-Itai (Ouch-Ouch) or severe bone damage disease in the local human population; and aluminium in domestic drinking water, Camelford, Cornwall, England (from the accidental addition of excess aluminium sulphate to water during the treatment of raw water).

An ever increasing problem is the effect of atmospheric pollution and acid rain on the aquatic environment. Relatively pure rain water has a pH of 5.5 but owing to SO_2 and NO_x emissions, the pH of rain can drop as low as 2–3. This increases the acidity of lakes and accelerates the leaching of trace elements from soils, which in turn affects both the metabolism of soil organisms and aquatic life. An example of this process is the increase in aluminium in lake waters due to the increased mobilisation from soils. Aluminium species are more soluble at lower solution pH, and this directly affects both water quality and fish life.

15.3.3   Soils and sediments

Trace element contamination of surface soils and sediments is slowly occurring on a global scale, in response to growing industrial and agricultural activities. In many cases soils act as natural buffers controlling the transport of trace element contaminants to the hydrosphere and biota. For example, pH and the organic content of soils play a major role in the complexing and subsequent dispersion and bioavailability of elemental ions. As such, the persistence of contaminants in soil is much longer than in the other components of the biosphere.

Global surface soil trace element contamination occurs at a slow rate and represents the long range transport of some pollutants, such as Pb, ^{137}Cs, ^{90}Sr. In contrast, local contamination occurs mainly around industrial centres, such as, metal smelting, chemical process factories, coal and fuel combustion plants, alongside motorways, and adjacent to domestic or industrial waste incinerators. Various agricultural practices, including the addition of fertilisers, pesticides or herbicides, sewage sludge, and waste water irrigation can add trace elements to soils. For example, Cd, As and Pb, are common trace elements in rock phosphate, which is widely used as superphosphate fertilisers. Similarly, the long-term application of sewage sludge to garden and farmland soils increases the levels of Cr, Ni, Cu, Zn, Cd, Hg and Pb.

The addition of trace elements to soils normally results in accumulation in the upper layers of the soil profile. This is primarily due to the soil organic horizon, which is rich in humic substances and organic acids that have a high affinity for elemental ions. In general, different soil types, their physical and chemical properties, pH, water content, the plant species and soil organisms associated with the soil will influence the degree of trace element contamination, the distribution within the soil profile and the bioavailability to plant roots.

Whilst the trace element contamination of soils is principally related to aerial sources, those involving the hydrosphere concentrate in sediments. The hydrous oxides of Al(III), Mn(IV) and Fe(III) readily complex trace elements in sediments. For many trace metals, Cr, Ni, Cu, Zn, Cd, Hg and Pb, sediment analysis is used to investigate the transport of elements, and the potential contamination of rivers, lakes, estuarine, and near shore aquatic environments. Moreover, many important chemical and biological processes occur in the water–sediment interface, including the biomethylation of trace metals, such as, As, Sn, Hg and Pb.

15.3.4 Plants

The trace element contamination of plants has a major impact on both the environmental cycling of nutrients and the quality of foodstuffs. Plants can accumulate trace elements, especially heavy metals, from soils, waters or air, and in themselves become elemental sources for animals and man. The normal mode of elemental uptake is via the soil solution and roots. However, in industrial mining and smelting areas, aerial deposition will result in significant elemental foliar uptake. This is also important for the application of fertilisers and the absorption of radionuclides released into the atmosphere (^{131}I, ^{134}Cs, ^{137}Cs, ^{90}Sr).

The impact of contaminated elemental uptake by plant roots is dependent upon many factors, including: the magnitude and chemical form of trace element(s) present; soil pH, moisture, aeration, temperature, organic matter and phosphate content; the presence or absence of competing ions; the plant species, rooting depth, age; and seasonal growth effects. Elemental uptake and toxicity are closely related, and the toxicity of one element will affect the uptake and toxicity of another element. Moreover, excessive elemental uptake will induce toxicity or deficiency effects on both the plant and plant-soil microorganisms. For example, the addition of high phosphate fertilisers or sewage sludges to surface soils reduces the biological activity of mycorrhizal fungi, which in turn imbalances the movement of nutrients from the soil solution to the plant.

An accumulation of elements in a plant can have major effects on key plant metabolic processes, such as respiration, photosynthesis, and fixation or assimilation of major nutrients. In some cases, the impact of pollution on the metabolism of plants can be used to monitor air-borne and soil contamination. Table 15.9 shows the excessive levels of elemental uptake by plants grown in non-contaminated and contaminated sites. The major problem of plant elemental contamination is the effect it has on the dietary and health requirements of animals and humans.

Table 15.9 Elemental uptake of plants grown in non-contaminated sites

Trace element	Site	Elemental range (mg kg^{-1}, dry weight)		
		Grass (leaves)	Grass (roots)	Lettuce (leaves)
Pb	Non-contaminated	0.8–3.4	1.7–6.8	0.2–1.1
	Contaminated:			
	Motorway	198–347	18–64	33–67
	Mining and ore processing	360–870	78–160	–
	Metal processing	350–3845	63–280	45–70
	Battery manufacture	90–240	43–112	27–42
	Sewage sludge	43–121	78–310	18–63
Cd	Non-contaminated	0.05–0.14	0.08–0.24	0.01–0.08
	Contaminated:			
	Motorway	1.7–16.7	1.2–4.6	0.4–6.7
	Mining and ore processing	0.7–5.6	1.3–9.8	–
	Metal processing	1.7–20.3	1.8–5.6	1.1–6.2
	Sewage sludge	0.5–8.7	5.6–48.3	1.1–4.3
Cu	Non-contaminated	3.5–14.5	4.1–18.2	3.0–12
	Contaminated:			
	Motorway	64–96	42–71	48–70
	Mining and ore processing	78–290	73–320	–
	Metal processing	35–78	21–75	24–92
	Sewage sludge	31–68	48–107	23–41
As	Non-contaminated	0.004–0.009	0.006–0.007	0.005–0.011
	Contaminated:			
	Fertiliser	0.07–0.23	0.11–0.37	0.01–0.14
Zn	Non-contaminated	15–67	21–72	14–60
	Contamined			
	Mining and ore processing	80–274	94–460	–
	Metal processing	130–380	180–690	90–270
	Sewage sludge	120–265	118–380	72–190

15.3.5 Animals and humans

The relationship between dietary status, health and exposure to elements through food, water and air is a major area of analytical activity. The threat of ever-increasing elemental pollution, especially for heavy metals, may be directly related to the increased incidence of various diseases, including asthma, birth abnormalities, impaired mental development, cancer and Alzheimer's disease. Although at present numerous trace elements are known to be essential for animals and humans (As, Cr, Co, Cu, F, I, Fe, Mn, Mo, Ni, Se, Si, Sn, V and Zn) under conditions of excessive dietary or environmental exposure, they are all capable of becoming toxic and impairing their respective biological functions, resulting in poor health status. Aluminium is considered as a non-essential trace element, but under conditions of environmental or biological stress (low pH, Fe^{3+} or $Si(OH)_2$ deficiency) it is believed to be toxic. Cadmium, Hg, Pb and thallium are known as toxic elements, and even at small increases in magnitude they can dramatically impair health.

Moreover, the impact of many trace elements on animal and human health is not only related to magnitude, but also chemical form, and the 'cocktail effect'. Many trace elements are more toxic in organic rather than inorganic forms; for example, CH_3Hg^+, $(CH_3)_2Hg$ are more toxic than Hg^{2+}. A 'cocktail of trace elements' can include interelement interactions whereby an excess of one will induce a deficiency of another element. Also, in some cases elements will compete for specific enzyme binding sites or carrier ligands, and thereby impair the absorption, transportation, or utilisation of other elements. This results in altered physiological activity and poor health status. For example, a dietary or environmental excess of lead will interfere with the absorption of Ca, Fe, Cu and Zn, and this may induce health problems, such as, osteoporosis (weak or brittle bones), anaemia, reproductive and growth depression. Table 15.10 shows some of the health effects associated with a deficiency or excess of trace elements.

Table 15.10 Health effects associated with a deficiency or excess of trace elements

Trace element	Health effect	
	Deficiency	Excess
Fe	Anaemia	Cirrhosis of liver, haemochromatosis
Cu	Anaemia and changes in ossification	Wilson's disease
Zn	Growth retardation, hypogonadism, mental lethargy, poor appetite, skin lesions	Nausea, anaemia, neutropenia
Se	Keshan disease, cardiomyopathy, white muscle disease.	Lassitude, skin depigmentation, hair loss
Cr	Impaired glucose tolerance	Eczema and linked to cancer
V	Reduction in red blood cell population, upset Fe metabolism, lack of growth of bones, teeth, cartilage	Kidney failure, linked to manic depression
Cd	–	Hypertension, renal damage, osteomalacia, anosmia (no sense of smell)
Pb	–	Impaired haemesynthesis, insomnia, headaches, irritability, impaired mental activity, reproductive and developmental problems

**5.4 Sampling and
sample preparation**

There is a need for continuous surveillance and monitoring of all the environmental compartments, including air, water, soil, sediments, plants, foodstuffs, animal and or human tissues and fluids, in order to evaluate the magnitude and impact of trace element contamination of the environment. The environmental analytical chemist is confronted with many questions before even the first sample is collected. For example, whether total elemental or chemical speciation data; single or multi-element analysis; trace element levels in the natural and/or polluted environment; samples to be investigated that will provide the most meaningful interpretation of the environmental problem; analytical method to be used, sensitivity, selectivity, precision, accuracy.

In recent years more attention has been focused on the sensitivity and detection limit of the analytical instrument, than the problems arising before the sample is introduced into the instrument. Environmental analytical studies are only of any value if serious attention is devoted to all aspects of the analytical sequence (see chapter 1). Sampling and sample preparation are very important in trace element analysis for a variety of reasons. The comparison between natural and contaminated sites means that the magnitude of various elements may vary between trace ($\mu g \, l^{-1}$, $mg \, l^{-1}$) to major ($mg \, ml^{-1}$) levels; or have different chemical speciation forms affecting the choice of the type of sampling, storage, and sample preparation methods. Environmental sampling normally involves the collection of small masses or volumes from a bulk material. The question must be asked: How representative and homogeneous is the sample collected? It is important when samples are being collected for trace or even ultratrace analysis that steps are incorporated to avoid sample contamination and analyte loss during sampling, storage and preparation. For example, liquids are normally stored in polyethylene bottles, but without any pretreatment, analytes can be absorbed on to the bottle surface or leached from the bottle material into the solution. This process is enhanced if liquids are unfiltered, whereby particulate matter will adhere to the surfaces and lead to trace element losses. Also, bacteria and algae growth in effluents and waste waters will scavenge trace elements from the water. This process is dependent on pH and temperature. More recently, trace element speciation studies have required a reassessment of traditional sampling procedures to prevent modification of the metal species between the real sample collected and that which is analysed.

15.4.1 Atmospheric samples

The main considerations are: size of sample (which is governed by the trace element levels to be measured, and sensitivity of analysis method), rate of sampling (relates to collection efficiency of sampling device), and duration of sampling (time of day, time intervals). Most sampling devices operate for medium- or long-interval samplings, so the results represent the average elemental concentration during that period. High flow rate devices can be used for short-interval sampling but this may cause detection problems.

Trace element atmospheric measurements are normally undertaken by the sampling of aerosols with particulates in the general size range 0.001–10.0 μm. Batch sampling is usually employed with collection of particulates by filtration, impaction, or electrostatic precipitation.

Filtration is carried out using either low (0.1–3 $m^3 h^{-1}$) or high (10–60 $m^3 h^{-1}$) volume samplers. Glass, quartz, fibre, or plastic fibre, and membrane filters are used. Some problems arise with trace element blank filter contamination (glass), poor mechanical strength (plastic), or high airflow resistance (membrane). Impaction collectors separate the aerosol into size fractions and provide good efficiency levels for the collection of particulates down to 2 μm in size. Electrostatic precipitators are very sophisticated collectors which will collect particles down to 0.001–0.01 μm.

The collection of volatile gaseous species which may be adsorbed on particulates (organolead and mercury species) require the use of a range of absorbants and extractants (iodine monochloride in hydrochloric acid with CCl_4 extraction for organolead), activated charcoal and graphite cup collectors, or vapour traps.

Atmospheric dry or wet deposition (dust particles or particulates in rainwater) can be achieved using polyethylene funnels mounted on a collection bottle. Major problems can occur through sample contamination, and losses during collection and storage. The addition of acid to prevent trace element losses through adsorption onto the bottle surface is recommended, but eliminates the further analysis of the particulate matter.

Air particulate filters can either be analysed directly (by neutron activation analysis, XRF, or laser ablation and probe methods) or after subsequent acid digestion (normally with HNO_3, HF, or $HClO_3$ mixtures). Chemical speciation information can be obtained by extraction schemes which differentiate elements into water-soluble, acetic acid-leachable, easily reducible, H_2O_2-oxidisable, moderately reducible, and residual fractions.

15.4.2 Water samples

In general water sampling requires a very close inspection of the required procedures because most trace elements to be measured in natural samples will be at very low or ultratrace levels (μg l^{-1} or ng l^{-1}). Sample contamination and analyte losses are potential problems.

Water samples can be collected from numerous sites, including domestic faucet outlets, surfaces of rivers, lake waters, estuarine and oceanic waters, rainwater, and varying types of effluents associated with industrial, mining process, landfill waste leachates, motorway run off, and sewage or water treatment works. The frequency and duration of sampling is important in order to obtain both a representative and reproducible sample. Composite sampling is popular in which individual samples taken over regular time intervals are combined.

Various container pre-treatment procedures should be undertaken before sample collection. Only polyethylene or Teflon® vessels should be used. Prior to use all new containers should be washed and stored in 10% HNO_3 (ultra-pure Aristar®) for 2 days, and rinsed with double distilled deonised water. All cleaned containers should be sealed in clean polypropylene bags until required. Following collection, acidification of the sample (normally with 2 ml 10% HNO_3 or 5 M HCl) will reduce or eliminate trace element adsorption and hydrolysis. However, acidification cannot be undertaken if speciation analysis is required. For such studies immediate freezing of the water sample ($-20°C$) is recommended.

Surface water collection is normally undertaken using polyethylene sampling bottles attached to a telescope arm. This device reduces contamination from dust or sediment during the immersion process. Depth samplers are used to collect samples at a specific depth. These devices contain mechanisms for opening and closing the stopper at the required depth.

If tap water is collected, the first water running from the tap must be avoided because there will be a high accumulation of trace elements stemming from the pipes, soldering and welds. The best method is to sample after the tap water has continuously run for at least 2 min. Collection of rain samples requires special containers to prevent dry deposition entering the sampler between periods of precipitation. The samples must be at least 2 m above ground to eliminate contamination by soil splashing.

All water samples should in theory be analysed immediately after collection. Whilst this is not practical, special stabilisation procedures are recommended. It is important to undertake all associated measurements (pH, temperature, conductivity) before solution stabilisation and storage. Many water pollution studies may involve the monitoring of other physical or chemical components (dissolved gases, particulate and dissolved matter, ammonia, phosphate, nitrate, nitrite, sulphate, hardness of water, BOD, COD, bacteria, microorganisms, etc.) which can routinely be undertaken by on-site field microanalysis kits.

Most water samples require filtration immediately following collection. This removes algae, bacteria, and particulate matter which can either contaminate or rapidly absorb trace elements from the water. Filtration with 0.5 μm membrane filters is recommended. Some ultratrace element analyses may require the water sample to be preconcentrated immediately after collection. This is also common when large sample volumes are collected in the field, such that the preconcentration step reduces the water sample volume required for transporting to the laboratory. Micro columns with Biorad™, Sep-pak™, Chelex™, containing 8-hydroxyquinoline and other chelating agents are recommended.

Stabilising agents are commonly added after pretreatment. Nitric, hydrochloric and sulphuric acids are widely used in amounts that lead to pH values between 1 and 3.5 in the stabilised samples. For total mercury analysis the addition of nitric acid and potassium dichromate is recommended. Stabilisation for organomercurial compounds is achieved with 1% sulphuric acid.

Before storage, all water sample collection bottles should be completely full. The presence of air may chemically or biologically alter the sample. This is particularly true for As(III) and Fe(II) speciation analysis. In order to prevent oxidation, acidification to pH 2, and deoxygenation with nitrogen is required.

All water samples should be stored in the dark, either by refrigeration (4°C), or by deep-freezing (− 20°C). The latter is reported to eliminate the leaching and/or sorption of trace elements from polyethylene bottles, however, the effect of thawing the sample may cause problems for Si and Al analysis. Long-term storage of frozen water samples may also be a problem for some trace metal species especially tributyltin.

15.4.3 Soils and sediments

The trace element concentrations in soils and sediments are normally much higher than those for water samples. However, the many precautionary steps relating to sample container preparation, and sampling of waters also apply to soils and sediments. In particular, the need for the samples to be representative of the sampled land is very important. A careful site inspection should be undertaken in order to establish the sampling strategy. Most studies use either a random or systematic square grid pattern sampling scheme. Choice depends upon the area to be sampled, number of samples, the amount of sample, topography and density of obstacles (houses, roads, fences, hedges, forests, rivers/lakes, etc.). The investigation of polluted areas normally requires a comparable non-contaminated area to also be sampled. It is important to assess whether differing geology or geochemistry, weather conditions, plant density, or topography do not influence the comparison between the two study sites.

Surface soil sampling should be carried out using a non-metal trowel. If a stainless steel tool is used, thorough washing with acetone and double distilled deionised water is required. The operator should also wear non-powdered disposable polyethylene gloves. Soil depth sampling is carried out using a 30–40 mm diameter auger or corer.

In many cases it is important to take sub-samples within a defined area in order to provide a representative composite sample. The total amount of mineral or organic soils collected should provide at least 500 g of sieved air dry soil.

Sampling of sediments requires a careful sampling strategy. Seasonal variation in climate water flow or depth conditions may influence both the sediment and associated biota. Sediment sampling is normally carried out using a tube corer or grab sampler. Composite sediment samples of 1 kg are common. It is important to avoid contamination from the metal surfaces of the sampler by sub-sampling from the middle of the bulk sample. Polycarbonate or perspex cores should be sealed underwater with polyethylene end caps.

The studying of sediment interstitial waters is becoming very important for evaluating the transport of trace elements from the sediment to the overlying water and vice versa. Centrifugation will provide separation of the sediment which can then be placed directly into a pre-acid washed polyethylene bottle.

Soil and sediment samples should be stored in sealed polyethylene containers at 40°C until arrival at the laboratory. Following the physical separation of debris and other large particles (stones, wood) samples should be separated into particle sizes by wet or dry sieving. Drying at 110°C is often practised, but this may result in the loss of volatile elements, such as Hg. Air drying is common (~24 h) but this will present problems for Fe speciation analysis. Most studies use a 63 μm sieve to remove sand and gravel. However, 20 μm or less sieve sizes can be used for fractionation studies. In all cases only polyethylene or nylon sieves should be used. All sieved and dried soils or sediments should be stored in sealed polyethylene containers at 40°C. Moist sediments are often used for metal speciation analysis. Such samples should be sealed under nitrogen in polyethylene containers and frozen.

The pre-analysis procedures for soils and sediments will be detailed in a later section. Before chemical analysis samples should be homogenised and dried (30–60°C) to a constant weight. Homogenisation is achieved by a grinding mill or agate pestle and mortar. A popular method of homogenising powdered material is to cone and quarter. The powdered sample is poured into a cone shaped heap, divided into four equal parts. Two opposite quarters are combined and re-coned. The process is repeated until the amount of sample is reduced to that required for analysis.

15.4.4 Plants

The collection of plant material may relate to both leaves and/or the root system. The aerial parts are obtained by cutting with acetone-distilled deionised water pre-cleaned secateurs at least 3 cm above soil level. Aerial parts must be stored separately from root samples for transport to the laboratory. During sample collection the analyst should wear disposable polyethylene gloves. All samples should be stored in polyethylene bags or pre-acid washed containers.

As with soils and sediment sampling, care must be taken to ensure that the plant is representative of the area being studied. Plant sampling regimes are similar to those previously mentioned (section 15.4.3) although plant population or density, especially for food crops is an important factor. Plant contamination from spraying, industrial or transportational particulate fallout or the application of fertiliser requires a carefully planned sampling programme. The collection and transfer of plant leaves or stems should not affect the surface particulate contamination. Composite sampling should provide 0.5 kg of fresh weight material.

Wet plant material should be transferred to the laboratory immediately after field collection. The sub-sampling or sectioning of samples must be carried out on a polyethylene sheet in a laminar-flow work station.

Washing is desirable for terrestrial plant material, in order to remove surface contamination, and in some studies to provide an evaluation of surface particulate contamination (contrasts between washed–unwashed material). All root material should be washed with double-distilled de-ionised water as any trace element contribution from soil particles will influence the plant material results.

Freeze drying (2–5 g sample sizes) is the preferred method for removing moisture from samples without trace element losses or contamination. Air oven drying at 100°C can be used, but very volatile species, such as elemental Hg and many organometals may be lost at this stage. Muffle ashing at 500–550°C is used to decompose most organic matter, but should be avoided in studies for Hg, As, Sn, Se, Pb, Ni and Cr. Plant ash should be homogenised with a polyethylene tool. Freeze dried plant is often powdered using an agate mortar and pestle, grinder or blender, but care should be taken to avoid any sources of metal contamination. Storage of freeze-dried or ashed plant samples poses no major problems. It is important to ensure that all material is homogenised and dried to a constant weight before acid digestion or solvent extraction is undertaken.

15.4.5 Biological tissues and fluids

Elemental analysis of biological tissues and fluids requires a careful examination of the study protocol before sample collection. Samples may be taken as part of a clinical monitoring programme of patients, evaluation of industrial or occupational exposure, general nutrition or dietary assessment, identification of a disease or an inherited disorder of human trace element metabolism, etc. Moreover, samples maybe required from neonatal, postnatal, children, adults, pregnant or severe and chronic disorder patients. Availability and selection criteria for both study and control or normal populations may influence both the types of samples, amount of sample, time of sampling, etc. The choice of sample is of fundamental importance is such studies. Often readily available material, such as, scalp hair, urine, or even blood serum or plasma, may not be the ideal material for studying a particular trace element or chemical, biochemical, nutritional, dietary toxiological, or metabolic problem.

The major concern in the collection of biological samples is contamination. Stainless steel blades, biopsy needles, metal scissors or tweezers should not be used. Blood samples should be taken using Teflon® or polyethylene catheters. Glass tubes or Vacutainers™ should be avoided. Urine collection requires mid-stream samples to be delivered into a covered acid-washed polyethylene container in order to prevent dust contamination.

Animal or human organ samples should be sampled using a quartz or tantalum blade, and not stainless steel which will result in high Mn, Cr and Ni levels. It is essential to wear disposable non-powdered polyethylene gloves, and all samples (including blood or urine collection tubes) should be stored in sealed polyethylene bags.

Sample preparation methods for biological tissues and fluids can be major sources of trace element contamination. In particular, sample homogenising is normally undertaken with a laboratory homogeniser or blender that has metal blades or inner walls. Hair samples require a washing procedure to remove exogenous elements. The recommended method is ultrasonication with 0.1% Triton™ X-100, filtration, washing with methanol or acetone followed by repeated double-distilled deionised water. Hair samples are dried under an infrared lamp with the material enclosed in a quartz container to prevent dust contamination.

Freeze drying is also widely used for blood, urine, milk and tissue samples, which are then stored. Most biological material is stored frozen, although whole blood samples can hemolyse. On thawing, the sample should be thoroughly mixed before an adequate sample is taken for acid digestion or solvent extraction. Routine digestion of biological samples is undertaken by dry ashing (muffle furnace at 500–550°C), wet digestion (open water bath or hot plate), pressurised decomposition with mineral acids in Teflon® bombs, low-temperature plasma ashing, or microwave digestion.

15.4.6   Sample digestion methods

The majority of trace element analytical techniques require the sample to be in solution form. There is no universal dissolution procedure for all types of samples. The most desirable features of such procedures are: (1) the ability to dissolve the sample completely (no insoluble residues); (2) reasonably quick and always safe; (3) no possible sources of sample loss through volatility, adsorption onto the walls of the vessel; and (4) elimination of sample contamination from the reagents used in the dissolution process. The majority of dissolution procedures involve dry ashing or wet digestion using one or a combination of concentrated mineral acids.

Muffle ashing at 500–550°C will decompose most organic matter, although problems can occur through volatilisation of Hg, As, Sn, Se, Pb, Ni and Cr. Wet digestion is often the preferred method for soils, sediments, biological tissues and blood. For trace and ultratrace element analysis, the reagent blank is very important. The mineral acids used in wet digestion procedures can be sources of many elements, especially Al, As, Mn, Cr, Ni and Zn. Only ultrapure or Arista® grade purity acids should be used. For ultratrace element analysis distillation of acids in a quartz sub-boiling still is necessary. Soil, sediment, and sludges can be readily digested using perchloric acid, or nitric-hydrofluoric acid mixtures (1:1 HNO_3/HF), with 0.1–0.5 g of sample attacked by 10 ml concentrated acid solution in a polypropylene squat beaker, heated by water or sand both placed in a special fumecupboard. Care must be taken for elements that form volatile chlorides or fluorides, and unstable nitrate complexes. Aqua regia (3:1 HCl/HNO_3) is often used for soil digestion. Biological tissues and fluids (0.25–1.0 g or 1–5 ml) can be digested in micro Kjeldahl digestion vessels, with controlled temperature heating mantles using nitric acid and H_2O_2. However, wet digestion using an open vessel always is subjected to possible element volatility problems. Pressurised decomposition (0.1–0.2 g or 0.5–1.0 ml) with nitric acid in Teflon® digestion bombs eliminates this problem and has the added feature of increased digestion efficiency through smaller acid volumes and pressure digestion.

Microwave digestion provides both a closed system method and shorter digestion times. Optimal conditions depend on the sample (weight, composition, volume of digestion reagents, reaction temperature, pressure, and time), and the digestion system (especially power ratings).

Both bombs and microwave digestion systems require strict attention to general safety rules in order to prevent explosive type reactions. In particular: (1) digestion of very fine powder samples must be restricted to less than 500 mg (ideally 100–200 mg) of sample, add only a maximum of 2 ml HNO_3, and start with the minimum amount of power and shortest possible time (microwave); (2) avoid the digestion of biological material containing more than 20% fat content; and (3) digestion of organic material generates a large amount of vapour, so after digestion be careful and never open a warm vessel.

A major limitation of existing sample digestion methods is that they are slow, labour intensive, and require strict attention by the analyst under the codes of laboratory Health and Safety, Good Laboratory Practice, or Quality Assurance (NAMAS, ISO 9000, BS5750, EN 29000). Also during the normal course of an analytical routine, it is necessary to undertake various procedures, such as weighing or volume measurements, sequential dilution of standard or digested sample solution, and the preparation of multielement calibration standard solutions. A basic systematic error can be introduced by incorrectly calibrated balances, pipettes or volumetric containers. It is always important to remember that most errors and poor trace element results occur before the sample is presented to the analytical instrument.

15.5 Methods of analysis

The ideal analytical technique for measuring trace elements in environmental samples must offer: (a) very low detection limits, (b) a wide linear dynamic range, (c) simple interference-free data, (d) qualitative, semi-quantitive and quantitative analysis, (e) possible simultaneous multielement capability, (f) simple sample preparation, and (g) high throughput and low cost per determination. In practice, whilst manufacturers or salesmen may claim that their technique is superior and ideal for water or sediment or rock analysis, in reality there is no universal analytical technique for environmental analysis. All the main contenders, including atomic absorption spectrometry (flame or electrothermal) (AAS, ETAAS), atomic fluorescence spectrometry (AFS), inductively coupled plasma optical (or atomic) emission spectrometry (ICP-OES or ICP-AES), neutron activation analysis (NAA), X-ray fluorescence (XRF), proton-induced X-ray emission (PIXE), spark source or isotope dilution mass spectrometry (SSMS, IDMS), electrochemical (anodic stripping voltammetry and polarography) or inductively coupled plasma mass spectrometry (ICP-MS) have their advantages and disadvantages (see chapters 7–10).

The main limitations for most of these analytical methods are sensitivity or precision problems due to interferences and sample matrix effects. In some cases these problems have been reduced or eliminated by the use of pre-analysis separation schemes, matrix-matching of reagent blanks, calibration standards and samples; the inclusion of instrument background correction (AAS) or the coupling of hydride generation (HG), electrothermal vaporisation (ETV), cold vapour (CV), flow injection (FI) ultrasonic nebulisation (US) or laser ablation (LA) devices to the trace element detector. Tables 15.11 and 15.12 summarise the general analytical features of the main methods used for environmental or biological sample trace element analysis.

Detection limits relate to the least amount of material the analyst can detect because they yield an instrument response significantly greater than a blank. Multielement techniques (ICP-OES, NAA, ICP-MS, XRF) usually provide significantly better detection limits when operated in the single element monitoring mode. Electrothermal vaporisation, cold vapour (Hg), ultrasonic nebulisation, and hydride generation (As, Se, Sn, Sb, Bi, Te, Pb) devices can also reduce the detection limits either from the separation of matrix interferences, increased sensitivity, or improved analyte transport before the production of free atoms or ions. Neutron activation analysis allows for multielement determinations which are basically nondestructive to the sample (instrumental or INAA). Radiochemical separation methods (RNAA) can enhance sensitivity and selectivity by preconcentrating the analyte or eliminating interferences. Electrothermal vaporisation devices (graphite rods, furnaces, cups, tantalum boats, metal filaments) enable the analysis of small sample volumes (5–10 μl), and improve analyte sensitivity by separating the element from the sample matrix. The use of chemical modifiers can improve the volatilisation of the many analytes, such that

Table 15.11 Comparison of the typical detection limits for the main methods used in environmental or biological trace element analysis

	AAS		ICP-OES		NAA		ICP-MS		XRF
	FAAS	ETAAS	PN	US(ET)	INAA	RNAA	PN	LA	
Al	xx[a]	xxxx	xx	xxx	xx	–	xxx	xx	–
V	xx	xxx	xx	xxx	xxx	xxxx	xxx	xx	xx
Cr	xx	xxxx	xx	xxx	xx	xxx	xxx	xx	xx
Mn	xx	xxxx	xxx	xxxx	xxx	xxxx	xxxx	xx	xx
Co	xx	xxxx	xx	xxx	xxx	xxxx	xxxx	xx	xx
Ni	xx	xxxx	xx	xxx	xx	xxx	xxx	xx	xx
Cu	xx	xxxx	xxx	xxxx	xx	xxx	xxxx	xx	xx
Zn	xxx	xxxx	xxx	xxx	xx	xxx	xxx	xx	xx
As	x[b]	xxx[b]	xx[b]	xxx[b]	xx	–	xxx[b]	–	xx
Se	x[b]	xxx[b]	xx[b]	xxx[b]	xx	–	xxx[b]	–	–
Rb	xx	xxxx	–	–	xx	–	xxxx	xx	xx
Mo	xx	xxxx	xx	xxx	xx	xxx	xxx	xx	–
Ag	xx	xxxx	xx	xxx	xx	–	xxxx	xx	–
Cd	xxx	xxxx	xx	xxx	xx	xxx	xxxx	xx	xx
Sn	x[b]	xxx[b]	xx[b]	xx[b]	xx	xxx	xxx[b]	xx	xx
Sb	x[b]	xxx[b]	xx[b]	xx[b]	xx	–	xxxx	xx	–
Hg	x[c]	xxx[c]	xx[c]	xx[c]	xxx	–	xxx[c]	–	–
Pb	xx	xxxx	xx	xxx	–	–	xxxx	xx	xx

[a] Detection limits: x, >1 mg dm^{-1}(mg kg^{-1}); xx, <1 mg dm^{-3}(mg kg^{-1}); xxx, <1 mg dm^{-3}(mg kg^{-1}); xxxx, <0.05 mg dm^{-3}(mg kg^{-1}).
[b] Improved by hydride generation (HG).
[c] Improved by cold vapour (CV).

coupled to an atomic absorption spectrophotometer, ETV enables single-element assays to be undertaken at levels typically below 1 µg l^{-1}. Flow injection (FI) on-line column preconcentration techniques provide for improved sensitivity and selectivity of atomic spectrometric methods. The main advantages of FI include a lower consumption of sample and reagents, with improved precision (typically 1–2% relative standard deviation), and the ability to analyse solutions with higher concentrations of dissolved solids.

The trace element analysis of solid samples normally requires ashing or digesting which may lead to the problems of loss and contamination. Laser ablation analysis of solids is becoming very popular, although both precision and detection limits are poorer due to homogeneity problems and the reduced signal produced through only a small sample mass (typically <1 mg) being ablated.

In recent years, the emphasis of trace element analysis techniques has diverted from the determination of total concentrations towards the measurement of elemental speciation. The need to evaluate the behaviour and fate of an element in the environment (e.g. its bioavailability, toxicity, and distribution) has demanded the development of coupling analytical techniques, so as to separate and detect the relevant elemental species (refer to chapter 13).

Table 15.12 Comparison of analytical performance characteristics of the main methods used in environmental or biological trace element analysis

	Single (S) or multi-element (M)	Elemental selectivity	Linear dynamic range	Sample volume (ml)	Matrix interferences	Max. matrix concentration (%)	Precision (%)	Accuracy	Cost
Flame AAS	S	Moderate	10^2	$1-10^2$	Large	>1	>10	Moderate	*
ET ASS	S	Good	10^3	$10^{-3}-10^{-1}$	Moderate	1	3–5	Good	**
ICP-OES	M	Moderate	10^3	1–10	Moderate	1	1–10	Moderate	**
NAA	M	Good	10^3	a	Moderate	<1	5–10	Good	****
ICP-MS	M	Good	10^8	1–10	Moderate	10^{-1}	<5	Good	***
XRF	M	Good	10^2	a	Moderate	<1	1–10	Good	**

a Solids preferred (mg).

In general, the choice of analytical method for making trace element measurements of environmental or biological samples depends not only on availability but on a variety of other reasons:

1. multi or single element determination;
2. required detection or determination limits;
3. required type of sample preparation;
4. possible matrix or interference effects;
5. total dissolved solids content;
6. liquid or solids analysis;
7. required levels of precision and accuracy;
8. one-off or routine analysis;
9. total or speciation analysis; and
10. throughput and cost per determination.

A review of the literature shows that ETAAS is the predominant method used, especially for natural waters and biological fluids (i.e. cases where both sensitivity and selectivity is required). Solids analysis (soils and sediments) is normally undertaken by NAA, XRF, and more recently LA-ICP-MS, although in most cases this requires an extensive research budget. Plants, and biological fluids and tissues are predominantly analysed by multielement methods, including NAA, ICP-OES, and ICP-MS. In particular, ICP-MS is becoming more widely used because of availability, its multielement capability over a linear dynamic range of six orders of magnitude and low detection limits. The ability to determine isotopic ratios or undertake isotope dilution studies on multiple or single elements at levels below $0.1~\mu g~l^{-1}$, with provisions of 0.2–1% relative standard deviations makes IC-MS an attractive method for environmental or biological elemental tracer or finger printing studies.

In conclusion, it is always important to remember that the trace element analysis of samples must be associated with an appropriate validation or quality control scheme (see chapters 1 and 2).

For the measurement of environmental or biological samples the use of certified reference materials, or inter-laboratory test comparisons must always be undertaken and published along with the results of the study (Table 15.13).

Table 15.13 Interanalytical method comparison for lead analysis

Reference material		Lead concentration (mg kg^{-1}, dry weight)			
		Certified	ICP-MS	ETAAS	ICP-OES
NIST-SRM 1577	Bovine liver	0.34 ± 0.08	0.32 ± 0.08	0.36 + −0.11	0.4 ± 0.1
NIES-CRM 6	Mussel	0.91 ± 0.04	0.89 ± 0.02	0.86 ± 0.04	1.0 ± 0.2
BCR-CRM 150	Milk powder	1.00 ± 0.04	0.97 ± 0.01	1.01 ± 0.04	1.1 ± 0.1
IAEA H9	Mixed human diet	0.15 ± 0.07	0.14 ± 0.01	0.14 ± 0.02	–
NRCC	Lobster	10.4 ± 1.9	9.74 ± 0.20	10.71 ± 0.91	12 ± 3

15.6.1   Lead

The widespread use of lead in our society makes it probably one of the most studied toxic trace elements. The principal source of environmental lead comes from motor vehicle emissions through the addition of lead-based antiknock compounds to petrol. Other sources include lead plumbing, glazed pottery, solder used in tin cans, old pewter, and lead-based paints.

Lead is found in rocks as galena (PbS) at concentrations of 0.1–10 mg kg^{-1}. During weathering Pb^{2+} can form carbonates and has the ability to be incorporated in clay minerals, in Fe and Mn oxides, and in organic matter. The impact of global Pb pollution means that most soils are likely to be enriched, especially in the high organic surface soils. Soil pH and organic matter content influences both the solubility and mobility of lead. A high soil pH may precipitate Pb as hydroxide, phosphate, or carbonate, as well as lead to the formation of Pb–organic complexes.

The forms of Pb in water vary with the type of water, the particle size (inorganic colloids), pH, and the presence of low soluble solids (phosphates, carbonates, sulphates and oxy-hydroxy species). The principal problems of lead in water are associated with mining and ore processing, industrial effluents, lead plumbing, and urban road or motorway run-off. Natural waters contain approximately 0.1–1 µg l^{-1} Pb. The European Community Environmental Quality Directive level (EC 80/778/EEC) is 50 µg l^{-1} for drinking water, and 100 µg l^{-1} Pb for the watering of livestock. Lead in tap water is a major problem in inner city areas where a large number of households have extensive lead plumbing, resulting in tap water levels >100 µg l^{-1}. Release of lead into natural water courses may pose a long-term environmental problem. Sediments act as sinks for lead, and recent evidence suggests that biomethylation takes place in the sediment–water interface producing the very toxic lead species $(CH_3)_3Pb^+$ and $(CH_3)_4Pb$.

Lead contamination of soils is a major problem for both man and animals. Normal surface soil lead levels are typically less than 40 mg kg^{-1}, but with increased values following contamination by metal mining (170–13 000 mg kg^{-1}), motorway soil (460–3600 mg kg^{-1}), urban gardens (270–450 mg kg^{-1}), and sewage sludged farmland (80–300 mg kg^{-1}).

Plants are able to take up Pb from soils to only a limited extent. A major problem is air-borne deposition of lead particulates onto the foliar surface of plants whereby it becomes a dietary source for animals or man. Normal plant foodstuffs have <2 mg kg^{-1} Pb (dry weight); with enhanced levels along side motorways (20–950 mg kg^{-1} Pb), battery works (34–600 mg kg^{-1} Pb), and metal-processing industrial sites (45–2714 mg kg^{-1} Pb). The United Kingdom standard for lead in food is 1 mg kg^{-1}, with the exception for baby food, which is 0.2 mg kg^{-1}. The World Health Organization has recommended that a tolerable intake of Pb per day for an adult is 430 µg.

Lead has been added to petrol since the early 1920s. Tetra-methyl and tetra-ethyl lead additives prevent the accumulation of large quantities of PbO within the combustion engine, and thereby prevent engine 'knocking'. The addition of lead scavengers, namely ethylene dibromide or dichloride, results in the emission of volatile lead halides $PbBr_2$, PbBrCl, Pb(OH)Br, $(PbO)_2PbBr_2$ into the environment via the exhaust gas. The European Community maximum permissible lead concentration in ambient air is $2 \, \mu g \, m^{-3}$, whilst concentrations of $>8 \, \mu g \, m^{-3}$ (motorways) and $\sim 2-3 \, \mu g \, m^{-3}$ (urban roads) are common. Lead in motor vehicle exhaust emissions consists principally of particle size $<0.5 \, \mu m$ in diameter, 90% of which are respirable.

Lead is readily absorbed through the gastrointestinal tract. In blood, 95% of the lead is in red blood cells (RBC) and 5% in the plasma. Around 70–90% of the lead assimilated goes into the bones, then liver and kidneys. Lead readily replaces calcium in bones. The symptoms of lead poisoning depend upon many factors including the magnitude and duration of lead exposure (dose), chemical form (organic is more toxic than inorganic), the age of the individual (children and the unborn are more susceptible) and the overall state of health (Ca, Fe, or Zn deficiency enhances the uptake of Pb). Young children exposed to mild forms of lead suffer effects to the central nervous system, shown by hypertension, irritability and impaired motor skills. There are some reports that mild Pb exposure will result in behavioural and educational abnormalities, and maybe even mental retardation.

Lead analysis of environmental or biological samples requires special considerations to be made during both collection and preparation stages. Lead in water appears to be primarily associated with particulates, so the type of filtration is important. The solubility products of common lead compounds found in the environment are reasonably low, so acidification of samples (especially below pH 6–4) will increase the solubility of various lead species. Many plant samples may have significant lead deposits on the foliar or root surfaces, so care must be taken during collection and storage. Contamination problems are also common during blood collection through lead contributions from heparin anticoagulants, glass containers, and laboratory dusts.

The principle methods of lead analysis are electrothermal atomic absorption spectrometry and inductively coupled plasma mass spectrometry. Interference problems may occur with NaCl and carbon content of the sample for both methods. Neutron activation analysis is not used in routine analysis as it does not have a principal lead gamma-ray. Organolead speciation measurements have been carried out using HPLC-ETAAS systems. Lead has four natural isotopes, Pb-204, Pb-206, Pb-207 and Pb-208. A recent analytical development has been the use of blood lead isotopic ratio measurements by ICP-MS in order to evaluate the impact of petrol lead on human blood and air in the city of Turin (Italy).

15.6.2 Aluminium

Aluminium is one of the main constituents of the earth's crust. Rocks commonly range from 0.45 to 10% Al. Weathering of primary rock minerals produces $Al(OH)^{2+}$ to $Al(OH)_6^{3-}$ species, which become the structural components of clay minerals. The Al content of soils is typically 1–5%. In aqueous solution Al^{3+} undergoes hydrolysis. The $Al(H_2O)_6^{3+}$ cation is stable only in acidic solution. At values more basic than pH 3, hydrolysis occurs with the appearance of $Al(H_2O)_5(OH)^{2+}$ and $Al(H_2O)_4(OH)_2^+$. The formation of $Al(OH)_3$ dictates the limiting solubility of Al^{3+} (~pH 6.5), and at more basic pH values $Al(OH)_4^-$ is formed. These species can interact with various other ligands, including oxygen, citrate, phosphate and silicic acid. The Al content of oceans is very low, typically 0.5 µg l^{-1}, although the presence of particulate matter dramatically increases the Al content. Deep sea clays and sediments can have levels as high as 95 000 mg kg^{-1} Al.

The principal release of Al into the environment is during chemical extraction and smelting from bauxite. Aluminium is widely used in a variety of commercial areas, including construction (car parts, aeroplanes, building sheets); packaging (household foil, food wrappings); coated beverage cans, kitchen utensils and appliances; and in food additives, antacid products, analgesics, and toiletries/cosmetics.

The Al content of plants is typically below 200 mg kg^{-1} dry weight. Most food and forage plants range from 2 to 70 mg kg^{-1} Al. Levels relate directly to soil type, plant factors, with increased Al availability under conditions of increased H$^+$ concentrations. Acid rain dramatically influences the release of Al^{3+} from soils into natural waters. Aluminium excess in plants is known to induce Ca deficiency or reduce Ca transport; and also involves interactions with the uptake of other nutrients, including P, Mg, K and N.

The Al content of most food stuffs lies in the range of 0.1–10 mg kg^{-1} Aluminium sulphate is also widely used in the treatment of raw water. The wide variety of food sources which contain natural Al levels, or have Al compounds added, makes it difficult to evaluate a typical daily intake, although values ranging from 5 to 100 mg have been reported.

The gastrointestinal tract is an effective barrier to Al absorption, which is usually only about 0.5% of intake. Greatly increased absorption is obtained in the presence of citrate, or in health conditions relating to decreased Fe, Ca and possibly Zn. Human blood serum contains <5 µg l^{-1} Al, although many published values may be in error due to the many sources of Al contamination during sample collection and preparation.

In recent years Al has been cited as a possible toxin in the pathogenesis of various diseases. Renal failure patients who received phosphate-binding Al gels for prolonged periods and who were dialysed using alum-treated water, often developed a fatal encephalopathy which was characterised by high brain Al levels. Patients suffering from senile dementia of the Alzheimer type also often have elevated brain Al levels.

Graphite furnace or electrothermal vaporisation atomic absorption spectrometry is the preferred method for measuring Al in biological samples. Neutron activation analysis has a short-lived gamma-ray, ^{28}Al ($t_{1/2} = 2.3$ min), although possible interferences can result from P and Si gamma-rays produced by a mixed neutron flux.

Inductively coupled plasma mass spectrometry has been used to measure low Al levels in biological fluids using ^{27}Al, although possible polyatomic interferences (^{26}Mg, ^{1}H, ^{12}C^{14}N^{1}H) can undermine the accuracy of the results if not properly taken into consideration. Chemical speciation measurements have been made of human plasma using ultra-filtration, gel filtration, and AAS, which showed that half the Al is bound to albumin and trasferrin, and the remainder is non-protein bound.

16 Environmental radiation and radioactivity

F.W. Fifield

16.1 Introduction

Ionising radiations are continuously present in the human environment. They may originate from *natural* or *artificial sources*. In the original genesis of the elements many radioactive species were produced and some with long half-lives, e.g. uranium, are still present today. These, together with their unstable decay products give rise to significant and measurable radiation doses (chapter 9, section 9.2.6). One impact of human activity has been to redistribute or concentrate naturally occurring radioactive substances, as with the mining and processing of uranium. The release of radon into the atmosphere from geological formations is also caused by mining or civil engineering unrelated to the extraction of radioactive materials. As a result the hazard posed by the ionising radiation emitted by the naturally occurring materials may be significantly increased, especially towards the personnel intimately involved in the operations. The second source of natural radiation is the flux of cosmic rays which bombard the earth, and originate largely from the sun. Few primary cosmic particles fully penetrate the atmosphere of the earth. Most interact with the gases in the upper atmosphere. One important product of the products of these interactions are the γ-rays, which have a long range, and contribute to the natural radiation background at the surface. Other products of the interactions are radioactive nuclides, especially ^{14}C and ^{3}H which become intermixed with the carbon and hydrogen in the environment, contribute to the natural radiation background, and additionally provide the basis for carbon and tritium dating.

Artificial sources of radioisotopes and radiation are largely associated with nuclear power or with medical and research uses. It is the medical use in diagnostic and therapeutic medicine which provides the greatest contribution to the radiation budget of the average individual, after natural background. Figure 16.1 shows the contributions from various sources to the average budget for UK citizens. Because the contribution shown from nuclear power is so small this does not mean that it is negligible for all people, as certain groups such as radiation workers or those exposed in accidents will be at higher risk than the general public.

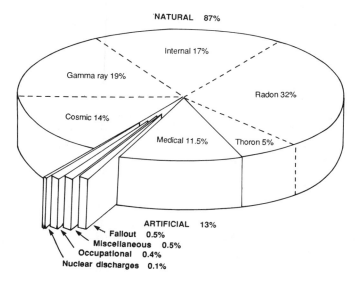

NATURAL   87%

Internal 17%

Gamma ray 19%

Radon 32%

Cosmic 14%

Medical 11.5%   Thoron 5%

ARTIFICIAL   13%
Fallout   0.5%
Miscellaneous   0.5%
Occupational   0.4%
Nuclear discharges   0.1%

Figure 16.1 The composition of the total radiation exposure of the UK population. Source: NRPB.

The measurement and assessment of environmental ionising radiations and radioactive materials for whatever reasons is a complex matter requiring the use of a diverse range of methods and instruments. Some of the more important aspects are considered in the following sections of this chapter. Key instrumentation has already been reviewed and discussed in chapter 9.

Damage caused by ionising radiation results from the energy that is transferred to matter on interaction with the radiation. The overall processes involved are complex, and remain only partially understood. As the name implies, the ionisation of atoms and molecules is the important phenomenon. In biological tissue the biomolecules themselves may suffer direct damage from the radiation, but given the high proportion of water present this is the site for a high proportion of the initial interactions. Ionisation of the water molecules initiates free radical reactions leading to significant biochemical changes in the tissue. Cellular and chromosome damage results. The latter may be quantified and used as an assessment of the radiation dose received. Where an intense exposure occurs, the level of damage may be high enough to produce significant physical damage in a so-called *radiation burn*. Intense exposure of the whole body, 100 Gy or more, leads to death. The effects of lower levels of exposure are less easy to assess. Damage to *somatic cells* can lead to mutation and the development of carcinomas. Damage to *germ cells* will have consequences for the offspring rather than for the parent.

16.2 The hazards of ionising radiations and their assessment

When radiation exposure results from an *external source*, the level may be measured, monitored and precautions taken to reduce the dose level if need be. Even when an accidental over exposure is involved, once the person is removed from the source the dose is reduced. If, however, radioactive material has become ingested into the body a more difficult situation arises, in part because the body may retain the isotope for a long period. For example ^{14}C, ^{90}Sr and ^{40}Ca will be retained more or less permanently, and U and Pu for tens of years. On the other hand because of the rapid turnover of water in the body much of any ^{3}H ingested will be excreted again quite quickly. It is more or less self-evident that exposure from *internal* sources can be long-term and in most cases impossible to reduce. The problem is compounded by the internal transport of some radionuclides which not only brings them close to parts of the body which are particularly sensitive to radiation damage but may deposit them locally. Particular examples are the concentration of iodine in the thyroid gland, and strontium and plutonium in the bone close to the bone marrow. Measurement of ingested radio-nuclides and the doses that they provide are very difficult problems. Where some degree of excretion occurs, monitoring of the urine may give some indication of the levels of ingestion. If the radionuclides emit γ-rays on decay then external counting is a possibility. In this context, sensitive γ-ray detectors placed close to the thyroid gland or the exterior of the throat can measure radioactive iodine isotopes. For nuclides more widely distributed through the body, *whole body counting* is a possibility, and is used for the routine screening of radiation workers. A whole body counter consists of an array of γ-radiation detectors (NaI) or semiconductor assembled in a tubular form and placed in close contact with the body over the greater part of its surface. Long counting times of several hours may be needed and sophisticated γ-ray spectrometry is required to separate the signals of ingested material from background ones. An important source of research information has been *postmortem analysis*. The proof of the significant increase in ^{90}Sr bone content, resulting from nuclear weapons fallout, came largely from the examination of stillborn foetuses.

16.3 Natural sources of radiation

In section 16.1 the distinction is made between radionuclides originating in the original genesis of the elements 4.6×10^9 years ago and those being produced continuously by cosmic ray interactions. It is convenient to consider natural radiation sources in these two classes.

16.3.1   Radionuclides of geological origin

Only radionuclides with very long half-lives will have persisted in significant amounts since their original genesis. Table 16.1 exemplifies these isotopes. To these must be added the nuclides in the decay sequences with ^{232}Th,

^{238}U and ^{235}U as their initial parent nuclides. These are illustrated by the ^{238}U series shown in Figure 16.2. Some of these decay products are of environmental significance. *Radon* which occurs in all three decay sequences is one such, and is considered more fully below.

Table 16.1 Some naturally occurring radionuclides having their origins in the original genesis of the elements

Nuclide	Half-life
Parent nuclides	
^{232}Th	1.4×10^{10} years
^{238}U	4.8×10^{9} years
^{235}U	7.0×10^{8} years
^{40}K	1.3×10^{9} years
^{87}Rb	4.7×10^{10} years
Examples of decay products (daughters)	
^{234}Th	24.1 days
^{234}Pa	1.2 min
^{226}Ra	16×10^{3} years
^{222}Rn	3.8 days
^{218}At	1.3 s
^{210}Pb	22 years
^{206}Tl	4.2 min

Uranium and thorium

Thorium is of limited industrial importance but may have a role as a nuclear fuel in the future. It occurs environmentally in significant deposits and is important as the initial parent in one of the decay sequences outlined above.

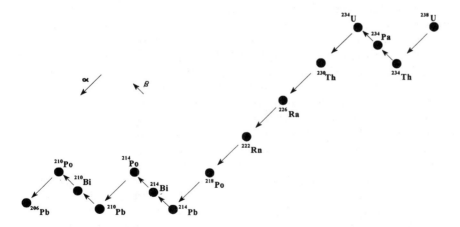

Figure 16.2 Decay sequence for ^{234}U. Note the formation of ^{222}Rn. Courtesy of N. Varley, Kingston University.

Uranium is an element of major industrial and environmental import-ance. It is the basis of nuclear power and the nuclear fuel cycle, and is increasingly used in high duty alloys, especially for armour piercing am-munition. It is thus mined, concentrated, processed and returned to the environment in large quantities. Naturally, it is widely distributed and is present at significant levels in many granitic strata. The main sources of extraction are, however, a limited number of deposits of high concentration in various parts of the world. In its natural state uranium contains two principal isotopes ^{235}U and ^{238}U. These isotopes have different but similar decay sequences and are illustrated by that for ^{238}U shown in Figure 16.2. The final product in each case is an isotope of lead. Isotopic ratios in lead deposits and their use in source identification are discussed in chapter 15, section 15.6.1. For use as a nuclear fuel or explosive the ^{235}U, which is fissile with thermal neutrons, is concentrated. The *depleted* starting material is rejected, and the uranium used in alloys generally comes from this source.

The determination of uranium in environmental samples is generally carried out at trace levels, and may often require the measurement of isotopic ratios in order to deduce its origins. Thus the principal stages in the analysis of samples for uranium will be a concentration step, chemical separation and preparation for final measurements including those of isotopic ratios. For an aqueous sample initial concentration may involve the use of a surface active precipitant such as hydrated ferrous oxide, which is then dissolved in a small volume of nitric acid. Readily soluble uranium, metal or salts, can be leached out from solid samples with concentrated nitric acid. Water insoluble minerals will need to be solubilised by fusion, e.g. with lithium borate, or by treatment with hydrofluoric acid. Chemical separation most commonly employs ion-exchange resins (chapter 5, section 5.3.4) Uranium and the actinides are characterised by their variable oxidation states and readiness with which they form complex ionic species which may be separated by ion-exchange chromatography. The uranium preferentially absorbed by the resin and subsequently eluted can be then measured on the basis of its radioactivity. Liquid scintillation counting is sensitive but provides only modest spectral resolution for the different α-particle energies. Where isotopic ratios are required to be measured accurately good quality α-spectra are required. These are obtained by electrolytically plating the uranium on to a stainless steel disc to provide a thin source, before measurement in an α-spectrometer. A typical spectrum is shown in Figure 16.3. This method is effective but time-consuming and the use of ICP-MS (chapter 7, section 7.3.3) is increasingly attractive as an alternative. Figure 16.4 illustrates its application to the measurement of uranium isotopes. Finally it should be noted that samples may be monitored for uranium by γ-ray spectrometry. This technique is not as sensitive as those discussed above but requires a minimum of sample preparation. An illustrative γ-spectrum from a uranium containing sample can be seen in Figure 16.5.

Radon and thoron. Radon and thoron (^{220}Rn) are intermediate products in the decay of uranium and thorium respectively. The nuclides are gaseous and their production is exemplified in Figure 16.2. Being gases, a proportion

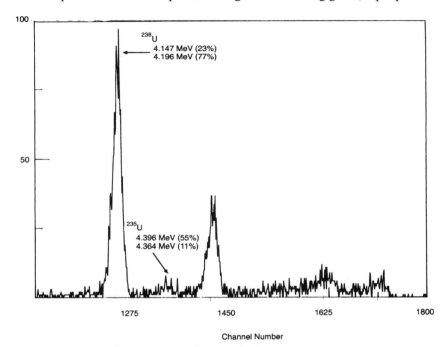

Figure 16.3 Low resolution α-spectrum showing the partial resolution of uranium isotypes. Courtesy of G.A. Wells, Kingston University.

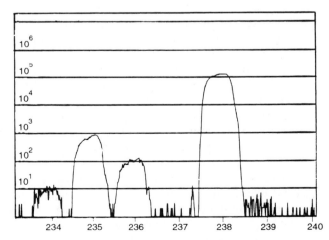

Figure 16.4 Uranium isotopes ^{234}U, ^{235}U, ^{236}U and ^{238}U, measured directly in U_3O_8 by laser ablation-ICP-MS. Source: VG Instruments Group.

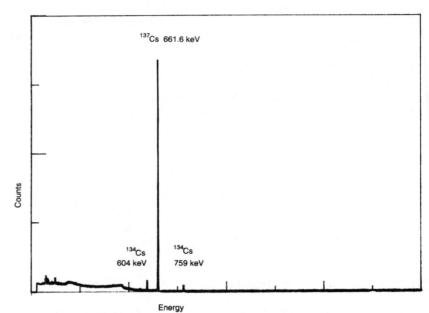

Figure 16.5 γ-Ray spectrum of a soil sample from the Chernobyl fallout area in Belarus, showing peaks for ^{134}Cs and ^{137}Cs. Courtesy of G.A. Wells, Kingston University.

of the radionuclides will emanate into the atmosphere from their solid precursors. Thus the atoms may be inhaled and undergo the next stage of decay inside the lungs, which not only experience the effects of the decay radiation but also become the deposition site for the resulting solid member of the radioactive series, polonium. Subsequent decay stages then occur in close proximity to the lung tissue. Radon and thoron thus present a perceived health risk which needs to be assessed. The level of the gases will vary geographically with the distribution of uranium and thorium as Figure 16.6 illustrates, being highest in areas with granitic strata. Gases emanating into the open air are rapidly dispersed so the serious hazard arises where the emanating gases are trapped and can concentrate. This happens in mineworkings and poorly ventilated buildings including domestic ones. Experimental deep drilling for *geothermal energy* extraction has sometimes led to the release of large quantities of radon.

Various techniques are used in the measurement of environmental radon. Where the gas is dissolved in water a straightforward measurement may be made using a liquid scintillation counter. Atmospheric measurements and the assessment of long-term doses for example in domestic housing present more complex problems. Some examples of the principles involved in some typical methods are:

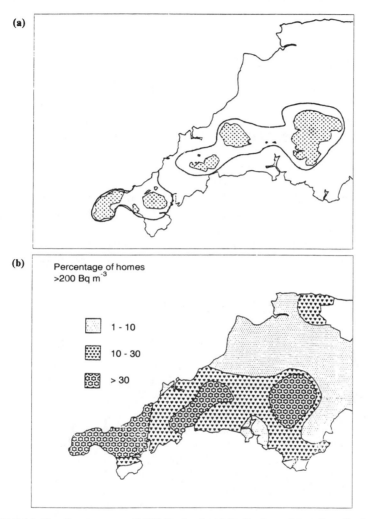

Figure 16.6 (a) Granite exposures of SW England and limit of uranium mineralisation and stream sediment uranium anomalies. (b) Estimated percentage of houses above the Action Level (after Ball and Miles, 1993). Courtesy of N. Varley, Kingston University.

1. absorption of radon on activated charcoal which is subsequently measured by liquid scintillation counting;
2. pumping the gas into an ionisation chamber for measurement; α-scintillation cells into which the sample is admitted; the interior surfaces are coated with ZnS as a scintillator and the cell is coupled to a photomultiplier;
3. thermoluminescent detectors which may be used for long-term assessments of exposure, e.g. inside buildings;

4. particle track film which may be used as an alternative method for long-term assessment.

16.3.2   Radionuclides resulting from cosmic rays

Carbon-14 is produced continuously in the upper atmosphere from nitrogen-14 as a result of the fast neutron flux from initial cosmic ray interactions (equation (16.1)) The rate of production is dependent upon cosmic ray intensity, thus being related to changes in sunspot activity.

$$^{14}_{6}C(n, p) \xrightarrow[t_{1/2} \sim 5.5 \times 10^3 y]{\beta^-} {}^{14}_{7}N \qquad (16.1)$$

The ^{14}C becomes incorporated into the CO_2 in the carbon cycle. All living organisms hence contain a proportion of ^{14}C which contributes to their overall radiation budget.

Tritium (^3_1H) is also produced in the upper atmosphere and mixes with the environmental hydrogen in water. It is also radioactive (equation (16.2)) and will contribute to natural radiation levels:

$$^3_1H \xrightarrow[t_{1/2} \sim 12.5 y]{\beta^-} {}^3_2He \qquad (16.2)$$

Both ^{14}C and ^3H can be used in dating procedures (see section 16.5). Environmental levels are now dependent also on artificial sources.

The β^- energies of ^{14}C and ^3H are ~0.2 meV and ~0.02 meV, respectively, and the decays are unaccompanied by γ-ray emission. Suitable instrumental procedures used for measuring such low energy, pure β^- emitters, have been discussed in chapter 9. Sample preparation is a key step for measurements on environmental samples. Early gaseous counting methods, in which ^{14}C was measured as CO_2 and ^3H as H_2 have been replaced by methods which use catalysts to convert the elements into liquid aromatic forms, with final measurements being made by liquid scintillation counting. The β^- energies are sufficiently different for spectral discrimination to be effective. Finally, it is becoming increasingly common for measurements at low levels to be made using mass spectrometry (chapter 8) as an alternative to radioactivity measurements. Progress is also being made in the use of NMR (chapter 8) for their determination in organic molecules.

16.4   Artificial sources of radiation

In section 16.1 the two primary areas of interest with regard to exposure to radiation from artificial sources have been identified, namely research and medical uses on the one hand and nuclear power, both civil and military on the other. Ironically it is the first which contributes most significantly to the average radiation budget but the second which has led to the production of very large amounts of radioactive materials.

16.4.1 Radiation and radioactivity in research and medicine

The unique characteristics of the radioactive decay process which remain unaffected by physical or chemical environments, taken together with the great sensitivity with which emitted radiation can be detected, make radioactive nuclides very valuable as *tracers*. In following the path of a chemical reaction a compound could for example have ^{14}C incorporated into it i.e. it is *labelled* or *tagged*. The route taken by the molecule can thus be traced by monitoring the radioactivity. *Diagnostic medicine* uses, for example, ^{131}I and ^{99m}Tc in monitoring the behaviour of the body. ^{131}I can be used in the study of thyroid function. ^{99m}Tc is preferentially absorbed at the site of carcinomas and can be used to detect and locate them. X-rays are extensively used in the study of the body where the varying abilities of materials to absorb them enable bone and tissue structure to be examined. X-rays and γ-rays are used to destroy cancerous cells in cancer therapy. Although patients are often exposed to quite high doses of radiation, this is done under control with a risk assessment being made prior to exposure.

Personnel working regularly with radioactive substances or X-ray generators need constant monitoring in order to protect them. In many cases this requires more than simple measurements of exposure. Where radioisotopes are used, the workplace must also be under constant surveillance to guard against contamination and ingestion. Measurement techniques as reviewed in chapter 9 will need to be used. With the exception of large installations such as γ-ray sources for cancer therapy, most radioisotopes are disposed of by environmental disposal within the terms of licences granted by the responsible government body. Disposal into the mains drainage, land fill sites and by incineration are all employed. Quantitative prior measurement of the disposals is required and comprehensive records are kept.

16.4.2 Nuclear power

The radionuclides involved in both the civil and military uses of nuclear power in general are similar, and in this sense they pose the same analytical problems. However, a significant difference exists with regard to the environmental distribution of the radioactive products. A *nuclear explosion* occurring in the atmosphere produces limited amounts of radioactive material but these are extensively drawn into the upper atmosphere by the accompanying thermal currents and distributed on a global pattern of *fallout*. If the explosion is in contact with the surface of the earth much surface material is converted into radioactive forms by the intense neutron flux accompanying the explosion. This material may be vaporised or drawn into the atmosphere in a particulate form with a net result of producing much higher levels of fallout containing a wide range of radionuclides.

Nuclear power generators contain the nuclear reactions within their reactor vessels. Under normal conditions the radioactive products are only released in significant quantities when the spent fuel in removed and reprocessed. Localised pollution may result from this process. One major nuclear accident in a power station has occurred from which widespread pollution has resulted. This happened at Chernobyl in the Ukraine in 1986. The pollution has been detected on a global scale and large areas of land in Europe remain significantly affected. The analytical problems are not merely associated with assessing the degree of contamination which has taken place as a consequence of fallout from either nuclear power stations or nuclear explosions. A matter of major concern is the degree and manner of the incorporation of the radionuclides into natural ecosystems. *Food webs* and *aquatic transport systems* are obvious examples. In this, of course, the nuclides are little different to their non-radioactive counterparts. The general comments on the chemistry and behaviour of trace elements made in chapter 15 apply also to the radioactive isotopes. One special feature concerning the handling of trace amounts of radionuclides is worthy of note. Significant amounts of ionising radiations may be emitted by very small amounts (μg or pg) of the nuclide. Chemical manipulation of such amounts may be difficult as they are not visible to the eye, and significant processing losses may arise from adsorption onto apparatus, precipitates or at solution interfaces. Where the ultimate measurement is of radiation emitted *inactive carriers* may be added in larger (mg) amounts in order to make the manipulations easier. Being non-radioactive they do not interfere with the final analytical measurement.

Nuclear power contributes to the level of environmental radiation in four ways.

1. emission of radiation by the power generating reactions;
2. the production of waste in the form of *fission products*;
3. the conversion of non radioactive isotopes into radioactive ones as a result of bombardment with neutrons originating in the power generating reactions;
4. the production or *breeding* of transuranic elements, e.g. Pu, Am.

In the first case the radiations are of consequence in the immediate vicinity of the reactor or explosion and do not have long-term implications for the global environment or the population as a whole. In a power station, local measurements of radiation are made to monitor the safety of the working environment. The second and third sources present similar analytical problems and may be considered together, but the final case has a number of special features deserving separate treatment.

Nuclear fission and breeding rections. The underpinning reaction for nuclear energy is the neutron induced fission of ^{235}U.

$$^{235}_{92}U(n, f) \rightarrow \underbrace{X + Y}_{\text{fission products}} + \sim 2.5n \tag{16.3}$$

(fission of Pu has a similar pattern)

Figure 16.7 shows the distribution of fission products. The neutrons produced react with further ^{235}U atoms to sustain the chain reaction and with other elements, primarily in (n,γ) or (n,p) to induce radioactivity in non-radioactive nuclides, e.g.

$$^{59}_{27}Co\,(n,\gamma)\,^{60}_{227}Co \xrightarrow{\beta^-/\gamma} {}^{60}_{28}Ni \tag{16.4}$$

Neutrons from the same source react with the non-fissile ^{238}U and initiate sequences leading to the formation of *neptunium, americium, plutonium* and *curium, e.g.*

$$^{238}_{92}U\,(n,\gamma)\,^{239}_{92}U \xrightarrow{\beta^-/\gamma} {}^{239}_{92}Np \xrightarrow{\beta^-/\gamma} {}^{239}_{94}Pu \tag{16.5}$$

The actinides heavier than Cm are made in more specific experiments; they have short half-lives and little environmental significance.

Fission products and induced radioactivity. In fission reactions the parent heavy nucleus splits into two *fission products*, represented by X and Y in equation (16.3). These parts are unlikely to be identical and the pattern of their production is shown in Figure 16.7. Obviously, induced radioactivity from neutron fluxes will not follow the same pattern. In this case the nuclear characteristics of the parent nuclide with regard to its tendency to undergo

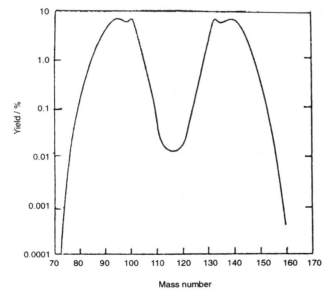

Figure 16.7 Fission yield of products from the fission of ^{235}U.

neutron capture, and the overall composition of irradiated material, will be the deciding factor. It will be seen from Figure 16.7 that some of the more common products fall in the range of elements some of which have biological functions, thus they become incorporated into food webs and may ultimately be concentrated and ingested into the human body by critical pathways. Others may become firmly attached to soil particles and remain, providing a high *ground dose* of radiation. In forest ecosystems it seems that fallout on to the trees is washed to the ground, taken up by roots and continuously cycled and concentrated within the trees by the fall, breakdown and reabsorption of leaf litter as nutrients. There are essentially two approaches to the determination of radionuclides in these groups. Where possible direct measurement using γ-ray spectrometry is employed. This avoids time-consuming chemical separation procedures. However, it has limited sensitivity and moreover can only be used where the analyte emits γ-rays with sufficient intensity and adequate spectral individuality. Thus concentration and separation of the analyte, employing appropriate combinations of precipitation, ion exchange chromatography and possibly solvent extraction are frequently required.

Plutonium and americium.   Of these two elements *americium* is of minor importance compared with *plutonium*. It is a by-product in nuclear reactors and is thus present in nuclear waste but is not manufactured or separated for any large scale uses. On the other hand many tons of plutonium have been separated for use as a nuclear fuel and as a nuclear explosive, and as such it is distributed for use on a global scale. Naturally, it exists in only minute amounts and is associated with uranium ores. Anthropogenic plutonium is environmentally distributed on a global scale as a result of fallout from nuclear explosions and concentrations above natural levels exist where nuclear waste has been released into the environment, e.g. in the Irish Sea. Plutonium is a highly toxic element and has been allocated a *maximum permissible body burden* of only 0.6 μg.

Analytical methods hence must have great sensitivity. For the determination of ingested material, whole body counting as described briefly in section 16.2 is used. The designated body burden will give a signal close to the sensitivity limit of this technique. Routine urine analysis is essential in monitoring the exposure of anyone working regularly with plutonium. To achieve the low detection levels needed for most environmental samples chemical concentration and separation by a combination of precipitation and ion-exchange chromatography is used. The final measurement is by α-spectrometry of the plutonium which has been electrochemically plated on to a stainless steel disc.

16.5 Radiogenic dating

In understanding environmental changes and events it is often important to know with some accuracy when they occurred. Some examples include:

1. the formation of rocks;
2. the date of occupation of an archaeological site;
3. the dating of animal remains;
4. the deposition of particulate matter from industrial emissions or a natural fire;
5. the time taken for water to move through a groundwater system.

Radiogenic dating has a contribution to make in all these areas. Whilst the principles are simple the application in specific cases, especially geological dating may be very complex. The kinetics of radioactive decay have been discussed in chapter 9. Reference to equation (9.6) will show that if the initial amount of a radionuclide is known and the amount left after decay for time *t* can be measured, provided λ is known the value of *t* can be calculated. The relationship can be used in environmental dating provided that the decay takes place in a *closed system* so that changes in the amount of the radioisotope result from decay alone. Indeed the essence of dating is to define the point at which the system became closed. The crystallisation of a molten rock, deposition of an element from solution or the percolation of rainwater into an aquifer are examples of natural systems which become closed. The half-life of the nuclide used in dating must match with the length of the lapsed time since the system became closed. On the one hand it must not be so long that the decay is not measurable, whilst on the other hand it must not decay to such a low level that the remaining atoms cannot be detected. Dating can generally be carried out back over a maximum of ten half-lives. Naturally the precision of the estimate of lapsed time will decrease with its length. Analytical methods used to determine the degree of decay that has taken place will generally need to be very sensitive and will need to be able to distinguish between different isotopes of the same element.

16.5.1 Geological dating

Key information in geological studies is often the date at which a particular rock formed. Geological time scales stretch over many millions of years and radionuclides with long half-lives are needed. *Uranium dating* is an important method which has been in use for some time. In liquid magma, elemental uranium behaves according to its own chemical and physical properties and is separate from its decay products. On solidification, the decay products are no longer able to move away so that the extent of the uranium decay may be estimated from the balance of uranium and decay products. This is also the case with uranium dissolved in seawater and deposited in the shells of shell fish and in coral which contribute to the formation of carbonate rocks. Some other decay sequences which are used in geological dating are summarised in Table 16.2.

Table 16.2 Some radionuclide systems used in geological dating

Decay	Half-life (years)
^{87}Rb/^{87}Sr	4.7×10^{10}
^{147}Sm/^{143}Nd	1.0×10^{12}
^{176}Lu/^{176}Hf	2.4×10^{10}
^{187}Re/^{187}Os	4.0×10^{12}
^{40}K/^{40}Ar	1.3×10^{9}
^{40}K/^{40}Ca	1.3×10^{9}
^{138}La/^{138}Ce	1.1×10^{11}
^{138}La/^{138}Ba	1.1×10^{11}

16.5.2   Carbon dating

The formation of ^{14}C from ^{14}N as a result of cosmic ray interactions in the upper atmosphere has been introduced in section 16.3.2. ^{14}C/^{12}C ratios in the atmosphere have remained roughly constant over time although perturbations produced by variations in cosmic ray intensity and in the last 200 years by the burning of fossil fuels (effectively without any ^{14}C content) must be noted. In its decay to ^{14}N, ^{14}C has a half-life of 5513 years and can be used for dating back for ca. 50 000 years. Living organisms incorporate carbon from carbon dioxide containing the ^{14}C:^{12}C ratio present in the atmosphere during their growing period. At the time of death, the incorporation will cease and the ratio will change. Measurement of the ^{14}C count can then be used to determine the date of death, enabling sites of human habitation to be dated. Careful calibration using artifacts or buildings of precisely documented origin and by the use of tree ring data has been necessary to deal with the perturbations in ratios noted above. Methods of measurement are outlined in section 16.3.2.

16.5.3   Tritium dating

A roughly constant proportion of tritium is maintained in the atmosphere as a result of its formation from cosmic ray effects. It is incorporated into the hydrosphere. Where water becomes part of a closed system, such as in groundwater or deep sea currents the principle of radiogenic dating may be applied. The half-life of tritium is 12.5 years and is useful over a 120-year period for dating. The movement of water can be studied by making measurements on borehole or spring samples. An interesting basis for studies has also been provided by the large amounts of tritium globally distributed as a result of thermo-nuclear explosions in the 1950s, which gave a marker date of significantly enhanced tritium activity. The measurement of tritium is discussed in section 16.3.2 and chapter 9. On the lighter side, tritium dating can also be applied to the bottling of wine or spirits.

16.5.4 Lead-210 dating

A recently developed technique with considerable potential for dating industrial activity utilises ^{210}Pb, whose formation and decay is shown in Figure 16.8. From this figure, it will be seen that ^{210}Pb is formed from ^{226}Ra via ^{222}Rn which is gaseous, subsequently decaying to ^{210}Pb. Any ^{210}Pb formed in the ground will be directly related to the amount of ^{226}Ra present and is said to be *supported*. ^{210}Pb formed in the atmosphere will be deposited with particulates. As this fraction has been formed from ^{210}Pb in the atmosphere it will not be related to the ^{226}Ra content of the earth and is said to be *unsupported*. This unsupported ^{210}Pb acts as a marker for the deposition of particulates. The amounts of particulates are closely related to the development of industrial activity in a region or to major natural fires. ^{210}Pb has a half-life of 22.3 years and can be used as a marker for dating over about 200 years, by the examination of its profiles in undisturbed cores from peat bogs, or sediments from lakes, estuaries or rivers. The ^{210}Pb content is best determined via the chemical separation and electroplating of ^{210}Po with a final determination using an α-particle spectrometer.

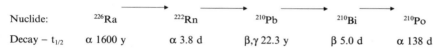

Nuclide:	^{226}Ra	^{222}Rn	^{210}Pb	^{210}Bi	^{210}Po
Decay – $t_{1/2}$	α 1600 y	α 3.8 d	β,γ 22.3 y	β 5.0 d	α 138 d

Figure 16.8 Decay sequence for ^{210}Pb.

16.5.5 Thermoluminescence or electron spin resonance dating

Certain materials, including bone and especially teeth have the ability to store the energy deposited in them by ionising radiations. The energy so stored reflects the *radiation dose* that has been experienced by the material. If the *dose rate* for a particular environment is known or can be determined, the exposure time can be calculated from a measurement of the energy stored. Two approaches to this measurement are possible. Firstly, heating the sample will lead to the emission of a pulse of UV radiation in a *thermo-luminescence process*. The size of this pulse can then be used to infer the amount of energy, stored.

An alternative measurement can be made on the basis of the nature of the energy storing process. In this, electrons are promoted to excited energy levels in an *unpaired form* (see chapter 3). The paramagnetic property that they then have can be detected and quantified by an *electron spin resonance* (ESR) spectrometer. This technique, also known as *electron paramagnetic resonance* (EPR) spectrometry, is also being pioneered for dose assessment in nuclear radiation accidents.

For dating purposes the radiation background dose rate, which comes largely from uranium and its decay products, is assessed by direct γ-ray measurement. The technique is effective over 100 000 years and can thus extend the dating of the deposition of animal remains considerably beyond that attainable by carbon dating.

Further reading

Ivanovich, M. and Harmon, R.S. (1982) *Uranium Series Disequilibrium: Applications to Environmental Problems.* Clarendon Press, Oxford.

Lowe, J.J. (Ed.) (1991) *Radiocarbon Dating: Recent Applications and Future Potential.* Cambridge.

Rollinson, H.R. (1993) *Using Geochemical Data: Evaluation, Presentation, and Interpretation.* Longman, UK.

Contaminated landsites 17
F.W. Fifield

17.1 Introduction

Contaminated landsites are among the commonest environmental problems that arise and their assessment depends heavily on chemical analysis. All anthropogenic activity gives rise to some form of pollution, and in the developed countries where land has been used and reused, sometimes over many centuries, the problems are particularly acute. It is the period over the last 200–300 years during which the industrial revolution, accompanied by major population growth, has gathered pace that has led to the greatest difficulties. Broadly speaking the sources of contamination fall into two groups. There are those which are due to specific activities such as manufacturing or mining, and those which concern the disposal of solid domestic waste.

From the analytical point of view it matters little which type of source is being assessed as any analysis is targeted on an identified problem. It is self-evident that chemical analysis will also play a key role in assessing the effectiveness of any steps taken to clean up a site following its assessment. With regard to regulations concerning landsites, e.g. for maximum levels of identified pollutants, the picture is unclear. Decisions on sites are taken according to national guidelines which do not necessarily contain quantitative limits. In Europe, Holland pioneered the setting of regulatory standards in its Soil Cleanup (Interim) Act 1983. The standards set in this document have attracted attention by civil engineering organisations across Europe. Apart from any technical problems presented by the making of relevant measurements, there are two other factors which need to be borne in mind. Proper assessment of a site may generate a large number of samples and commercial pressures can mean that results are required in the shortest possible time.

In reaching decisions about the suitability of sites for redevelopment a comprehensive *risk analysis* is required. It must also be recognised that the consequences of contamination for human health are not the only ones to be considered. Damage to building materials may be brought about by substances not otherwise thought to be harmful. High levels of sulphates and chlorides probably represent the best known examples here.

17.2 The nature of contaminated sites

17.2.1 Origins of contamination

A comprehensive list of types and characteristics of contaminated landsites would be very long, and within the context of this text of limited value. It is, however, useful to comment, briefly, upon some important classes.

1. Industrial sites which have been used for a specific and limited purpose, e.g. smelting a particular ore, or manufacturing a limited range of products. Potentially hazardous contaminants are readily identified in these circumstances.

2. Industrial sites which have been exposed to a wide range of materials. Prominent amongst these are sites once used for producing town gas from the coking of coal, which often had associated chemical products plants, and oil refineries. Some chemicals storage sites also come into this category. Assessment here may present complex analytical problems.

3. Sites on which significant amounts of air-borne products have originated, e.g. from fossil fuelled power stations or metal smelting works. The contamination in these circumstances may be spread over wide areas.

4. Mine workings, both deep and opencast produce very large spoil heaps. These can contain hazardous minerals and may also undergo reactions, e.g. atmospheric oxidation in which further undesirable products arise.

5. Landfill disposal sites have been used for a very long time as a way of disposing of both domestic and industrial waste. The former currently represents a major problem. Prior to the 1970s selection of suitable sites and the control of the dumping processes were often rather haphazard. Modern sites are more closely controlled but the assessment and monitoring of older ones often presents very considerable problems.

Table 17.1 Some important types of pollutants/contaminants associated with various types of site

Contaminant/pollutant	Type of site
Toxic elements e.g. Cd, Pb, As, Hg, Cu, Ni, Zn	Mines, metal works Electroplating works General engineering Scrap yards Waste disposal sites
Oils, tars, aromatic and polyaromatic hydrocarbons	Gas works Power stations Chemical works Oil refineries
Methane	Landfill sites Filled dock basins
Aggressive substances e.g. sulphates, chlorides, acids	Slag heaps Waste tips
Asbestos	General industrial buildings Waste disposal sites

6. Agricultural land where sewage sludge has been used as a fertiliser can show enhanced concentrations of metals which can then become involved in biological uptake, thereby entering food chains.

Table 17.1 shows some typical pollutants which are associated with identified uses of landsites. The principal hazards linked to the pollutants are also summarised.

17.2.2   Physical characteristics of landsites

Key elements in affecting the environmental impact of contaminated landsites are their geographical locations and the underlying geology. The environmental transport of contaminants will be significantly affected by both. *Aquatic transport* is a principal route by which contaminants are spread beyond the area initially affected. Transport may be in *solution*, as *suspended solids* or by *sorption onto suspended solids*. The former is a potential mode for transport both by surface run off into streams and rivers, and via groundwater. Suspended solids on the other hand are important only for surface runoff.

It is important to remember that aquatic transport by surface runoff can be both rapid and extend over long distances. Groundwater may be affected at a slower rate with the resultant contamination also moving slowly as a result of the low rate of movement of groundwater itself. The corollary to this of course is that even when the source is removed, any contamination will take a corresponding long period to flush out from the system. In the United Kingdom, nitrate contamination in aquifers is still present 40–50 years after its penetration. A further feature associated with groundwater contamination concerns those sources of contamination which are well below the surface. Mine workings obviously come into this category as do many landfill sites. In both cases the source can be very close to the aquifer itself leading to rapid and intense contamination. The purifying effect of sorption onto soil particles as water seeps through the surface layers before penetrating into the aquifers is also lost.

Air-borne transport of materials from a contaminated landsite can be by gases, vapours and particulate material. Clearly the prevailing winds are important considerations here, and in these, gases and vapours will ultimately be dispersed. However, in the shorter term they may migrate through underground pathways and collect to a dangerous degree in enclosed spaces. Methane, long recognised as an explosive hazard in coal mining (fire-damp), is also a product of the breakdown of organic material in landfill sites and is known to collect in the foundations and basements of nearby buildings. Carbon dioxide presents no hazard of explosion but can lead to asphyxiation in confined spaces. Radon, discussed in chapter 16, is a parallel problem but is not significantly associated with the majority of contaminated landsites. Particulate material from a point source of contamination can be spread over a wide area by winds. The land area contaminated will be accordingly extended. Figure 17.1 illustrates this with data from a lead smelting plant in the north of England.

17.3 Assessments The overall aim of an assessment will be to reach conclusions concerning the suitability of a site for future use. In order to do this it will be necessary to:

1. determine the nature of any contamination;
2. assess the degree and extent of contamination;
3. decide if clean-up of the site is needed;
4. plan for future monitoring of the site.

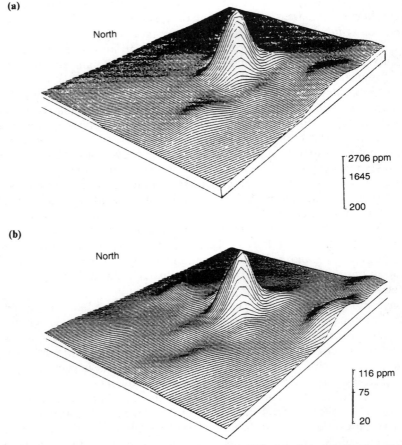

Figure 17.1 Surface pollution over a 15 km square around a lead smelting complex, investigated by the analysis of lichen samples. The close similarity between the distribution of (a) lead and (b) copper is clear. It should not be assumed that all metals show the same pattern.

Analytical measurements will be extensively required in this for both qualitative and quantitative purposes. The essence of this book is to provide the background to the techniques and methods needed and used in the environmental context. They are extensively covered in other chapters. Chapters 14, 15, 18 and 19 discuss, respectively, some of the practical aspects of the analysis of atmospheric samples, the determination of trace elements, the analysis of water and the determination of trace amounts of organic substances. At this point it is sufficient to note the complexity of the problems of landsites as summarised in section 17.2 and to recognise that a very wide range of techniques and methods will be required. It is, however, important to consider here the overall nature of the investigation of a landsite, and the important aspects of the sampling programme required to support it.

17.3.1 Investigational plan

A plan can be formulated in three stages.

1. an investigation of the physical nature and history of the site use;
2. a preliminary analytical survey;
3. full survey and monitoring programme.

It is extremely important to start with a study of the records of the site. Details of previous use will give good indications of likely contaminants, although complete reliance must not be placed on records alone. Key environmental pollutants outside a historical list should always be considered (Table 17.1). As has been indicated in section 17.2.2, the general physical and geographical nature of the site is important in defining potential problems. On the basis of this, a preliminary survey can be planned. Samples would be taken to represent parts of the site where contaminants might be expected to be present in the greatest concentrations, or places to which they might have been transported. Such a survey would be expected to include examination of surface soil samples and core samples at depth, covering the extent of the site to which contamination might have spread. Core samples at various depths will be required. Runoff must be examined by analysis of water taken from nearby streams, rivers and lakes. Sampling should include sediments and vegetation from the surface aquatic system. It is particularly important to examine regions downstream of the site. Surface soil sampling should be backed up by using appropriate biological indicator organisms as discussed in chapter 12. Groundwater can be sampled at any existing boreholes and springs, although it will often be necessary to sink additional sampling boreholes. After a preliminary survey, it may be decided that no further investigations are required. However, if any indications of significant contamination are found, a fuller survey will be required.

In a full survey, in addition to careful examination of samples from the points discussed above, systematic sampling on a grid pattern is required. As this must cover the whole site in detail, a large number of samples can be generated, leading to a time-consuming and costly survey. Gaseous emissions will need to be measured either by portable equipment on-site, by trapping devices in which gaseous samples are removed to a laboratory for analysis, or by thermal desorption cartridges as described in chapter 14.

Final site assessment. The final assessment will seek to place the site in one of three categories with regard to contaminants:

1. *at background levels* with the site requiring no further investigation;
2. *at enhanced levels* requiring further investigation;
3. *significantly contaminated* and requiring clean-up.

The data in Table 17.2 which are taken from The Dutch Soil Cleanup (Interim) Act provide a basis for such a classification with regard to some contaminants.

Table 17.2 Indicative contamination levels[a] for the classification of landsites (data from The Dutch Soil Cleanup (Interim) Act

Contaminant	Soil (mg kg^{-1} of dry soil)			Groundwater (μg dm^{-3})		
	A	B	C	A	B	C
Metals						
Cr (chromium)	100	250	800	20	50	200
Co (cobalt)	20	50	300	20	50	200
Ni (nickel)	50	100	500	20	50	200
Cu (copper)	50	100	500	20	50	200
Zn (zinc)	200	500	3000	50	200	800
As (arsenic)	20	30	50	10	30	100
Mo (molybdenum)	10	40	200	5	20	100
Cd (cadmium)	1	5	20	1	2.5	10
Sn (tin)	20	50	300	10	30	150
Ba (barium)	200	400	2000	50	100	500
Hg (mercury)	0.5	2	10	0.2	0.5	2
Pb (lead)	50	150	600	20	50	200
Inorganic substances						
NH$_4$ (expressed as N)	–	–	–	200	1000	3000
F (total)	200	400	2000	300	1200	4000
CN (free)	1	10	100	5	30	100
CN (total-complex)	5	50	500	10	50	200
S (total)	2	20	200	10	100	300
Br (total)	20	50	300	100	500	2000
PO$_4$ (expressed as P)	–	–	–	50	200	700
Aromatic substances						
Benzene	0.01	0.5	5	0.2	1	5
Ethylbenzene	0.05	5	50	0.5	20	60
Toluene	0.05	3	30	0.5	15	50
Xylenes	0.05	5	50	0.5	20	60
Phenols	0.02	1	10	0.5	15	50
Aromatic substances (total)	0.1	7	70	1	30	100

Table 17.2 *Cont'd*

Contaminant	Soil (mg kg^{-1} of dry soil)			Groundwater (μg dm^{-3})		
	A	B	C	A	B	C
Polycyclic hydrocarbons						
Naphthalene	0.1	5	50	0.2	7	30
Anthracene	0.1	10	100	0.1	2	10
Phenanthrene	0.1	10	100	0.1	2	10
Fluoranthene	0.1	10	100	0.02	1	5
Pyrene	0.1	10	100	0.02	1	5
1,2-Benzopyrene	0.05	1	10	0.01	0.2	1
Polycyclic hydrocarbons (total)	1	20	200	0.2	10	40
Chlorinated hydrocarbons						
Aliphatic (individual)	0.1	5	50	1	10	50
Aliphatic (total)	0.1	7	70	1	15	70
Chlorobenzenes (individual)	0.05	1	10	0.02	0.5	2
Chlorobenzenes (total)	0.05	2	20	0.02	1	5
Chlorophenols (individual)	0.01	0.5	5	0.01	0.3	1.5
Chlorophenols (total)	0.01	1	10	0.01	0.5	2
Chlorinated PAHs (total)	0.05	1	10	0.01	0.2	1
PCBs (total)	0.05	1	10	0.01	0.2	1
EOCI (total)	0.1	8	80	1	15	70
Pesticides						
Organic chlorinated (individual)	0.1	0.5	5	0.05	0.2	1
Organic chlorinated (total)	0.1	1	10	0.1	0.5	2
Pesticides (total)	0.1	2	20	0.1	1	5
Miscellaneous						
Tetrahydrofuran	0.1	4	40	0.5	20	60
Pyridine	0.1	2	20	0.5	10	30
Tetrahydrothiophene	0.1	5	50	0.5	20	60
Cyclohexanes	0.1	6	60	0.5	15	50
Styrene	0.1	5	50	0.5	20	60
Gasoline	20	100	800	10	40	150
Mineral oil	100	1000	5000	20	200	600

[a] Indicative levels: A, background values not requiring further investigation; B, further investigations necessary; C, clean-up required.

Where a long-term monitoring programme is decided upon it will require regular visits to the site for sample collection and laboratory analysis. The use of instrument packages, permanently in place, providing a constant record of the parameters being measured, should be considered. Modern developments in instrument design have made this possible in many cases.

Further reading

Dutch Soil Cleanup (Interim) Act (1983) Dutch Ministry of Housing, Planning and Environment.
Interdepartmental Committee on Remediation of Contaminated Land (ICRL) (1983) Circulars.
    N.B. Circular 59/83 Department of the Environment (UK).
UK Water Supply (Water Quality) (1989) Regulations.
N. Nicolson (1993) *An Introduction to Drinking Water Quality* The Institution of Water and
    Environmental Management.

18 The analysis of water

F.W. Fifield

18.1 Introduction

Water is a substance with a number of unique properties which have great environmental significance. It is self-evidently an essential for life and many chemical analyses are carried out to ensure that water of appropriate *quality* is supplied for human consumption. The water quality standards have legal backing and specify key characteristics for the water, dependent upon its intended use. Water intended for use as drinking water will be under the tightest standards whilst that on bathing beaches, for example, would be controlled at a less stringent level. Table 18.1 exemplifies standards for drinking water adopted in the UK. From this table it can be seen that a wide range of tests and analyses is required to ensure compliance with the standards. In conjunction with the specified tests and analyses a formal sampling protocol is also documented.

The geosphere, atmosphere and biosphere all contain large quantities of water where it is prominently involved in global cycles. Its excellent solvent properties mean that it is commonly involved in the transport of both natural and pollutant materials. Water is often the ultimate *sink* for a pollutant. Aquatic organisms are often prolific and play important roles in food chains. Indeed, the organisms may in themselves provide tools for the investigation of the characteristics of a particular body of water. This principle has been discussed in chapter 12. In assessing the characteristics of an environmental body of water it may also be appropriate to carry out analysis on associated materials such as sediments and biota.

In general terms the analytical methods used in the assessment of water characteristics are very much the same irrespective of the ultimate purpose of the analysis. However, there are sometimes special requirements, such as detection limits or selectivity which may dictate the use of one method rather than another. The need to consider the overall analytical process and its aims have been strongly emphasised in chapter 1. In the following sections of this chapter key measurements are identified and discussed with regard to both the aims of and need for them on the one hand and the technical nature of the methods used on the other. From the point of view of chemical analysis water can be regarded as a relatively simple matrix although the high salt content of seawater can produce complications. It should also be noted that even in fresh water, the speciation of elements must be carefully considered.

Table 18.1 Extract from The Water Supply (Water Quality) Regulations 1989 for The United Kingdom. Schedule 2: prescribed concentrations or values (Table A)

Item parameter	Units of measurement	Concentration of value (maximum unless otherwise stated)
1. Colour	mg/l Pt/Co scale	20
2. Turbidity (including suspended solids)	Formazin turbidity units	4
3. Odour (including hydrogen sulphide)	Dilution number	3 at 25°C
4. Taste	Dilution number	3 at 25°C
5. Temperature	°C	25
6. Hydrogen ion	pH value	9.5 5.5 (minimum)
7. Sulphate	mg SO_4 l^{-1}	250
8. Magnesium	mg Mg l^{-1}	50
9. Sodium	mg Na l^{-1}	150
10. Potassium	mg K l^{-1}	12
11. Dry residues	mg l^{-1}	1500 (after drying at 180°C)
12. Nitrate	mg No_3 l^{-1}	50
13. Nitrite	mg NO_2 l^{-1}	0.1
14. Ammonium (ammonia and ammonium-ions)	mg N l^{-1}	0.5
15. Kjeldahl nitrogen	mg N l^{-1}	1
16. Oxidisability (permanganate value)	mg O_2 l^{-1}	5
17. Total organic carbon	mg C l^{-1}	No significant increase over that normally observed
18. Dissolved or emulsified hydrocarbons (after extraction with petroleum ether); mineral oils	µg l^{-1}	10
19. Phenols	µg C_6H_5OH l^{-1}	0.5
20. Surfactants	µg l^{-1} (as lauryl sulphate)	200
21. Aluminium	µg Al l^{-1}	200
22. Iron	µg Fe l^{-1}	200
23. Manganese	µg Mn l^{-1}	50
24. Copper	µg Cu l^{-1}	3000
25. Zinc	µg Zn l^{-1}	5000
26. Phosphorus	µg P l^{-1}	2200
27. Fluoride	µ F l^{-1}	1500
28. Silver	µg Ag l^{-1}	10

18.2 pH, acidity and alkalinity

The pH represents the acidic or basic nature of water on a scale of 0–14 (see Figure 18.1), from strongly acidic to strongly basic, being neutral at pH 7. The acceptable range for domestic water supplies is from pH 5.5 to pH 9.5. Environmentally, water is considered to be most satisfactory when close to pH 7. Pure water in atmospheric equilibrium has a pH of 5.5, unpolluted soft water in the range 5.5–7.1, hard water 7.9–8.9, and seawater 8.1–9.1. Outside this range its productivity becomes increasingly limited although where higher or lower pH values have been established for a long time,

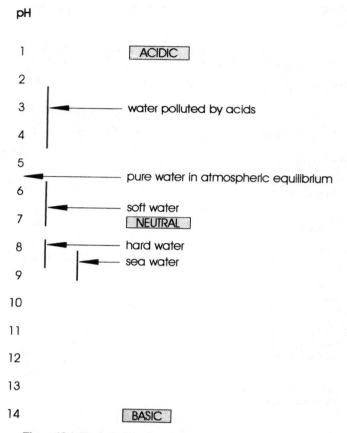

Figure 18.1 Typical pH values for environmental water.

localised ecosystems may have developed. The pH reflects the total concentration of hydrogen ions irrespective of their sources which may be natural or artificial (Figure 18.1). Three methods of pH measurement are commonly used. Test sticks of papers impregnated with pH sensitive dyes or dye solution are simply and easily used. For laboratory measurements and for greater precision and accuracy, pH electrodes are to be preferred. These, based upon the glass electrode discussed in chapter 10, are intrinsically the most satisfactory and can be configured in simple robust forms for use in field measurements.

The term *acidity* when applied to water has a specialised meaning and is defined as the capacity of the water to neutralise OH^-. It has clear implications in assessing the ability of water to accommodate alkaline waste or treatment of water for purposes such as boiler feed. The acidity can result from both strong and weak acids of natural or anthropogenic origin. Any strong acids present are usually referred to as *free mineral acids*. Distinction between the total acidity and that due to strong acids alone can be made by titration with a base such as sodium hydroxide, employing different indicators. Strong acids are estimated by titration to the methyl orange end point at pH 4.3 and total acidity by titration to the phenolphthalein end point at pH 8.2 (chapter 4). Alternatively the end points may be sensed electrochemically (chapter 10).

In a similar way *alkalinity*, the capacity of the water to absorb hydrogen ions can be assessed by titration with an acid such as HCl. Natural alkalinity mostly arises from bicarbonate, carbonate and hydroxide ions (equations (18.1), (18.2) and (18.3)) with minor contributions from other sources.

$$HCO_3^- + H^+ = CO_2 + H_2O \qquad (18.1)$$

$$CO_3^{2-} + H^+ = HCO_3^- \qquad (18.2)$$

$$OH^- + H^+ = H_2O \qquad (18.3)$$

The alkalinity due to stronger bases, e.g. hydroxyl ions is determined by titration to pH 8.3 (the phenolphthalein end point) and *total alkalinity* by titration to pH 4.3 (the methylorange end point).

18.3 Dissolved oxygen and oxygen demand

A characteristic of major importance in many aquatic systems is the amount of *dissolved oxygen* present. This will depend in the first instance upon the saturation concentration, which is determined by the ambient temperature and pressure, the latter being affected by height above sea level (see Table 18.2) Saturation will be most closely approached overnight in turbulent conditions or in the presence of oxygenating plants. A measure of *supersaturation* may sometimes occur.

Table 18.2 Solubility of oxygen in water at 760 mm pressure as a function of temperature

T (°C)	O_2 solubility (ppm)
10	11.28
15	10.07
20	9.08
25	8.26
30	7.57
35	6.98
40	6.47

Apart from a rise in temperature, dissolved oxygen levels can be depleted by the biochemical breakdown of organic materials present in the water. This latter can be greatly increased by discharges from industrial and agricultural sources. The *oxygen demand* of the organic substances can be assessed by measuring the amount of oxygen present before and after the incubation of a sample for 5 days at 20°C. This gives the so called *biochemical oxygen demand*, or BOD. The advantage of BOD measurement is that it parallels the natural processes closely, although it is a lengthy measurement to complete. A more rapid estimate may be obtained by measuring the amount of chemical oxidation produced by acid potassium permanganate in a *chemical oxygen demand*, or COD test.

The determination of dissolved oxygen may be made by using a *dissolved oxygen electrode*, a version of which is based on the Clark cell. In this a pair of silver electrodes in an electrolyte are separated from the sample by a gas permeable membrane. As oxygen diffuses through to the electrodes it is reduced according to equation (18.4):

$$O_2 + 2H_2O + 4e^- = 4OH^- \tag{18.4}$$

Accordingly a current will flow which is proportional to the rate of diffusion and thus in turn to the concentration of oxygen in the sample (see chapter 10).

Alternatively the long established *Winkler method* can be used. This is a titrimetric method based upon an initial reaction between the dissolved oxygen and freshly precipitated manganous oxide, to produce higher oxidation states of the manganese. On subsequent acidification in the presence of iodide ions, iodine is liberated which is then titrated with sodium thiosulphate. The key equations in this analysis are equations (18.5)–(18.8).

$$Mn^{2+} + 2OH^- = MnO \cdot H_2O \tag{18.5}$$

$$MnO \cdot H_2O + \tfrac{1}{2}O_2 = MnO_2 \cdot H_2O \tag{18.6}$$

$$MnO_2H_2 + 2I + 4H^+ = Mn^{2+} + I_2 + 3H_2O \tag{18.7}$$

$$I_2 + 2S_2O_2^{2-} = S_4O_6^{2-} + 2I^- \tag{18.8}$$

from which it can be seen that 1 mol $S_2O_3^{2-} \equiv 0.25$ mol O_2.

A routine way of representing the oxygen demand is by means of a *permanganate value* (PV) or *permanganate index* (PI). In this measurement an excess of $KMnO_4$ is added to the sample, and after a specified reaction time under specified conditions, e.g. 10 min at boiling point, the unreacted portion is determined by the liberation of iodine from added potassium iodide, which is then titrated with sodium thiosulphate. Alternatively direct reaction with sodium oxalate may be employed. The permanganate value is then expressed in terms of the oxygen equivalent of the permanganate used up in the reaction. In the initial oxidation Mn(VII) is reduced to Mn(II).

$$MnO_4^- + 8H^+ + 5e^- = Mn^{2+} + 4H_2O \tag{18.9}$$

From which it can be seen that 1 mol $MnO_4 = 5$ mol e^-. The equivalent in oxygen can be seen from equation (18.10)

$$O_2 + 2e^- = O_2^{2-} \tag{18.10}$$

It follows that 1 mol $KMnO_4 = 5/2$ mol O_2. Finally, the PV is expressed as mg O_2 dm^{-3} of the sample.

Both BOD and COD give indications of the oxidisability and oxygen demand of water samples. Neither, however, measures the total organic loading of the water. When this is required a determination of the *total organic carbon* (TOC) is made. Methods vary and routine analysis is by quantitative oxidation to carbon dioxide after acidification to remove interference from carbonates or bicarbonates. Oxidation in a gas stream passing through a heated tube or wet oxidation with potassium peroxodisulphate have both been used. The latter is less convenient but more sensitive and can be used at levels below 1 mg dm^{-3}. The carbon dioxide produced is measured either by conductivity after absorption in solution or by catalytic conversion to methane which is then passed to a flame ionisation detector as used in gas chromatography. The TOC measurement is the most objective of those concerned with oxygen demand and as such is gaining in favour.

18.4 Total organic carbon

The determination of metals in water presents one of the more straightforward analytical problems, unless the water is very heavily polluted or contains a high level of suspended solids. Atomic spectroscopic methods are most widely used with FES, AAS, ICP-AES and ICP-MS all finding application, and being used appropriately to meet the specificity, sensitivity and multielement demands of the sample. None of these methods is well suited to measurements in the field and in these circumstances methods based upon the use of colour reagents and simple photometry are employed. Electrochemical techniques such as polarography or anodic stripping voltammetry may be employed where ultrasensitive analysis or speciation is required. The determination of metals is more fully covered in chapter 15.

18.5 Metals

Water is an excellent solvent for ionic substances which dissociate into ions on going into solution (chapter 3). The overall concentration of dissolved solids is a major factor in determining the characteristics of the water. It is easy to see this in the big differences exhibited by fresh water and saline environments. At the same time, the concentration of individual species, even at low levels can be of great importance. For example, productivity is much influenced by small amounts of nitrate and phosphate ions whilst small amounts of chromium are extremely toxic towards micro-organisms. There is thus a need both for analysis, which indicates the total concentration of dissolved substances, and for selective methods, which can determine individual species with good sensitivity. The former is readily achieved by *electrical conductivity* measurements, but the latter requires either selective individual reactions or chemical separation of the species prior to measurement. Some of the important analyses are reviewed in the following paragraphs.

18.6 Dissolved salts

Water hardness is a term used to describe the properties of some dissolved substances which lead to the deposition of metal carbonates (lime scale) on boiling and reactions with soap or detergents leading to scum formation rendering them ineffective. The main source of hardness is calcium carbonate which is dissolved readily by water containing carbon dioxide, i.e.

$$CaCO_3(s) + CO_2(aq) + H_2O = Ca^{2+} + 2HCO_3^- \qquad (18.11)$$

This reaction is reversed when the CO_2 is removed by boiling and the $CaCo_3$ deposits. Other minerals containing sulphate and metals such as magnesium and small amounts of iron, zinc or aluminium also contribute. Only carbonates can be deposited by boiling and their contribution is known as *temporary hardness* and the remainder as *permanent hardness*. Although all species can be separately determined normally only *total hardness* is regularly measured. A complexometric titration using EDTA is used for this employing a buffered solution at pH 10 to ensure that all metals remain in solution at the high pH required for stoichiometric reaction (chapter 4, section 4.2.3). For simplicity the hardness is expressed as mg Ca dm^{-3} as if all is due to calcium (Table 18.3).

Table 18.3 An approximate classification of water hardness

$CaCO_3$ mg dm^{-3}	
0–50	Soft
50–100	Moderately soft
100–150	Slightly hard
150–200	Moderately hard
200–300	Hard
>300	Very hard

Anionic species in water are most comprehensively determined by chromatographic separations followed by non-selective measurement such as conductivity. In chapter 5, the principles of ion exchange chromatography have been reviewed. Methods based on this technique are very effective and are well established. Figure 18.2 shows a typical chromatogram for a natural water sample. Rapid developments are currently taking place in capillary electrophoresis and it is a technique that is likely to expand in terms of future use. Figure 18.3 illustrates the analysis of a water sample by CE. Both techniques require complex and relatively expensive laboratory equipment and are generally unsuited for use in the field. On the other hand, they deliver high quality analyses and can be fully automated.

Field analysis kits often use methods based upon colorimetry. Colour reagents and others are packaged for use with simple portable equipment and rapid on-site measurements for species such as nitrate, sulphate, chloride, nitrate and phosphate can be made. Alternatively ion selective electrodes can be used for nitrate, fluoride and chloride. A general purpose

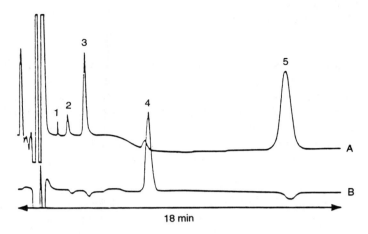

Figure 18.2 Analysis of a groundwater sample by ion chromatography. (A) Electrochemical detection; (B) UV. 1, F^-; 2, HCP_3^-; 3, Cl^-, 4, NO_3^-; 5, SO_4^{2-}.

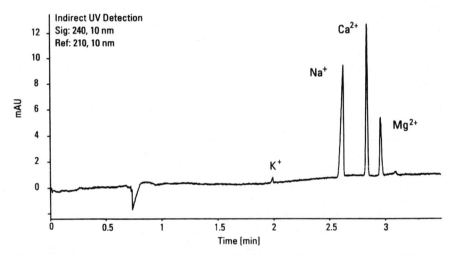

Figure 18.3 Inorganic cations in water determined by capillary electrophoresis. Teinacher mineral water. Concentration according to bottle lable: Na 0.12 g/l, Ca 0.12 g/l, Mg 0.03 g/l.

meter can be used for these analytes and for pH and dissolved oxygen electrodes as well.

18.7 Trace organics

In chapter 19 the methodology used in the determination of trace amounts of organic compounds in environmental samples is discussed, and comments pertinent to the analysis of water samples are included. Reference to this later chapter should therefore be made. Special aspects of sampling in water assessments are explained in section 18.9.

18.8 Radioactivity and radionuclides in water

Water is a sink for many pollutants and will collect radionuclides from fall out and run off. Natural radionuclides will also be transported by water. A rather special case is that of radon gas which emanates from rock formations. Mining and civil engineering operations can lead to the release of radon into the water, which would otherwise remain immobilised in the strata. The measurement of radionuclides from both natural and artificial sources is discussed in chapter 16.

18.9 Water surveys and sampling

Where water quality and characteristics are defined by regulation, the questions of what measurements to make, the methods to use, and the sampling protocols are clearly recorded. When carrying out surveys of environmental water for investigational or research purposes the situation is less clear cut. Most of the analyses required will be drawn from the list discussed in this chapter but obviously may also involve determination of specific substances and more physical measurements such as turbidity. Clearly defined aims for the survey or investigation must be established before the appropriate methods can be selected.

There are some general points that need to be consistently borne in mind when establishing a sampling plan.

1. Although readily homogenised, aqueous solutions in environmental circumstances are not always so. The possibility of horizontal and depth variations must always be considered.
2. Temporal variations are common. These may result for example from pollution discharges at varying times or simply from seasonal variations due to rainfall or temperature. Nitrate fertilisers for example are applied at specific times in the agricultural cycle.
3. Organic substances do not generally have a high water solubility and may well be more closely associated with humic substances and sediments. Analyses of these materials must always be considered as an option.
4. Natural processes lead to the accumulation of many metal ions on hydrous oxides of Al, Fe and Mn which form part of the sediments. Analysis of the sediment rather than the water, may well be a better indicator of metals present.
5. Where sediments remain undisturbed, as in lakes and ponds, the analysis of coves can give an historical record of pollution. This may be extended to bioindicators. Much work has been done on pollen types in the investigation of the vegetational history of surrounding areas.
6. The analysis of bioindicators, showing bioaccumulation and bio-magnification, as discussed in chapter 12, is also a valuable option.

7. Careful consideration must always be given to the need, in some cases, to make measurements shortly after sampling, e.g. dissolved oxygen and pH.

8. Remote sensing, using instrument packages suspended in the water is undergoing rapid development. It may provide the appropriate approach for long-term surveys or measurements at difficult depths and locations.

Further reading

Nicolson, N. (1993) *An Introduction to Drinking Water Quality* The Institute of Water and Environmental Mangement.

Reeve, R.N. (1994) Environmental Analysis, (ACOL Series) Wiley, Chichester.

19 The determination of trace amounts of organic compounds

C.J. Welch

19.1 Introduction

This chapter discusses elements of the analysis of trace organic materials, many of which are considered pollutants in the environment. What constitutes the environment, and what constitutes an organic pollutant will depend on the scope of the investigation. Environmental analysis may be limited to the surounding environment, where we may investigate the presence of unnatural or unwanted organic materials in water, soil and air. However the term 'environment' should be universal and include living systems both plant and animal that occupy that space. Indeed, it is very often via these living vectors that we discover the presence of organic pollutants in the environment as a whole, the presence of a trace organic material often only made known after it has had an effect on the ecosystem. The investigation of organic materials in living systems is usually undertaken by biochemists and toxicologists, however if biological systems are to be considered part of the environment, then it is important that the environmental analyst becomes expert in some of the methods used in toxicological investigation. To answer the second question, namely, what constitutes an organic pollutant, is equally as difficult and ambiguous. By definition a pollutant or poison is a material that has some harmful effect on a living system. Some confusion arises, however, when we consider that all materials, even water are potentially toxic to living systems, and it is simply a matter of degree or concentration that separates the noxious from the innocuous. At what level a biological response is initiated, and at what concentration that response becomes detrimental is a question that has still to be answered in many cases. When one considers the number of organic materials, and the diversity of life, the number of possible combinations becomes so large as to evade resolution. A general guideline is that if a material has a damaging effect at median or low concentrations, then it can be considered to be toxic, however what constitutes a median or low concentration will depend on chemical activity, and the biological system under investigation. Having clouded the horizon, we will now clarify matters by classifying the many organic materials that have been shown to be environmentally hazardous, and discuss how we can conduct chemical analysis for the presence of these materials.

If all organic compounds can be considered to be toxic at some concentration then classification will greatly simplify our task, for compounds of comparable chemical character should lend themselves well to similar methods of analysis. Organic materials that fall into one of the following categories are of most interest, however the list is by no means exhaustive, and as with all classifications, the boundaries are artificial, and should be viewed as convenient guidelines rather than rigid delineation. Environmental organic pollutants include aliphatic hydrocarbons, aromatic hydrocarbons, aldehydes, alcohols and phenols, anilines and benzidines, chlorinated solvents, polychlorinated biphenyls (PCBs), pesticides (encompassing insecticides and herbicides), and products used in the polymer industry such as phthalates and adipates. A detailed account of the analytical protocols used for the analysis of compounds from each category is beyond the scope of this chapter, however a few important examples have been chosen for more detailed discussion.

The analysis of organic materials can be divided into two major classifications, the first being the identification of bulk materials, and the second the analysis of trace organic materials. The analysis of a bulk material is usually conducted on a single homogeneous component of known composition, and the procedure is one of identification, quantitation not usually being required. Techniques such as nuclear magnetic resonance (NMR) and infrared (IR) spectroscopy which have been traditionally used as aids in identification rather than for quantitation are most applicable.

Trace organic analysis is required where the components of interest are present at very low concentrations and contained within a bulk matrix that is often of an undetermined and heterogeneous composition. Consequently the methods used for the analysis of trace analytes are not as easily defined as for bulk organic materials. Physical separation (extraction and partition) of the analyte from the matrix is usually required before analysis can begin. Once achieved, further physical separation may be needed to remove interfering components from the extract that may have been co-extracted with the analytes. Low levels of co-extractives are tolerated provided they do not interfere with analyte determination, but if significant levels are present they will have to be removed, commonly via simple column chromatography using adsorbent packing materials such as silica gel or alumina. The presence of co-extractives interferes not only with analyte determination, but contributes to the fouling of chromatographic columns and detectors, and hence adversely affects the sensitivity and precision of analysis.

Only after the analytes have been isolated can analysis proper begin. Where determination of a single analyte is required, direct spectroscopic methods of detection are favoured as they can simply and quickly provide a means of positively identifying the analyte as well as enabling quantitation.

Where several analytes have been isolated as a mixture, spectroscopic and non-spectroscopic physical detectors are used in conjunction with a variety of chromatographic techniques to enable quantitation. The combination of detectors and chromatographic techniques such as those commonly used for high performance liquid chromatography or gas chromatography, are particularly suited to the analyses of multiresidue samples. For example, several homologous aromatic hydrocarbons may have been isolated from a sample of contaminated drinking water; however only by chromatographically separating the homologs from each other is it possible to determine the amount of each individual component in the extract.

Methods commonly used for trace organic analysis can be assigned to one of three subdivisions; sample preparation, chromatography and detection.

19.2 Sample preparation   Sample preparation is a necessary prerequisite for almost all trace organic analysis. The method of sample preparation is dictated by the physical characteristics of the sample matrix and analytes, and the phrase 'like dissolves like' best describes the approach to successful analyte separation. Simply stated, sample preparation will ideally, with a high degree of efficiency, transfer only analytes of interest from the sample matrix into a solvent of known and uniform composition. To achieve this, a solvent with some chemical and physical characteristics in common with the analyte is often used. Hence, hydrophilic analytes such as organic acids are extracted using polar solvents like ethyl acetate and dimethyl ether (Lee and Thurmon, 1993), while lipophilic analytes such as neutral lipids, e.g. triglycerides require non-polar solvents like hexane and n-heptane (Christie, 1982).

The method of sample preparation must also be compatible with chromatographic and associated detection systems that will be used later for analysis. For example, the use of dichlorobenzene as a solvent for extracting chlorinated pesticide residues from a variety of raw materials is commonplace. These residues are usually volatile at moderately elevated temperatures, hence separation by gas chromatography and quantitation using an electron capture detector, which is highly sensitive to the presence of chlorinated hydrocarbons, offers an ideal solution. The presence of a chlorinated solvent, however, would completely mask the presence of halogenated trace analytes, so that sample preparation must include a step where dichloromethane is evaporated and replaced by a non-chlorinated solvent, such as hexane, prior to analysis.

In general, most protocols for sample preparation include several common steps which can be summarised as follows:

1. maceration;
2. dissolution;

3. extraction;
4. partition;
5. concentration.

19.2.1 Maceration, dissolution and extraction

Any sample for analysis can be considered as containing two basic elements; a small amount of analyte which is of interest, and the remaining bulk of the material which comprises the sample matrix and other minor components which are of no analytical interest. The sole purpose of the first two steps is to homogenise the sample in order to facilitate a uniform and efficient extraction of analytes from the sample matrix prior to partition. There are exceptions to this premise, where a degree of matrix integrity (heterogeneity) needs to be maintained. For example, if a fruit crop such as oranges were suspected of having been exposed to pesticides during the spraying of an adjacent field just prior to harvesting, then it may be assumed that only the skin of the fruit may be contaminated, while the inner flesh remains untainted. Under such circumstances it may be deemed necessary to analyse the skin and flesh separately, so that sample preparation would include separating the peel and flesh of the fruit for individual analyses.

For the purpose of sample preparation, samples can be assigned to one of two categories, namely those that are hydrophilic and will associate readily with aqueous or polar organic solvents, and those that are lipophilic and will associate with non-polar solvents.

If the sample is hydrophilic and a dry solid, then preparation can be achieved simply by crushing and grinding it to a powder. Wet mineral samples such as soils are most easily handled by drying in a vacuum oven prior to grinding. Mechanical disruption serves to create smaller sample particles of somewhat uniform size and generate a larger sample surface area, as well as ensuring any encapsulated analytes are liberated. The addition of an aqueous buffer after grinding results in the formation of a slurry where the matrix is fully solvated and ensures the homogeneity of the sample. Rehydration is an important prerequisite to extraction, where transfer of the analytes from the matrix to a miscible solvent such as acetone or methanol is mediated. Water miscible solvents are used because they are able to break up the sample matrix and better dissolve the more polar analytes which would otherwise remain associated with the matrix during partition. For extraction to occur efficiently and reproducibly, contact between the matrix and the extracting solvent must be as complete as possible, and is most fully attained when the matrix is fully solvated.

Matrices that contain a high proportion of water, such as those of an organic origin, i.e. vegetable and animal samples, are usually prepared by chopping or grinding (maceration), with the addition of a minimum volume of mild aqueous buffer to make a homogenised paste. The addition of water miscible solvents again assists in the liberation of analytes from the matrix. Initially non-polar solvents such as hexane were used to extract pesticide residues of DDT and DDE from crop samples. Later acetonitrile was used as an extracting solvent for organochlorine (OC) and organophosphorus (OP) residues. The change to a water miscible solvent had a two-fold effect; first the more polar OC and OP residues had a higher affinity for acetonitrile than hexane resulting in higher recoveries, and second, the non-polar 'fatty' co-extractives remained with the matrix as they had a lower solubility in acetonitrile, yielding cleaner sample extracts. Lipophilic matrices such as oils, lipids, and synthetic polymers make up a smaller proportion of samples when compared to the vast array of those that are hydrophilic. Extraction of analytes is achieved by solvating the sample with non-polar solvents such as hexane or *n*-heptane, however the co-extraction of some fat residues is not uncommon, and sample clean-up using adsorption chromatography on alumina columns, or gel permeation chromatography is often required.

19.2.2   Partition

The process of partition is often confused with that of extraction which as stated earlier is the transfer of an analyte from the matrix to a miscible organic solvent. Partition is specifically the transfer of an analyte or analytes from one solvent to another, both usually having very different polarities. The extent to which partition is achieved is measured by the fraction of analyte as a whole that is transferred, where 100% efficiency represents a transfer of all the analyte from the extractant to the partitioning solvent. In practice, this can never be achieved, although ideally it is approached. Analyte partition between two immiscible liquids is more correctly called liquid–liquid partition, and is mediated by the transfer of the analyte from one liquid, the extractant, to another, the partition solvent. Because the partitioning solvent is not miscible with the extracting solvent, an interface exists where the two phases meet. This interface is a physical barrier which bars the exchange of solvent molecules, but permits the migration of solute (the analyte) in both directions. The extent to which the analyte is associated with a particular phase describes the dynamic equilibrium of the system, and is called the equilibrium or distribution coefficient. Other factors such as the surface area of the interface will affect the kinetics of the transfer, i.e. the speed at which a steady state equilibrium is arrived at, but ultimately the efficiency of partition will be controlled by thermodynamic factors, i.e. the physical interaction between the analyte, extractant and partitioning solvent.

Liquid–liquid extractions are conducted using a separatory funnel. The solvated sample is introduced after filtering to remove particulate matter, and then the immiscible solvent is added. The stoppered funnel is inverted several times to allow mixing of the contents, the tap should be opened briefly and then closed on the last invertion to allow venting of built-up vapour pressure within the funnel. Vigorous shaking is to be discouraged as this will only lead to the formation of an emulsion, which is difficult to separate, and excess vapour pressure which may lead to spillage from the funnel. If an intractible emulsion is formed on gentle invertion, as sometimes occurs when an aqueous layer is extracted by a solvent such as dichloromethane, breakdown of the emulsion can be accelerated by the addition of a few millilitres of methanol and gentle mixing. This appears to function as an surfactant, reducing surface tension at the liquid–liquid interface, and dispersing the emulsion.

Partition can also be achieved between a solid and a liquid phase, as occurs in solid phase extraction (SPE). SPE is actually a combination of the extraction and partition processes described earlier. The solvated sample is applied to a solid phase that acts as an agent for extraction, selectively transferring the analyte from the sample to the solid phase. This step is likened to that of the transfer of analyte from the sample to a water miscible solvent. Once the analyte has been bound by the solid phase, i.e. extracted, then partition is required to transfer the bound analyte from the solid phase, to an aqueous buffer or organic solvent. This is often called the elution step, as the process is analogous to eluting components from a chromatography column. In fact as we will appreciate later, SPE is very like some forms of column chromatography, and a number of the same chemical phases are used in their construction. With SPE, the elution step transfers the analyte from the solid phase to a liquid solvent, and hence constitutes solid–liquid partition. The reader may have noticed what at first may seem to be a contradiction. To achieve liquid–liquid partition it was imperative that an immiscible solvent, for example dichloromethane or hexane was used to create a physical interface across which water miscible solvents could not cross, but which permitted analyte migration. Solid phase extractions often use eluting solvents that are miscible with the sample solution, so the question arises how is partition achieved. This dilemma is readily resolved when we realise that the solid phase performs not only the function of extraction agent but serves as a physical interface during elution (solid–liquid partition).

Most solid phase cartridges have a similar construction, comprising a funnelled barrel manufactured from polyethylene which is partially filled by a solid porous core. The core itself is composed of two distinct components; a porous support which is ideally as inert as possible and made from inorganic materials such as silica gel, or organic matrices such as the macroporous styrene–divinyl benzene copolymers (XAD resins). To facilitate extraction and provide an interface for partition, the support is coated with a chemically distinct liquid phase which is typically covalently bound via spacing groups to the support. This provides the chemical surface to which the analytes are selectively attracted.

Selective sorbents include silica, which is used primarily as a normal phase adsorbent, and bonded methyl (C2), octyl (C8), octadecyl (C18), cyano (CN), amino (NH$_2$), diol, phenyl, cyclohexyl and ion-exchange phases such as phenylsulphonic acid. SPE cartridges of various volumes are shown in Figure 19.1.

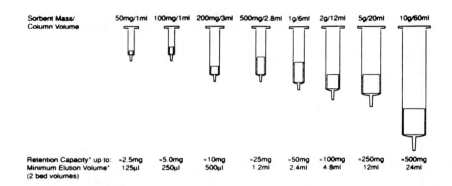

Sorbent Mass/ Column Volume: 50mg/1ml 100mg/1ml 200mg/3ml 500mg/2.8ml 1g/6ml 2g/12ml 5g/20ml 10g/60ml

Retention Capacity* up to: ≈2.5mg ≈5.0mg ≈10mg ≈25mg ≈50mg ≈100mg ≈250mg ≈500mg
Minimum Elution Volume* (2 bed volumes): 125μl 250μl 500μl 1.2ml 2.4ml 4.8ml 12ml 24ml

Figure 19.1 Capacity and elution characteristics of Bond Elut columns. Approximate values: capacity (mg) = 5% of sorbent mass; bed volume = 120 μl/100 mg of sorbent.

A typical protocol for the use of SPE cartridges is as follows:

1. wet the cartridge with methanol to solvate the sorbent;
2. follow by a conditioning wash with water or buffer to remove residues that may remain after cartridge manufacture;
3. load the sample solution;
4. wash the cartridge with a suitable solvent mixture to remove impurities;
5. selectively elute the co-extracted analyte with an appropriate eluant.

The first step in any SPE protocol involves wetting the cartridge with a water miscible solvent such as methanol. This serves to reduce any hydrophobic interactions that may occur between the support and the water solvated sample. Solvents such as acetone should not be used as wetting agents as they have a propensity to strip the liquid phase from the support. If this occurs, the selectivity of the cartridge is altered, and extraction and elution would be achieved most probably via adsorbtive interactions, if at all.

After wetting, the cartridge is washed with a mild buffer, and the sample is loaded. Often the sample is highly particulate, and may contain components such as lipids and proteins which will restrict the flow of buffers and eluting agents through a cartridge. A method which has proven successful in overcoming this problem is to add a small layer of dry celite to the top of the column before sample addition. This serves to prevent particulates from clogging the top of the support, allowing the free passage of solvents into the column, but which itself does not appear to adsorb analytes. SPE cartridges which incorporate an upper layer of celite into the support have recently been made commercially available.

To assist in loading the sample, and pulling elutants through the solid phase, a vacuum is applied by means of manifold to the cartridge outlet. Modern manifolds are constructed of transparent materials such as tempered glass or perspex and can hold sample bottles or test-tubes to collect elutants.

Once the sample/analyte has been loaded onto the solid phase, the cartridge is washed again with a mild buffer. The aim of the second wash is to elute co-extractants from the cartridge that may have been bound with the analytes on sample addition. Sample preparations that use extraction followed by liquid–liquid partition are not suited to clean-up by this method, and often require steps after partition to ensure the removal of co-extractants. The ease by which many co-extractants can be removed from analytes using solid phase extraction is often overlooked when choosing a method for sample preparation. Sample clean-up using this method can be viewed as a two-stage partition that selectively elutes co-extractants before analytes. This stepwise partition necessitates that the co-extractants and analytes have different affinities for the solid phase; for example, weaker solvent such as aqueous buffers can be used to elute weakly associated co-extractants before a stronger solvent like methanol is applied to elute the analytes. If the co-extractants display a higher affinity for the solid phase than the analytes then it is likely that judicial selection of a suitable solvent will selectively elute the analyte, leaving the co-extractants bound to the solid phase. Problems are encountered when co-extractants and analytes have very similar affinities for the solid phase, which can only be overcome by changing the selectivity of the solid phase, i.e. changing the type of liquid phase used for partition. An example is afforded by the application of SPE for the extraction of triazine herbicides from environmental water samples using C18 cartridges (Nash, 1990). Octyl bonded silica supports are used extensively because they have the ability to extract a wide range of trace organic pollutants. However, this universality can be a disadvantage, resulting in the transfer of interfering co-extractants as well as analytes of interest. By exchanging the C18 cartridge for one with a liquid phase that incorporated a strong cation exchanger like phenylsulphonic acid, it was found that at low pH the protonated forms of simazine, atrazine and propazine could be selectively extracted, excluding the interfering compounds previously seen (Land, 1994).

19.2.3 Concentration of the analyte

Little time is expended in most texts in discussing the concentration of analytes during sample preparation, and one can only presume that it is a straightforward procedure, usually achieved by gently heating the mixture to evaporate excess solvent. As the last step in sample preparation it should be viewed as the most critical, since mistakes made at this late stage will involve much wastage of time and materials if the process must be started again. In addition, as this is a concentration step, errors introduced during sample handling will be magnified by the time we are ready to start quantitation. A common mistake is to reduce the volume of solvent as quickly as possible by the application of too much heat. Here, the danger is that some alteration or decomposition of the analyte will occur in solution, which in some circumstances may be catalysed by the presence of impurities. For example, extracts containing tetracycline antibiotics will oxidise at elevated temperatures in the presence of some trace metal ions (Welch, 1991). More importantly, high temperatures may evaporate volatile analytes making quantitative recovery impossible. It is better to use a lower temperature, at reduced pressure if possible, to expedite rapid solvent removal. Access to the use of a rotary evaporator is ideal, particularly where large volumes of solvent are to be removed. For most practical purposes, solvents with boiling points of 100°C or above, including water, are considered difficult to evaporate. The higher temperatures afforded by heating the sample on a hot plate will be necessary to remove these solvents, but care must be taken when reducing flammable solvents, as their flash point may be exceeded. Once the solution has been reduced to a small volume, ca. 5 cm^3, it is better to remove the sample from heat altogether. Below this volume, it is too easy to allow the mixture to evaporate to dryness, by which time it is certain that all but the most non-volatile of components would have been at least partially volatilised. A more prudent method is to direct a stream of dry nitrogen gas onto the surface of the extract, which may safely be allowed to reduce to dryness without loss of analytes.

Before analysis, a known volume of a suitable solvent will be required to reconstitute the sample. This method of sample reduction and reconstitution has the added advantage of enabling the removal of a solvent that is known to interfere with analysis, but has proven efficacious as an agent for partition, e.g. the combination of dichloromethane and the electron capture detector mentioned earlier. The United States Environmental Protection Agency (EPA) have published methods for the concentration of organic materials from water samples (EPA, 1980). Volatile organics are purged from the water and trapped on solid adsorbents (Bellar and Lichtenberg, 1974) (GC-MS method 624, or GC method 601, 602 and 603). The EPA recommends a different adsorbent trap for each of the four methods. After the organics are adsorbed they are released from the solid phase by heat desorption onto the GC column. Non-volatile organics are preconcentrated by solvent extraction of the water sample, and reduced to minimal volume prior to analysis. Details of the extraction techniques and clean-up procedures are given in the Federal Register (EPA, 1979).

Following sample preparation, isolation of individual analytes from a mixture can be achieved by highly efficient column chromatography. In its simplest form, the stationary phase may be a particulate solid such as diatomaceous earth or silica gel, and separation occurs via adsorption where the analyte interacts directly with the support; this is commonly called normal phase chromatography. Alternatively, if the support has a liquid phase coated or bonded to its surface, then analyte separation occurs via partition; this type of separation is said to be reverse phase.

19.3.1 *Gas chromatography*

Typically, GC columns are of a very small diameter (about 0.25 mm i.d.) and quite long (25–50 m). Shorter (2–3 m), wider bore (3 mm) columns are used in older systems, where the column is packed with a solid support, such as diatomaceous earth or some other zeolite, which may or may not be coated with a liquid phase. Packed columns, however, are not suited to providing the higher resolutions required by modern analyses, and hence the use of long, small diameter capillary columns is the norm. Contemporary GC analysis of trace organic materials uses open tubular capillary columns where a liquid phase has been bound to the internal wall. Columns of this manufacture are called wall coated open tubular (WCOT) columns. Modern capillary columns are constructed of fused silica coated externally with a temperature resistant polyimide. Two other forms of open tubular column are available; surface coated open tubular (SCOT) and porous layer open tubular (PLOT) columns, however these have specialised uses and are not commonly used for trace organic analysis. Both forms incorporate a porous particulate layer such as alumina or a zeolite deposited on the inner wall of the column, which may be used simply as a molecular seive (PLOT), or may be coated with a liquid phase for partition (SCOT). Most commonly used are liquid phases of the OV-1 type, which have methyl groups attached to a silicone backbone. Polarity is gradually increased by substituting methyl groups with unsaturated moieties, such as those of the OV-17 type (methylphenyl silicone), or with groups containing electronegative elements, for example OV-210 (trifluoropropyl-methyl silicone). Having decided a suitable liquid phase, several other practical considerations must be taken into account before choosing a column. If a mass selective detector (MSD) is to be used, then a conventional capillary column must be used. However, if an FID detector, or one of the more selective detectors such as ECD or the flame thermionic detector (NPD; nitrogen phosphorus detector), is to be used then the use of a shorter (15 m), large bore capillary column (0.5 mm, megabore) can be contemplated. The advantage of the megabore column is one purely of practical convenience, as a conventional capillary column will always provide greater separation efficiencies. The larger size of the megabore column makes it possible to coat a thicker layer of liquid phase onto its inner wall. This serves to greatly increase the capacity of the column for analyte, allowing large volumes of sample to be injected directly on column without overloading. As a practical guide, column overloading is easily recognised; the chromatograms exhibit peak fronting, as distinct from tailing which arises usually as a result of undesirable analyte adsorption. More importantly megabore columns are highly tolerant to the presence of co-extractives, so that impurities that would severely impair the operation of a conventional capillary column do not interfere with analysis.

19.3.2   High-performance liquid chromatography

High-performance liquid chromatography (HPLC) is almost exclusively performed in the reverse phase mode (RP-HPLC), using stationary phases of silica gel derivatised with silylated aliphatic hydrocarbons such as octasilyl (C-8) and octadecylsilyl (ODS) moieties. These are used in conjunction with mobile phases that are predominantly aqueous, ideally containing 20–30% organic modifier, such as ethanol or acetonitrile. HPLC is of most utility where the organic analytes have a large molecular mass, and hence have low volatility, and where the molecule is polar or contains many polar substituents. Very polar organic materials and those that are ionised are not accommodated well by conventional RP-HPLC, and are either run in the presence of a counter-ion (ion pair RP-HPLC), or on columns that operate in a normal phase mode, e.g. on aminopropyl columns.

19.3.3   Quantitation

It is necessary to demonstrate that quantitation of any organic pollutant is accurate and precise, and this is achieved by calibration; most commonly against a set of internal standards, but where this is not possible external standards may be substituted. Prior to any calibration, the analyst must demonstrate that the system is working by injecting a known mixture of the components to be analysed. A blank containing all reagents used is subject to the extraction and clean-up procedures and is analysed to confirm the absence of interferences. Calibration standards of concentrations thought to encompass the expected concentration of the compound in the sample are prepared, and used to establish the sensitivity, minimum detectable limit (MDL) and linear range of the detector for each component. A standard curve is plotted daily to ensure that precision and reproducibility is maintained. If doubt exists as to the identity of a peak, the confirmation must be made by mass spectrometry.

The aim of a screening in multiresidue analysis is to determine which environmental sample submitted may be contaminated or not. The analysis of any sample begins by weighing a representative portion, followed by clean-up and injection of the cleaned extracts onto a chromatograph. For example, pesticides present as contaminants are characterised by their retention data and their response factor where selective detectors are used. Peaks found in the chromatogram arise from different sources: pesticide residues, environmental contaminants that are not pesticides, co-extractants from the sample matrix, trace contaminants in solvents, and interfering components that may have eluted from a fouled chromatograph. Other sources may be packing material, most probably via column bleed of the liquid phase from older columns, or cross-contaminants. Cross-contamination is avoided by observing meticulous cleaning and drying regimes when preparing glassware. A general guide is that all glassware should be washed before use using a mild detergent, rinsed with tap water, rinsed again with distilled water, and finally rinsed with reagent grade acetone or methanol before drying. Trace contamination caused by chemicals and solvents can easily be identified during the analysis of sample blanks. The reliability of a screening run depends on careful control of the chromatographic system with respect to stability of retention time and response factors. A primary cause of deterioration in GC systems is contamination of the injection port and the top of the column by non-volatile deposits that may catalyse thermal degradation or give rise to adsorption. These effects can reduce the overall response to analytes, produce tailing peaks or in the most extreme cases result in complete adsorption, and an absence of analyte peaks. The latter case rarely occurs completely, and a significant reduction in detector sensitivity over time is usually indicative of the need to dismantle and clean the chromatograph. Seriously contaminated columns can be rejuvenated by removal from the chromatograph and passing hexane through the column to wash adsorbates from it. It is advisable to pass a stream of nitrogen through the column to dry it after washing, as the liquid phase becomes saturated with solvent and may bleed significantly at normal operating temperatures if this is not removed. Polar solvents such as acetone must not be used for washing as they will strip the liquid phase from the column. Fouling of the top of the column can best be remedied by breaking 5 cm from the top after scoring the column with a glass cutter. The loss of such a small length column is of no consequence, but provides a new surface for future sample loading. Often this practice is combined with reversing the direction of flow through the column, by reversing the column in the chromatograph.

Removing a portion of column is not an option where HPLC is used, and fouling of the column front must be avoided. Washing the column with a mobile phase that has a high proportion of organic modifier is usually sufficient to remove some adsorbed materials. However, good sample clean-up is a prerequisite for most HPLC analyses, although the injection

19.4 Screening analysis

of 'dirty samples' directly on column has been documented where a guard column has been used. A guard column has essentially the same specifications as the analytical column, but is much shorter (typically less than 10 cm) and is used in front of, and in tandem with the analytical column. Its purpose is not to effect separation, although is does contribute in this respect, but to act primarily as a barrier to contaminants such as co-extractants, and protect the analytical column from fouling. When chromatographic efficiency degenerates, then replacement with a new guard column is usually all that is required to restore the system.

19.5 GC applications: pesticide analysis

The analysis of pesticides residues by GC is well documented, and has been developed to enable the detection of the 500–600 different compounds used worldwide. The wide range of chemical structures within this classification provides an insight into how chemical structure dictates analytical methodology. Pesticides can be considered as including organochlorine pesticides (OCs), organophosphorus pesticides (OPs), carbamates, phenols, triazines, triazoles, pyrethrins, pyrethroids, imidazoles and mercaptoimides. No single analytical method or combination is applicable to all possible residue combinations. Analytical methods for pesticide residues generally require a procedure for extracting the residues, clean-up procedures to separate residues from co-extracted compounds, a technique to measure the residues and a technique to confirm their identity. Such analytical methods are referred to as multiresidue methods; they must provide reliable identification and quantitation of a large number of compounds at very low concentrations.

The invention of the electron capture detector by Lovelock in 1960 (Karasek and Denney, 1974) provided the first selective detector with extremely high sensitivity for halogenated compounds. In the decade of the late 1950s to 1960s chlorinated hydrocarbons represented the class of insecticide most used. In the mid-1960s the invention of the alkali flame ionisation detector (AFID) and the flame photometric detector (FPD) allowed selective detection of organophosphates and nitrogen containing compounds.

Gas chromatography used in conjunction with selective electron-capture and nitrogen phosphorus detectors allow the detection of contaminants at trace level concentrations in the lower ppb range. Contemporary analyses utilise GC with mass spectrometry (GC-MS). GC-MS is most probably the most powerful tool available for the analysis of trace organic materials because of its inherent high selectivity and good sensitivity. It combines a high-performance separation method with a high-performance measuring technique. Packed column GC is not suited to coupling to a mass spectrometer (MS) because of the considerable pressure difference at the interface. Open tubular columns as discussed earlier, have little back-pressure and can be linked to a MS simply via an open-split coupling. The interface enables rapid column changes to be made without losing the vacuum in the MS, and is versatile with respect to column types (WCOT, SCOT, PLOT) and mobile phase (gas) flow rates. By introducing a large helium flow via a scavenger gas line at the interface, it is possible to protect the ion source from undesired fractions such as solvent peaks, and matrix compounds that may deteriorate sensitivity in the ion source.

The mass spectrometer is undoubtedly the most specific detector available; this is achieved by bombarding the analyte molecules with high energy electrons (70 eV) to fragment them. The fragments will have a range of molecular masses, and the distribution of mass is characteristic of a particular structure. This is called the fragmentation pattern, and can be used as a fingerprint to positively identify the analyte present. By searching a library of many thousands of fingerprints held in computer memory, it is possible to identify an unknown structure by comparing its fingerprint with similar patterns within a few minutes.

An alternative method still commonly used to identify multiresidue samples is that of dual column GC. The method uses the comparison of analyte retention times with those of known standards to identify unknown residues. Using a single column is almost impossible as the retention time of individual components will change with small variations in chromatographic conditions. However, provided the conditions can be held relatively constant, the dual column technique does work. The requirements are that both columns are run in unison on the same chromatograph, and that only one injection port is connected (usually via an effluent splitter) to both columns, although dual detectors are used. The columns must have different selectivities, i.e. columns with different liquid phases, for example OV-17 and OV-210 are commonly used. The OV-17 column is less polar than the OV-210 counterpart, which means that the elution order of components in a multiresidue extract will differ for the two columns. By cross referencing the elution of components on one column with those of the other, it is possible to identify components based on their elution times.

Summarising the reported applications of capillary columns in pesticide residue analysis from the beginning to the present, it can be stated that only a few stationary phases are used consistently. Non-polar dimethyl poly-silicones (OV-1, DB1) or those with 5% diphenyl substitution (DB-5) are best suited for screening analysis, applying the more polar phases OV-1701 and OV-17 for confirmation. Although capillary column GC is widely used for pesticide screening, wide bore capillary columns are often used, as they allow open tubular columns to be used in older chromatographs that were designed originally to take packed columns. An advantage of the megabore column over conventional capillary columns is that it allows larger amounts of material to be injected on column, so that splitless injection of a 1-μl sample is possible. In Figure 19.2 we can see that chromatogram (a), using a conventional capillary column (0.25 mm i.d.) affords separation of 16 chlorinated pesticides within 21 min, which requires a split injection

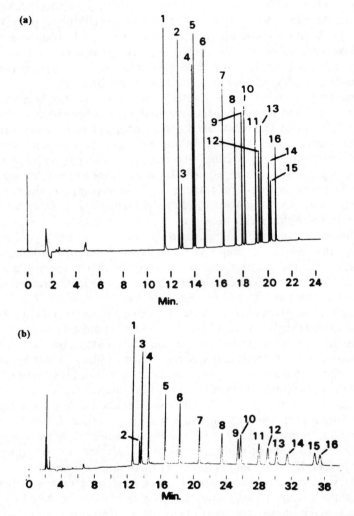

Figure 19.2 (a) Chlorinated pesticides resolved on a SPB-608 capillary column. SPB-608 fused silica capillary-column, 30 m × 0.25 mm i.d., film thickness, 0.25 μm; column temperature, hold 4 min at 150°C, then to 290°C at 8°C/min and hold 5 min; linear velocity, 25 cm/s. He, set at 290°C (a halogenated gas such as chloromethane or chlorotrifloromethane is used to determine linear velocity by ECD); make-up gas, N_2; 60 ml/min.; det., ECD, range, 10^{-11}; atten., ×16; sample, 0.8 μl decane containing 160 pg each pesticide. (1) α-BHC; (2) Lindane (γ-BHC); (3) β-BHC; (4) Heptachlor; (5) δ-BHC; (6) Aldrin; (7) Heptachlor epoxide; (8) Endosulfan I; (9) 4,4'-DDE; (10) Dieldrin; (11) Endrin; (12) 4,4'-DDD; (13) Endosulfan II; (14) 4,4'-DDT; (15) Endrin aldehyde; (16) Endosulfan sulfate. (b) Chlorinated pesticides resolved on a 0.75 mm i.d. SPB-5 capillary column. SPB-5 column, 60 m × 0.75 mm i.d.; film thickness, 1.0 μm; column temperature, 120°C to 215°C at 8°C/min and hold; flow rate, 15 ml/min; He, set at 120°C; det., ECD; sens., 10 × 64 (Varian 3700); sample, 1.0 μl n-decane containing 200 pg each pesticide, direct injection. (1) α-BHC; (2) γ-BHC; (3) Lindane (β-BHC); (4) δ-BHC; (5) Heptachlor; (6) Aldrin; (7) Heptachlor epoxide; (8) Endosulfan I; (9) Dieldrin; (10) 4,4'-DDE; (11) Endrin; (12) Endosulfan II; (13) 4,4'-DDD; (14) Endrin aldehyde; (15) Endosulfan sulfate; (16) 4,4'-DDT.

technique. Chromatogram (b) shows the separation of the same pesticide suite using a 0.75 mm i.d. megabore column. The 1-µl sample is injected directly on column, but the wider bore means that zone dispersion is increased, reducing component resolution, and increasing elution times. Sensitivity is often not impaired as more sample can be accommodated by the column.

Although the first applications of HPLC tended to be for the separation and quantitation of relatively large samples, there has been a progressive move towards using the technique for trace analysis. The term trace analysis is somewhat ambiguous, for it can apply to the detection and estimation of low levels of materials in large samples (e.g. polycyclic aromatic hydrocarbons in water) as well as to the determination of low levels of material in small samples (e.g. polychlorinated biphenyls (PCBs) in a food sample), and a different analytical approach is needed to meet the challenge posed by these two different problems. For example, in the first instance the compounds of interest could perhaps be concentrated by solid phase extraction or adsorption, where many column litres of water are passed over the stationary phase, in which case the sample enrichment process has produced a level of analyte that does not require the detector to operate close to its minimum detectable limit (MDL). On the other hand, the second problem is more likely to require a sensitive detection procedure, for even with extraction and concentration, such samples are unlikely to yield high levels of the analyte. In general, fluorescence (FL), electrochemical (EC) or ultraviolet (UV) detectors are most likely to be used where the quantity of sample for analysis is small. An important criterion is to select a detector for the task that will produce a large response for the analyte and a low response for other co-extractives (i.e. selectivity is required as well as sensitivity). In this respect UV absorbance is the least selective of the three detection methods and it will often be found that the limiting factor in an analysis is not the sensitivity of the detector, but how effectively the components of interest can be separated from other UV active solutes. A good example is that supported by the analysis of polycyclic aromatic hydrocarbons and phenols present in drinking water. Extraction and partition of these analytes produces a multiresidue sample that is dominated chromatographically by the strong UV absorbance of the conjugated species. A more selective chromatogram can be obtained by using electrochemical detection.

19.6 HPLC: trace analysis

19.7.1 *Aromatic hydrocarbons*

19.7 HPLC applications

Polycyclic aromatic hydrocarbons (PAHs) are a diverse group of compounds that constantly appear as trace environmental contaminants mainly as an adjunct to the use of fossil fuels. Their release into the environment may occur via one of many avenues, but they commonly appear as a result of fuel combustion, for example in the effluent from power station smoke stacks and exhaust from automobile engines, or raw materials processing where for example oil is used in the manufacture of synthetic polymers.

Polycyclic aromatic hydrocarbons are not all inherently toxic, but the oxidised derivatives such as phenols, quinones and *trans*-dihydrodiols are, so that bioaccumulation and subsequent metabolism (oxidation) of the reduced forms represents a hazard that necessitates environmental monitoring. Reverse phase HPLC is particularly well suited to the analysis of mixtures of lipophilic compounds and their metabolites, since polar metabolites have shorter retention times than the native forms, facilitating identification (see Figure 19.3).

The analysis of PAHs in air-borne particulate matter is often conducted using HPLC combined with fluorescence (FL) detection. Sample collection begins with the passage of a large volume of air through glass fibre filters. Adsorbed PAHs are partitioned using Soxhlet-extraction for 8 h with cyclohexane, and the subsequent extract is concentrated using a Kuderna-Danish or rotary evaporator. The sample is separated into fractions by

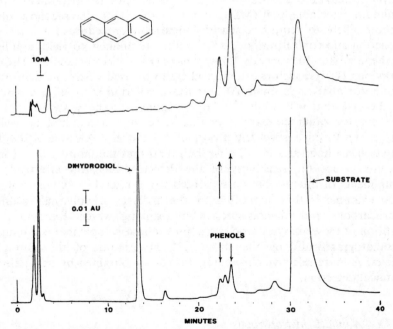

Figure 19.3 EC (top) and UV (bottom) chromatograms of the ethyl acetate soluble metabolites of anthracene. Components include: anthracene-1,2-dihydrodial (13.8 min), anthracene-1- and -2-ol (22–34 min), the substrate (31.0, UV), and an unknown (31.5 min, EC). Source: McMurtrey, K.D. and Welch, C.J. (1986) *J. Liquid Chromatogr*, **2**, 2749.

simple column chromatography on silica gel using hexane–chloroform as the mobile phase, which are quantitatively analysed by reverse phase HPLC with FL detection (HPLC-FL). Here excitation wavelengths of 250–280 nm and emission wavelengths of 370–390 nm are usually used. Lipophilic liquid phases such as octadecylsilyl-silica gel (ODS) are highly selective for most PAHs but require gradient elution to facilitate separations within an acceptable range of capacity factors ($k = 2$–4).

HPLC-FL has been applied to the analysis of PAHs in water samples, which are partitioned with immiscible solvents such as cyclohexane or dichloromethane. The sample can be reduced either by use of a Kuderna-Danish or rotary evaporator. Sample clean-up is achieved using an alumina column and a 1:1 (v/v) cyclohexane–benzene mobile phase. The eluant is then concentrated and analysed by reverse phase gradient elution HPLC.

HPLC analysis of PAHs in coal tar pitch and other bituminous products utilises a TLC clean-up technique. The sample is extracted with tetrahydro-furan, and spotted onto a TLC plate and developed with toluene, where the adsorbent is a mixture of silica gel and the organo-clay Bentone 34. The appropriate fractions are then scraped off the plate and the PAHs desorbed using hexane, before HPLC analysis with UV detection. Oxidised deriva-tives of PAHs can be selectively detected using EC (amperometric) detec-tion, which are highly electroactive and are reduced to their native forms by the working potential (ca. 0.7–1 V). When EC detection is used in conjunc-tion (tandem) with a UV detection, it has been shown that native and oxidised form of PAHs and polynitro aromatic hydrocarbons (PNAHs) such as the acridines can be readily identified (McMurtrey and Welch, 1986).

19.8 GC/HPLC applications

19.8.1 Trace organic analysis in water

The diversity of trace organic compounds that have been found to contami-nate water supplies has necessitated the development of many procedures for analysis. Statutory regulation of drinking water quality has meant that the ability to identify contaminants in µg/l (ppm) or even ng/l (ppt) concentrations is required. Organic materials found in water can be placed into one of three classifications; volatile, semi-volatile and non-volatile compounds. GC has been used for the analysis of volatile and some semi-volatile compounds, however it is estimated that only 5% of the organic contaminants normally present in water are volatile enough for analysis by GC. HPLC is ideally suited to the analysis of water samples, as a minimum of sample clean-up is required, and samples may be injected directly on column as aqueous solutions. HPLC is the method of choice for the analysis of semi-volatile and non-volatile compounds.

The EPA has published a series of bulletins (EPA, 1982), for GC and HPLC analysis (numbers 601–613), and GC/MS analysis (numbers 624 and

625 for volatiles and non-volatiles analysis, respectively) of 113 organic priority pollutants in wastewater. The methods categorise these pollutants into 13 classes and contain a protocol for the analysis of each class, thus simplifying the procedure.

Direct analysis for trace components in water is often not practical, as the detection of species at very low concentrations would require an extremely sensitive means of detection, and would most certainly yield ambiguous results due to the presence of interfering compounds. A concentration step prior to analysis is generally required, where the concentrate must be enriched by at least a hundred- and preferably a thousand-fold. For semi- and non-volatile organic compounds, liquid/liquid partition or solid phase adsorbants (SPE) are used extensively. In liquid/liquid partition, typically 1 dm^{-3} of water is partitioned with two or three 100–150 cm^3 portions of a solvent such as dichloromethane. The combined extracts are concentrated by reduction to 1 cm^3 or less by evaporating the solvent, using a Kuderna-Danish evaporator and a hot water bath. SPE techniques are frequently used to trap and concentrate semi- and non-volatile organic materials from water samples. The method of SPE extraction has been elaborated on earlier, but briefly stated involves passing a large volume of the water sample through the solid phase bed, and subsequently eluting retained materials with a low volume of a suitable organic solvent. Numerous sorbents have been studied, among which the XAD resins such as Amberlite XAD-2 have proven most effective. The organic materials adsorbed onto porous polymers can be fractionated into acidic, basic and neutral pollutants by eluting the polymer with dilute (0.05 M) sodium hydroxide, 1 M HCl and ether, respectively. Where GC is to be used for final analysis, analytes concentrated by SPE can be eluted by heating the extraction cartridge. If the cartridge has been mounted in line with the gas inlet to a chromatograph, then eluted analytes are swept directly onto the GC column for separation. Combined with mass spectrometry for identification, this purge and trap technique is a very powerful tool for the analysis of trace volatiles and semi-volatile organic materials in water. Where HPLC combined with UV or FL detection is to be used for analysis, transfer of the extract to solvents such as methanol or acetonitrile affords solvent compatibility with the mobile phase and detectors. Standard methods of analysis for contaminants such as phenols and PAHs in drinking water have been published by the EPA (1984). The extensive use of phenols in manufacturing and their generation as by-products means that industries must routinely monitor their discharge into the water supply. In the USA this monitoring is undertaken by the EPA while in the UK it is presently the responsibility of agencies such as the National Rivers Authority (NRA). A colorimetric method using 4-amino antipyrene is the most widely used method for routine analysis (Rand et al., 1976). However, this procedure lacks specificity and para-substituted phenols cannot be determined. Methods using GC combined with electron capture detection require that the phenols are derivatised with a halogenated derivatising agent (McKague, 1981). Direct detection by HPLC with UV detection has been reported (Sparacino and Minick, 1980), and HPLC with amperometric detection affords a rapid and accurate method of phenols analysis, while being simple enough to be used for screening purposes. The chromatogram in Figure 19.4 show the detection of phenol in wastewater from oil refinery effluent. Sample preparation involved acidification of the sample to pH 3.0 with hydrochloric acid, centrifugation to remove particulates, and dilution with citrate buffer if phenol concentrations exceeded 1.5 ppm. Detection was achieved using an applied potential of 900 mV, and the minimum detectable limit was shown to be better than 20 ppb.

Acidic and basic organic compounds can be separated from water by ion chromatography. Ion-exchange resins with an XAD structure are modified to carry charged groups to separate ionic nonvolatile compounds from the sample matrix. Where the group is an anion exchanger such as a tertiary amine, organic acids can be selectively separated from the water, then eluted with hydrochloric acid in methanol solution, prior to analysis by HPLC.

19.8.2   Analysis of organic materials in soil and sediments

While there are no specific guidelines for the analysis of trace organics in soil, any of the common methods of sample preparation can be used including

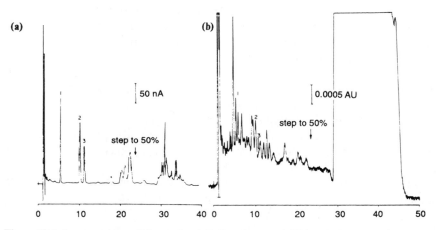

Figure 19.4 A comparison of electrochemical detection at +0.90V and ultraviolet detection at 254 nm. Mobile phase: 15% CH_3CN in 0.2 M sodium perchlorate/0.005 M sodium citrate buffer (pH 5.00). Stationary phase: biophase ODS 5 μm column (250 × 4.6 mm). Detector; LC-4A or LC-4B electronic controller with a glassy carbon working electrode. Applied potential: +900 mV versus Ag/AgCl. Flow rate: 2 ml/min. Temperature: ambient. (a) LCEC: 1, phenol; 2, *m,p*-cresol; 3, *o*-cresol; (b) LCUV.

headspace (GC) analysis and solvent extraction. Pesticides and polychlorinated biphenyls (PCBs) can be extracted with iso-octane, while semivolatiles like aromatic hydrocarbons are extractable with dichloromethane.

Methods used for the analysis of volatile organic materials in water can be applied to the analysis of volatiles in wet sediments and soils. Analysis of trace organic compounds in soil samples is greatly complicated by the sample matrix, and sample clean-up is very much dictated by the water and humic content of the soil. The analysis of semi- and non-volatile organics in dry or wet soil is relatively easier. Partition of analytes is often achieved via Soxhlet extraction with a non-polar solvent such as hexane. While analytical recoveries are common, the main problem with this technique is that high proportions of co-extractives are recovered, especially if the sample has a high humus content.

19.8.3   Analysis of amines

The analysis of amines found in environmental samples provides an interesting topic for discussion for two reasons. Firstly, the extensive use of these materials as raw materials or intermediates in the pharmaceutical, pesticide, polymer and dyestuffs industries, and hence the likelihood of an environmental presence, and secondly that they pose several analytical problems. The need to follow the progression of these analytes into the environment stems from their inherent carcinogenic or mutagenic behaviour and that they are readily bioaccumulated. Conversion of organic amines to nitrosamines on reaction with nitriles presents still further hazard.

The determination of amines in water can be approached via several routes, namely, direct gas chromatography of aqueous samples, concentration of amines on an adsorbent followed by desorption, separation and detection, and precolumn derivatisation of the amines prior to analysis.

Because amines are polar (primary > secondary > tertiary), they are difficult to remove from water and require highly polar solvents to effect extraction. To circumvent this problem, ion pairing agents such as sodium octyl sulphate or tetrabutylammonium phosphate can be added which associate with the analyte, effectively raising the hydrophobic character so that solvent extraction is facilitated. Alternatively ion-exchange SPE can be utilised to effect extraction.

Under basic conditions, alumina can be used as a weak cation exchanger suitable for the isolation and separation of amines. At low pH amines are fully protonate, and hence may be separated using conventional HPLC stationary phases like ODS. However, the tendency of these compounds to be adsorbed by the packing material makes their behaviour less than ideal, and the addition of ion pairing agents is often required to effect satisfactory separation (Sparacino and Minick, 1980). Aromatic amines are easily detected by the use of UV detection at 254 nm, but aliphatic amines have a low response factor, and are more readily detected using EC (amperometric) detection.

Primary and secondary amines can be derivatised with benzoyl chloride in sodium bicarbonate to form *N*-alkyl and *N,N*-dialkyl benzamides, which absorb strongly in the UV. The derivatives are stable and chromatograph well on a reverse phase column. Some derivatising reagents such as fluoro-acetates and *m*-toluamides will react with water as well as the amines, hence it is necessary to extract the amines into organic solvents prior to derivatisation.

Amines are difficult to work with by GC because the chromatograms display tailing peaks caused by the adsorption of the sample onto the support or internal column surface. As the amine basicity increases so does the tailing; consequently primary aliphatic amines and polyfunctional amines are most difficult to analyse. Aromatic amines, which are not as basic, pose less of a problem, while tertiary amines are least difficult, and can be analysed easily by HPLC under acidic conditions. GC of these compounds will almost always require derivatisation to precede separation, as the native amines will oxidise and/or irreversibly adsorb to the column. Pentafluorobenzaldehyde (Karoum *et al.*, 1975) is a derivatising agent selective for primary agents which chemically reacts under aqueous conditions to form imines which can be extracted with hexane, prior to GC. The introduction of a halogenated (fluorine) group in the molecule enables the use of the ECD affording the determination of very low concentrations of analyte (Rhoades and Johnson, 1970). Clarke *et al.* (1966) have reported a comparison of ECD responses for several derivatives of benzylamine. Heptafluorobutyrate gives the best ECD response, but pentafluoropropion-ate was favoured as it was possible to elute the derivative using lower column temperatures.

References

Bellar, T.A. and Lichtenberg, J.J. (1974) *J. Am. Water Works Assoc.* **66**, 739–744.
Christie, W.W. (1982) in *Lipid Analysis*, 2nd edition, Pergamon Press, Oxford.
Clarke, D.D., Wilk, S. and Gitlow, S.E. (1966) *J. Gas Chromatogr.*, **4**, 310.
EPA (1979) *Federal Register* **44** (233), Dec 3.
EPA (1980) *Federal Register* **45** (98), May 19.
EPA (1982) *Method of Organic Chemical Analysis of Municipal and Industrial Wastewater*, US EPA EMSL, Cincinnati, OH 45268 USA, EPA-600/4-83-057.
EPA (1984) *Federal Register* **49**, 112.
Karasek, F.W. and Denney, D.W. (1974) *Anal. Chem.* **46**, 1312.
Karoum, P., Gillin, J.C., Wyatt, R.J. and Costa, E. (1975) *Biomed. Mass Spectrom.* **2**, 183.
Land, C.C.J. (1994) *LC GC Int.* **7**, 215.
Lee, B.Y. and Thurmon, T.F. (1993) *LC GC Int.* **6**, 42.
McKague, A.B. (1981) *J. Chromatogr.* **208**, 287.
McMurtrey, K.D. and Welch, C.J. (1986) *J. Liquid Chromatogr.* **2**, 2749.
Nash, R.G. (1990) *J. Assoc. Off. Anal. Chem.* **73**.
Rand, M.C., Greenberg, A.E. and Taras, M.J. (1976) *Standard Methods for the Examination of Water and Wastewater*, 14th edition, American Public Health Association, Washington, DC.
Rhoades, J.W. and Johnson D.E. (1970) *J. Chromatogr. Sci.* **8**, 616.
Sparacino, C.M. and Minick, D.J. (1980) *Environmental Sci. Technol.* **14**, 880.
Welch, C.J. (1991) Dissertation Abstracts, University of Southern Mississippi, USA.

Glossary

Some of the more important terms and symbols which will be used in this book are defined below. More exhaustive lists may be found in the references at the end of Chapter 1.

Accuracy
The closeness of an experimental measurement or result to the true or accepted value.

Acid
A species which donates protons (Lowry–Brønsted) or accepts electrons (Lewis).

Acid rain
Rain with a pH less than 5.6 due to the presence of acids produced by natural or human causes.

Aerobic conditions
Environmental conditions in which oxygen is present.

Anaerobic conditions
Environmental conditions in which oxygen is absent.

Analyte
Constituent of the sample which is to be identified by qualitative analysis or measured quantitatively.

Anion
Negatively charged chemical species.

Aromatic compounds
Planar cyclic organic compounds in which there is a delocalized bonding electron system, e.g. benzene, naphthalene, anthracene, pyridine. Polynuclear aromatic hydrocarbons (PAH) are severe chemical pollutants.

Assay
A highly accurate determination, e.g. of valuable metal in a mineral ore, or of the purity of a material.

Atmosphere
The gaseous part of the environment.

Autotrophs
Organisms that can produce organic molecules from inorganic sources.

Background
That proportion of the measurement which arises from sources other than the analyte itself.

Base
A species which accepts protons (Lowry–Brønsted) or donates electrons (Lewis).

Biodegradable
A material which may be broken down by natural biological processes.

Biomagnification
The increase in concentration of a chemical that occurs in a living organism because the rate of excretion is lower than the rate of ingestion.

Biomass   Biologically derived material, both living and dead.

Biosphere   The part of the environment containing all living matter.

Biota   All living matter.

Blank   A measurement in which the sample is replaced by a simulated matrix, the conditions otherwise being identical to those for sample analysis. The blank can be used to correct for background effects.

Buffer   A solution which allows only a small change when ions are added to it. For example a pH buffer solution should allow only a very small change in pH when acid or base are added.

Calibration   A procedure which allows the response of an instrument to be related to the concentration, or other property of the sample by first measuring the response with a known standard. Calibration of apparatus may also be carried out using independent methods, for example, using electrical measurements.

Catalyst   A substance which increases the rate of a chemical reaction.

Cation   A positively charged species.

Cell   The fundamental unit of structure of living organisms.

CFCs   Chlorofluorocarbons, e.g. CCl_2F_2.

Cleanup   A stage in an analytical method in which the analyte is separated from the bulk of the matrix as a preliminary to the analysis of the sample. Commonly needed in environmental analysis because of the complex matrices that are encountered.

Complex ion or compound   A central atom or ion (e.g. a metal ion such as Co^{3+}) co-ordinated to a definite number of species (ligands) which may be ions or molecules (e.g. NH_3). For example, $[Co(NH_3)_6]Cl_3$.

Concentration   The amount of sample present per unit mass or volume of solvent. Many alternative methods of expressing concentrations must be used, and some are given in Table G.1.

Constituent   A component of a sample. It will be major if $> 10\%$, trace if between 0.01 and 0.0001%.

Table G.1   The units of concentration

Units	Name and symbol
moles of solute per dm^3	molarity, M
moles of solute per kg	molality, m
grams of solute per dm^3	$g\,dm^{-3}$
parts per million/parts per billion	ppm/ppb
percent by weight	% w/w
percent by volume	% v/v

Users of materials such as food, chemicals. **Consumers**

Bond between atoms formed by sharing of electrons. **Covalent bond**

A pathway via which a receptor can be exposed to a pollutant and its effects in **Critical pathway**
a concentrated or enhanced way. Typically involves the concentration of a
pollutant by physical, chemical and biological processes and delivery to the
receptor via its food chain.

A geometric structure in which the atoms, ions or groups are regularly **Crystal lattice**
positioned on a set of lattice points. The lattice is built of repeating unit cells
of lattice points.

The smallest amount or concentration of an analyte that can be detected by a **Detection limit**
given procedure with a given degree of confidence.

A quantitative measure of an analyte with an accuracy much better than 10% **Determination**
of the amount present.

The observation of the interaction of organisms with the environment. **Ecology**

The ability of an atom to attract electrons to itself. **Electronegativity**

A change which occurs with the absorption of heat from the surroundings. **Endothermic change**

A measure of the capacity to do work. The SI unit of energy is the Joule, and **Energy**
the energy may be of many forms, mechanical, potential, chemical, solar or
electrical. For chemical reactions, the change may be described in terms of the
internal energy, U, or the enthalpy, H, or the free energy, G, depending on the
conditions.

A thermodynamic property related to the disorder of the system or **Entropy**
surroundings.

The surroundings of a system or organism. **Environment**

A protein-based molecule which can act as a catalyst, often very specifically **Enzyme**
for a biochemical reaction.

The final state of a chemical reaction. The equilibrium constant will show the **Equilibrium**
relationship between the final concentrations (or pressures) of the reactants
and products at a particular temperature.

The amount of substance which produces, reacts with or can be indirectly **Equivalent**
equated to one mole of H^+ ions. The term is obsolete.

The difference between the measured value and the true value. **Error**

A semi-quantitative measure of the amount of analyte with an accuracy no **Estimation**
better than 10% of the amount present.

The enrichment of a body of water by nutrients. This is often due to the **Eutrophication**
release of nitrates or phosphates from fertilizers or effluent.

Exothermic change   A change which occurs with the release of heat.

Fault   A fracture in the Earth's crust along which there has been displacement.

Fermentation   An enzyme-controlled reaction causing breakdown of large organic molecules, e.g. starch to alcohol.

FGD   Flue Gas Desulphurisation. The removal of sulphur compounds from the gaseous effluent, particularly at power plants.

Food chain/web   The related species which produce and consume food from other materials or species.

Free radical   A chemical species containing an unpaired electron.

GCM   Global Circulation Model. A model of global climate trends.

Germ cells   Biological cells in an organism such as those in ova and sperm the characteristics of which are transmitted to the offspring.

Global warming   The increase in the global average temperature.

Greenhouse Effect   The effect on the heat balance of the Earth due to the difference in energy received and re-radiated. This amount depends on the amounts of 'greenhouse gases' such as CO_2 in the atmosphere.

Heterotrophs   Organisms that cannot use inorganic molecules to synthesize organic molecules, but must obtain them from other organics.

Hydrosphere   The part of the environment containing liquid water.

Inorganic   Formerly, compounds *not* produced by living organisms. Generally now used for compounds of elements other than carbon.

Interference   An effect which obscures or alters the behaviour of an analyte in the analysis.

Internal standard   A substance added to sample and standards in a fixed and known amount. This standard is then subject to the same conditions as the sample and should give a better accuracy.

Ion pair   Oppositely charged ions associated in solution.

Isomorphous   Two crystal lattices whose structures are the same if an atom or ion from one may replace one from the other without altering the geometry.

Isotopes   Atoms that have the same atomic number but different atomic mass due to the number of neutrons in the nucleus.

Kinetics   Reaction kinetics is the study of the rates and mechanisms of chemical reactions. The kinetic theory of gases models the behaviour of assemblies of gas molecules.

LD_{50}   A measure of toxicity as the lethal dose which kills 50% of a population.

Lithosphere   The part of the environment containing rocks.

The upper part of the Earth's inner structure.	**Mantle**
Treatment of a sample with a reagent to prevent interference from other constituents.	**Masking**
The remainder of the sample of which the analyte forms a part.	**Matrix**
The arithmetic average of a set of values.	**Mean**
The complete description of the instructions for a particular analysis.	**Method**
The conversion of other species (e.g. NH_3) to nitrites and nitrates. The reverse is denitrification.	**Nitrification**
A species having no charge or dipole. In general, non-polar solvents (e.g. benzene, tetrachloromethane) dissolve non-polar compounds, while polar solvents (e.g. water) dissolve ionic and polar compounds.	**Non-polar**
A distribution of results or values which follows a Gaussian curve.	**Normal distribution**
Formerly, compounds produced by living organisms. Now generally the chemistry of carbon compounds.	**Organic**
The removal of electrons from a species.	**Oxidation**
Polychlorinated biphenyls. Present a disposal hazard.	**PCBs**
A measure of the acidity of a solution defined by:	**pH**

$$pH = -\log[H^+]$$

where $[H^+]$ is the molar concentration of hydrogen ions.

A homogeneous region of a system. It may be gas, liquid or solid.	**Phase**
The breaking of chemical bonds due to absorption of electromagnetic radiation.	**Photodissociation**
The production of organic molecules from CO_2 and H_2O by plants under the action of solar radiation.	**Photosynthesis**
Having a charge or dipole.	**Polar**
Any change in a material, species or environment on land, in water or in air that may cause acute or chronic detriment or lower the quality of life directly or indirectly. The subject may be divided to indicate the type or area of pollution, e.g. chemical, thermal, energy or noise pollution, or water or air pollution.	**Pollution**
A substance of high molecular mass generally made from smaller molecules (monomers) by chemical reaction.	**Polymer**
The random or indeterminate error associated with the result. It is measured by the standard deviation.	**Precision**

Primary standard   A substance whose purity and stability are very well established and which is suitable for making calibrations or comparisons with other standards.

Producers   Organisms that produce organic molecules and are then used as a food source.

Productivity   The ability of an ecosystem to support life. Commonly applied to aquatic systems where pollution with nutrients leads to high productivity and a deterioration in the water quality.

Radiation   Electromagnetic waves of all wavelengths. Ionizing radiations produced by nuclear reactions.

Radiation budget   The make-up and nature of the radiation to which an individual is exposed.

Table G.2   Physical quantities and units including SI and CGS

Physical quantity	SI		CGS	
	Unit	Symbol	Unit	Symbol
length, l	metre	m	centimetre	cm
mass, m	kilogram	kg	gram	g
time, t	second	s	second	s
energy, E	joule	J	erg	–
			electron volt	eV
			calorie	cal
thermodynamic temperature, T	kelvin	K	kelvin	K
amount of substance, n	mole	mol	mole	mol
force, F	newton	N	dyne	–
volume, V	cubic metre	m^3	cubic centimetre	cm^3 (ml)
	cubic decimetre	dm^3	litre	l
electric current, I or i	ampere	A	ampere	A
electric potential difference, E	volt	V	volt	V
electric resistance, R	ohm	Ω	ohm	Ω
electric conductance, G	siemens	S	mho	Ω^{-1}
quantity of electricity, Q	coulomb	C	coulomb	C
electric capacitance, C	farad	F	farad	F
frequency, ν	hertz	Hz	cycles per second	cps
wavenumber, σ $(\bar{\nu})$			reciprocal centimetre	cm^{-1}
wavelength, λ	metre	m	centimetre	cm
	millimetre	mm	millimetre	mm
	micrometre	μm	micron	μ
	nanometre	nm	millimicron	$m\mu$
			Angstrom	Å
magnetic flux density, B	tesla	T	gauss	G
disintegration rate	curie	Ci	curie	Ci
	becquerel	Bq		
nuclear cross-sectional area	barn	b	barn	b

Source: Fifield, F.W. and Kealey, D. (1995) *Principles and Practice of Analytical Chemistry*, 4th edn, Blackie A&P, Glasgow. p. 12.

The removal of gases and dust from *inside clouds* by incorporation in rain.   **Rain out**

A chemical used to produce a specific reaction.   **Reagent**

An organism that is affected by a pollutant.   **Receptor**

A reaction involving the transfer of electrons.   **Redox reaction**

The potential of an electrode at which a reduction occurs compared with the   **Reduction potential**
standard hydrogen electrode. For example, the zinc electrode has a *standard*
reduction electrode potential, or half-cell emf of $-0.76\,V$ corresponding to
the reduction reaction $Zn^{2+} + 2e = Zn$. Older texts often use the *oxidation*
potential with opposite sign, corresponding to the reverse reaction.

The process of generating energy used by living organisms where inorganic or   **Respiration**
organic compounds are oxidized by inorganic species which are themselves
reduced. If oxygen is the species reduced, it is aerobic respiration.

Ribonucleic acid. Important in the storage and transmission of genetic   **RNA**
information. Also deoxyribonucleic acid (DNA).

A portion of substance about which analytical information is required, or a   **Sample**
selection of data for statistical purposes.

The change in response per unit signal, or the ability of a method to detect or   **Sensitivity**
determine an analyte.

The system of units now used for reporting all data. The most useful ones are   **SI units**
given in Table G.2.

The number of figures which indicate the precision of a measurement.   **Significant figures**

The ultimate site of a pollutant.   **Sink**

A mixture of smoke and fog, caused either by smoke particles in suspension in   **Smog**
water-based fog, or by photochemical production of harmful compounds
from reaction of oxides of nitrogen and volatile organic compounds.

Biological cells in an organism that are not transmitted to offspring. Thus any   **Somatic cells**
changes such as mutation are not passed on.

The uptake of a substance (e.g. a gas) onto the surface of a substrate   **Sorption**
(*adsorption*) or into its bulk (*absorption*).

A pure substance whose quantitative behaviour and reactions are known.   **Standard**
The term may also be used for a pure analyte or a solution containing an
accurately known amount which may be used for calibrations.

A mathematical parameter useful in comparing the precisions of measure-   **Standard deviation**
ments. A full discussion will be found in Chapter 2.

The determination of the concentration of an analyte solution from its   **Standardization**
reaction with a standard.

Stereochemical	The different arrangements of atoms or groups in space. This arrangement can affect physical and chemical properties, and two or more forms of the same compound with different arrangements in space are stereoisomers.
Stratosphere	The upper part of the atmosphere.
Synergism	The increased effect on one compound caused by the presence of another compound which is greater than the sum of the two when used separately.
Technique	The principle upon which a group of methods is based.
TEF	Toxic Equivalent Factor. The 'toxic equivalent' TEQ of hazardous organic compounds compared with the worst dioxin.
TLV	Threshold Limit Value. The maximum allowable level of toxic vapour thought to be safe.
Trace analysis	Determination of very small quantities or concentrations well below 0.1%.
Transpiration	The evaporation of water from surfaces, especially leaves of plants.
Troposphere	The layer of the atmosphere extending about 7 miles (11.3 km) upwards from the Earth's surface.
Weathering	The effects of exposure to natural conditions.

Index

accuracy 11
acid–base *see also* titrimetry
  equilibria 57
  Lowry-Brønsted theory 68
acid rain 263
activation energy 62
active biomonitoring 269
activity 196
adsorption 282
aerosols 315, 317
alpha particle spectrometry 190
alpha radiation 181
aluminium, determination 350
americium 364
amines, determination at trace levels
  406–7
amperometric sensors 218
amperometry 218
analysis *see* analytical chemistry
analysis of variance (ANOVA) 18, 23
analytical chemistry
  atmospheric 292
  biological materials 271
  classification of techniques 6
  definition 3
  measurements 8
  method selection 6
  methods 6
  qualitative 4
  quantitative 4
  structural 4
  techniques 6
anode 197
anodic stripping voltammetry (ASV) 218,
  284
aromatic hydrocarbons
  determination at trace levels 401–3
atmosphere 291
  composition 292
  indoor 292
  workplace 292
atomic absorption spectrometry (AAS)
  137–42 *see also* atomic spectrometry
  detection limits 143–4

electrothermal methods 140–2
  flame methods 139–40
  interferences and corrections 138–39
  uses 142
atomic energy levels 39
atomic number 35
atomic orbitals 37, 41
atomic spectra 35, 36
atomic spectrometry
  absorption 121
  detection limits 143–4
  emission 121
  fluorescence 121
  introduction 121
  mass 121
  plasma 124
  techniques 122
  uses 142
atomic theory 32
aufbau principle 38
auxiliary electrodes 213
auxochromes 147–8
average score per taxon (ASPT) 256

band spectra, molecular 146
Becquerel 187
Beer–Lambert Law 115–16, 146
best fit 26 *see also* graph plotting
bioaccumulators 251, 268
bioassays 251, 272
bioavailability 279
biochemical oxygen demand (BOD) *see*
  water
bioindicators 251, 255
biological
  indicators 249
  monitoring 3
  Monitoring Working Party Score
    (BMWP) 256–7
  surveillance methods 249
biomagnification 268
biomonitoring 250
biotic indices 255
Boltzmann equation and distribution 40

Bragg equation   136
buffers
    pH   201
    total ionic strength adjustment (TISAB)
        197

calibration
    Curie point for TG   230
    curves   5, 26
    process   5
calomel electrode   202
carbon electrode   215
carbon-14   360
carbon-14 dating   366
carbon monoxide   306
cathode   197
cements, DTA of   237
Chandler Biotic Score (CBI)   256
chemical bond
    covalent   41
    dative   41
    ionic   40
chemical flame   123–4
chemical oxygen demand (COD)   see
    water
chemical principles   32
chemical shift   see nuclear magnetic
    resonance spectrometry
Chernobyl accident   362
chlorofluorocarbons (CFCs)   314
chromophores   147–8
chromatography
    chromatograms, characteristics of   87
    column   87
    diffusion term   89
    efficiency   88
    electrophoresis   103
    gas (GC)   98–102
    gas–liquid (GLC)   98–102, 293
    gel permeation (GPC)   102–3
    high performance liquid chromatography
        (HPLC)   90–3
    introduction   84
    ion exchange   93
    mass transfer term   89
    paper (PC)   96–8
    peak width   89
    plate height or HETP (height equivalent
        to a theoretical plate)   89
    plate number   86
    resolution, R_s 89
    retardation factor, R_f 90
    retention volume (time)   86
    sorption mechanisms   84
    supercritical fluid (SFC)   102
    thin-layer (TLC)   90, 96–8
Clark oxygen electrode   208, 218
clays, DTA of   237, 246

coefficient of variation (CV)   see standard
    deviation
column chromatography   see
    chromatography
combustible gases   314
community structure studies   251
computer modelling of speciation   279
complexes
    complexation   61
    chelate effect   73
    formation constants   72, 73
    molecular spectra   149
    organo-metallic   47
    transition metal   47
Compton effect and scatter   184
conductance   197
conductometry   197, 222
confidence
    interval   15
    levels   15
    limits   15
constant composition, law of   32
contaminated landsite   369–75
    air-borne transport   371
    aquatic transport of pollutants   371
    assessment   372
    classification   374–5
    gases   371
    introduction   369
    investigations   373
    monitoring   375
    origins of contamination   370
    physical characteristics   371
    sampling   373
    types of pollutant   370–1
    vapours   371
control charts   26, 29   see also Shewhart
    and cusum charts
controlled potential techniques   210
correlation   26, 27
    coefficient, r, Pearson's   27
correlation spectrometry   295, 298
coulometry   222
coupled techniques   287
Craig counter-current principle   85
Curie point calibration of TG   230
cusum charts   30   see also control charts
cyclic voltammetry (CV)   215
cyclone sampler   318

data
    analytical   10
    assessment   9, 10, 20
    environmental   10
    interpretation   9, 10
    range of   12
    reliability   21
    spread of   12

Debye–Hückel theory of solutions   56
detection limit   4
detector organisms   264
dialysis   286
differential absorption spectrometry
  (DOAS)   297
differential LIDAR (DIAL)   297
differential pulse techniques   216
differential scanning calorimetry (DSC)
  233
differential thermal analysis (DTA)   233
diffusion   306
  layer   197
diode array detectors   92
dissolved oxygen   see water
distillation   103–4
distribution constant   55
distribution curves
  chi-squared   20
  frequency   12, 18
  normal (Gaussian)   19
  Student t 19
distribution profiles   86
distribution ratio   84
diversity indices   252
double-beam spectrometer   118–19
double junction reference electrode   203
dropping mercury electrode (DME)   213
dusts   317
dynamic calibration systems   306
dynamic mechanical analysis (DMA)
  240
dynamic mechanical thermal analysis
  (DMTA)   see dynamic mechanical
  analysis

efficiency of extraction   84
ethylene diamine tetraacetic acid (EDTA)
  74, 75
electric double layer   197
electrochemical
  cell   197
  detection   219
  half-cell reactions   59
  methods   196
  reactions   59
  sensors   299
  series   199
electrode potential   197, 200
electrodeless discharge lamp (EDL)   138
electrogravimetry   223
electromagnetic radiation
  absorption   115
  interaction with chemical species   113
  nature   109–11
  spectrum   112
electromotive force (EMF)   60, 197
electron, nature of   36

electron probe microanalysis   129, 133–5
electron spin resonance dating   see
  thermoluminescence dating
electrophoresis   103, 286
emission measurements   292, 303
  sampling   315
energy levels
  atomic   105–6
  molecular   105–8, 147
  selection rules   105, 121
  transitions   105–8
enthalpy
  of formation   51
  of reaction   51
entropy   52
environmental science   3
enzyme electrodes   219
equilibrium
  acid–base   57
  constant   54
  free energy and   54
  phase   55
error
  absolute   11
  bias   16
  constant   16
  determinate   16
  nature and origin   15
  normal error curve   17   see also
    distribution curves
  proportional   16
  relative   11
  systematic   16
evolved gas analysis (EGA)   242
exhausts   303
extractive sampling   302

F 22
  tables   23
  F-test   22, 23, 26
faecal contamination   266
fission products   362–4
flame emission spectrometry (FES)
  123
flow injection analysis (FIA)   219
flue gas treatment   247
fluoride electrode   206
Fourier-transform instruments   119–20
  spectrometry   295
free energy
  and equilibrium   54
  Gibbs   53
fulvic acid, complexation by   283
fundamental particles   33

Galvanic cells   198
gamma radiation   184–5
gamma ray spectrometry   190

gas analysis   293
    calibration   305
gas chromatography (GC)   *see*
    chromatography
gas detector tubes   301
gas ionization detectors   188–9
gas laws   49
gas–liquid chromatography (GLC)   *see*
    chromatography
gas sensing electrodes   208, 301
GC–MS   *see* mass spectrometry
Geiger–Müller counters   189
gel permeation chromatography (GPC)
    see chromatography
geological dating   365–6
geological materials   246
glass electrode   206
glass transition of polymers   237, 240
graphical methods   26
    least squares   29
    polynomial   29
    regression   29
gravimetry
    dust sampling   318
    organic precipitation reagents   79
    precipitation   78
    principles   77
    procedures   78
Gray   187

half-cell reactions   59, 198
head space analysis   104
Heizenberg uncertainty principle   36
high performance capillary electrophoresis
    (HPCE)   287
high performance liquid chromatography
    (HPLC)   *see* chromatography
histological indicators   261
hollow cathode lamp (HCL)   137–8
Hooke's law   154
humic acid, complexation by   283
Hund's rules   38
hydrocarbons   313

ICP torch   125
Ilkovic equation   211
Index of Air Purity (IAP)   260
indicator electrodes   197, 203
inductively coupled plasma–atomic
    emission spectrometry (ICP–AES)
    125–8
inductively coupled plasma–mass
    spectrometry (ICP–MS)   128–9
infrared atmosphere analysis   293
infrared spectrometry   152–60   *see also*
    molecular spectrometry
    applications in environmental analysis
        159, 293

combination with separation techniques
    158
group frequency correlation   156–8
infrared absorption   154–8
polyatomic molecules   156
sampling   152–4
in situ monitoring   302
intrusive microorganisms   266
ion exchange chromatography   93–6
ion exchange resins   95
ion selective electrodes (ISE)   204, 287
ion selective field effect transistors
    (ISFETs)   206
ionic activity   196
ionic reactions   63
ionic solvation   196
ionic strength   196
ionizing radiation   *see* radioactivity
ions, in solution   56
isokinetic sampling   316
isotopic ratios, in geochemistry and
    pollution studies   129

Jaccard coefficient   252

K_a 68

lasers   106
laser microprobe   127
lattice energy   40
LC–MS   *see* mass spectrometry
lead-210 dating   367
lead, determination   348–9
lichens   260
ligand   48
ligand field   48
light detection and ranging (LIDAR)   297
liquid
    junction potential   202
    –liquid extraction   289
    membrane electrodes   207
liquid scintillation counting
    192–4

macrophyte index   258
Margalef index   252
mass spectrometry   170–9
    applications   176
    base peak   172
    chemical ionization (CI)   171–2
    double focusing   172
    electron impact ionization (EI)   170
    fragmentation   174
    fragment ion   170
    GC–MS   177
    instrumentation   170–2
    isotopic composition   173
    LC–MS   178–9

molecular ion   170
nitrogen rule   173–4
primary ion   171–2
quadrupole   172
secondary ion   172
single focusing   172
mass transfer   87, 88
mean   13
comparison of   24   *see also* t-test
mechanical moduli   240
median   14
melting points   235
membrane electrodes   204
potential   205
metal complexes, in UV/VIS spectrometry
149
methane   313
microbiological
monitoring   265
surveys   251
microorganisms   265
mobile phase   86   *see also*
chromatography
molecules
bonding in   42
dissociation   108–9
energy levels   48–50
mode   13
orbitals   42–5
polyatomic   45
molecular spectrometry   *see also* under
individual techniques
introduction   105, 145
infrared (IR)   152–60
mass (MS)   170
nuclear magnetic resonance (NMR)
160–70
Raman   107
spectrometric processes   146
ultraviolet (UV)   146–52
visible (VIS)   146–52
monitoring   3, 251
morphological indicators   261
multivariate analysis   254

Nernst equation   199
Nernst partition (distribution) law   83
Nikolsky Eisenman equation   205
nitrogen oxides   307, 314
nitrogen rule   *see* mass spectrometry
non-dispersive
infrared (NDIR)   294
ultraviolet   294, 307
non-methane hydrocarbons   313
non-methane organics   313
nuclear fission   362–4
nuclear magnetic resonance spectrometry
(NMR)   160–70

anisotropic shielding   165
applications   170
chemical shift   163–5
FT–NMR   162
instrumentation   162
magnetic properties of nuclei   161
nuclear spin   161
peak areas   165
spin decoupling   169
spin–spin coupling   166–9
solvents   163
spin–lattice relaxation   162
spin–spin relaxation   162
quantum numbers   161
nuclear spectrometry   188

optically transparent thin layer
electrochemical cells   213
orbitals   47   *see also* atomic and molecular
anti-bonding   47
bonding   47
non-bonding   47
organic compounds, determination of trace
amounts   386–407
amines   406–7
aromatic hydrocarbons, by HPLC
401–3
biological response   386
concentration of analytes   393–5
dissolution   389
extraction   389
gas chromatography   395   *see also*
Chapter 5
HPLC   396, 401   *see also*
Chapter 5
introduction   386
Kuderna–Danish evaporator   404
liquid extraction   390
liquid–liquid extraction   390   *see also*
solvent extraction, Chapter 5
maceration   389
partition   390–3
pesticide analysis by gas
chromatography   398–401
quantitation   396
rotary evaporator   394
sample preparation   388–9
screening analysis   397–8
sediments, determination in
405–6
soils, determination in   405–6
solid–liquid extraction (SPE)   391–3
see also Chapter 5
solid phase cartridges   391–3, 404
solvent evaporation   394
water, determination in   403–5
organic nitrates   314
organic scintillators   192

outliers   21   *see also* Q-test
oxidation–reduction reactions   59   *see also* redox reactions
oxidative stability of polymers   238
oxygen demand   *see* water
oxygen sensors   310–11
ozone   308

paper chromatography   96–8   *see also* chromatography
particle track film detection   194
particulates   315, 317
passive biomonitoring   269
Pauli's principle   38
pellistors   315
periodic table   34
permeation methods   306
peroxyacetyl nitrate (PAN)   314
pesticides, determination by GC   398–401
pH   58, 200, 281
   meter   200
photoacoustic spectrometry (PAS)   299
photochemical reactions   64
Pinkham and Pearson index   252
pIon   200
pK_a 69
Planck's constant   36
plasma spectrometry   124
plutonium   364
pOH   58
polarography   213
pollutant mapping   260
polymer analysis   236
polymer degradation   231
population   11
potentiometric biosensors   208
   techniques   197
   titrations   209
precision   12
   comparisons   22   *see also* F-test
proportional counter   189
pulse height analysis   188
pulsed fluorescence analyser   309
purity determination   236, 248

quantum numbers   37
Q-test   21   *see also* outliers

R_f, Retardation factor   90
radiation dose   180, 187
radioactivity   33
   alpha radiation   181
   beta radiation   181–4
   detection and measurement   187–95
   energy changes in decay processes   181
   gamma radiation   184–5
   ionizing radiations   180–86

internal conversion   185
introduction   180
kinetics   63, 186–7
negatron   *see* beta radiation
positron   *see* beta radiation
separation of radionuclides   195
units   187
radioactivity and ionizing radiation in the environment   352–68
   artificial sources   360
   breeding reactions   362–3
   cellular effects   353
   external exposure   354
   internal exposure   354
   introduction   352–3
   medical uses   361
   nuclear power   361–2
   nuclear fission   362–4
   radionuclides resulting from cosmic rays   360
   radiogenic dating   364–8
   radionuclides of geological origin   354–5
   radon   355, 357–60
   research uses   361
   sources   352–3
   whole body counting/monitoring   354
radiogenic dating   364–8
radionuclide   *see* radioactivity
radon   *see* radioactivity in the environment
reaction kinetics   61, 63   *see also* equilibrium
recycling   246
redox electrodes   204
reference electrodes   197, 201, 213
remote sensing   303
replicates   12
residues   247
River Invertebrate Prediction and Classification System (RIVPACS)   265
Rydberg constant   36

safety   9
salt bridge   198
sample
   analytical   11
   sampling   7
   statistical   11
Saprobic (Saprobien) index   255
saturated calomel electrode (SCE)   202
scatter plots   28
selectivity coefficients   205
semiconductor detector   189
sentinel organisms   264
separation factor   84
separation techniques, general introduction   81
sequential measurements   127
sewage fungus   265

Shannon index 252
sharpline radiation and sources 137–8
Shewhart charts 29 *see also* control charts
similarity index 252
Simpson index 252
simultaneous measurements 127
    thermal analysis (STA) 242
    thermogravimetry–mass spectrometry (TG–MS) 244
sodium iodide detectors 191–2
soils, thermal analysis of 246
solid phase extraction (SPE) 93, 391–3
solid state reactions 64
solubility 56
    products 282
solvent effects, on molecular spectra 148
solvent extraction
    summary of phase systems 82
    separation of mixtures 82, 83
    use in environmental analysis 81, 82
Sorensen index 252
soxhlet extraction 82, 83
spark erosion 127
speciation 279
    computer modelling 279
    coupled techniques 288
    experimental 284
species abundance patterns 253
spectral resolution 117
spectrometric methods 293
spectrometry, general
    introduction 105
    instrumentation 106–17
    techniques 111
sputtering 138
stability constant 280
standard deviation 12, 14
    pooled 15
    relative (RSD) 14
standard electrode potentials 61, 199
standard hydrogen electrode (SHE) 199
stationary phase 86 *see also* chromatography
stripping film detection 194–5
stripping voltammetry 214
sub-isokinetic sampling 316
sulphur dioxide 262, 309
supercritical fluid chromatography (SFC) *see* chromatography
super-isokinetic sampling 316
supporting electrolyte 213
surveillance 3

t (Student) 19
    tables 24

t-tests 24, 25
thermal desorption cartridge 99
thermal methods
    atmosphere effects 226
    definitions 225
    for mixtures 232
    rate of heating 227
    samples 226
thermodynamics
    first law of 51
    second law of 52
thermogravimetry (TG) 228
thermogravimetric analysis (TGA) *see* thermogravimetry
thermoluminescence dating 367–8
thermomechanical analysis (TMA) 239
thermomicroscopy 245
thin-layer chromatography (TLC) 96–8 *see also* chromatography
thin layer amperometric detection 221
thorium 355
thoron 357
titrimetry
    acid-base 68 *see also* acid–base
    complexometric 71–5 *see also* complexation
    end-points and detection 67, 71, 75, 76
    potentiometric 209
    redox 75–7
    redox reagents 77
    strong and weak acids 69, 70
    titration curves (acid–base) 70
    types of titration 67
    visual indicators (acid–base) 71
    visual indicators (complexation) 75
total ionic strength adjustment buffer (TISAB) 197, 201
total organic carbon, (TOC) *see* water
trace elements
    air pollution 328–30
    analytical methods 344–7
    animal and human concentrations 334–5
    aquatic presence and pollution 327, 330–2
    chemical form 325–8
    contamination and pollution 328
    definition 320–1
    environmental 320–1
    examples 348–50
    food chain incorporation 322
    health effects 335
    movement in the environment 322
    natural levels 323–5
    root uptake and levels in plants 327, 333–4

trace elements *contd*
  sample digestion 342–3
  sampling and sample preparation
    336–43
    atmosphere 336
    biological materials 341–2
    plants 340–1
    soils and sediments 339–40
    water 337–8
  soil and sediment levels and pollution
    326, 327–33
  tissue, content of 327
Trent Biotic Index (TBI) 256
troposphere 291
tritium 360
tritium dating 366
true result 11
tunable diode laser spectrometer (TDL)
  296
two-way indicator species analysis
  (TWINSPAN) 254

ultra violet spectrometry 146–52
  applications 149
  fluorescence methods 149–51
  combination with separation techniques
    151–2
uranium 356
uranium dating 365

v-mask 30   *see also* cusum charts
Valence Shell Electron Pair Repulsion
  Theory (VSEPR) 45, 46
validation
  methods 9
  standards 9
van Deemter equation 87
van't Hoff equation (isochore) 55
vaporisation studies 247
variance 15
visible spectrometry 146–152
  applications 149
voltammetric techniques 210, 220

water
  acidity 378
  alkalinity 379
  analysis of 376–85
  biochemical oxygen demand (BOD)
    380
  chemical oxygen demand (COD) 380
  conductivity 381
  dissociation of 57
  dissolved oxygen 379
  dissolved salts 381
  hardness 382
  ionization constant 58
  metals determination 381   *see also*
    Chapter 15
  organics, traces 383–4   *see also* Chapter
    19
  oxygen demand 379
  permanganate value/index 377, 380
  pH 377–8
  Quality index (WQI) 259
  quality regulations 377
  radioactivity 384   *see also* radioactivity
    and radiation in the environment
  sampling 384–5
  surveys 384–5
  total organic carbon (TOC) 381
  Winkler method 380
wave equation 36, 37
Winkler method   *see* water
working range 26

X-ray
  analysis 245
  emission and absorption 131–3
  emission techniques 129
  energy dispersive analysis (EDX) 131
  spectra 35
  spectrometry 190
  wavelength dispersive analysis (WDX)
    131
X-ray fluorescence spectrometry (XRFS)
  135–7